AF575119

BAYESIAN METHODS FOR MANAGEMENT AND BUSINESS

BAYESIAN METHODS FOR MANAGEMENT AND BUSINESS

Pragmatic Solutions for Real Problems

EUGENE D. HAHN
Department of Information and Decision Systems
Salisbury University
Salisbury, MD

Published by John Wiley & Sons, Inc., Hoboken, New Jersey
Published simultaneously in Canada

Library of Congress Cataloging-in-Publication Data:

Hahn, Eugene D.
Bayesian methods for management and business : pragmatic solutions for real problems / Eugene D. Hahn, Department of Information and Decision Systems, Salisbury University Salisbury, MD.
pages cm
Includes bibliographical references and index.
ISBN 978-1-118-63755-5 (hardback)
1. Management–Statistical methods. 2. Commercial statistics. 3. Bayesian statistical decision theory.
I. Title.
HD30.215.H34 2014
650.01′519542–dc23

2014011434

10 9 8 7 6 5 4 3 2 1

To Gene, Nora, David, Tara, and Nok.
Thank you for all that you do and have done. Also, thanks to Jeff Kottemann for comments early in the book's development.

CONTENTS

PREFACE

The use of Bayesian statistics has exploded over the past two decades. Researchers in many disciplines have used Bayesian statistics to reveal new insights and understand difficult problems. However, there are few discussions of Bayesian statistics that are focused on its practical applications to business and management. As a result, people with an orientation toward business and management problems were left to bridge the theory/application gap themselves with little guidance. In this book, we show how Bayesian statistics can help generate insights into business and management data. The book features a practical orientation with *In Practice* sections that go into detail the use of Bayesian approaches with actual business data.

On a personal note, I have spent countless hours to make this book, as much as possible, one that blends concepts and intuition along with mathematics. There is math in this book like in all statistically oriented books. However, it is my belief that **an understanding of every formula is not required** for getting something out of this book. As an analogy, you may have gone to a foreign country where you did not understand every word but were still able to get around. Here we will see that modern Bayesian methods use Monte Carlo Markov chain (MCMC) methods to perform much of the more difficult math involving calculus. Whenever you see the abbreviation MCMC, you can also think of it as the abbreviation for "Makes Calculus More Convenient" or "Makes Computation More Convenient". MCMC makes these things more convenient because computer simulations handle the job for us. It is good to have a conceptual understanding of what an integration is for the few occasions that we discuss it in this book. But to summarize, an integral here is typically used like a weighted averaging technique to average out something that has variability. We are all very familiar with using summary numbers like average monthly sales to fill in for sales numbers that vary every month. When we integrate a metric over a probability distribution, we are doing something very similar. If you can remember this working definition, you will have a working understanding of the few formulas that mention calculus explicitly. The greatest amount of math appears in Chapters 3 and 4. Feel free to skip these entirely on a first reading and come back later at your leisure.

Instead of math, I think **you should emphasize learning about modeling**. A Bayesian model has three components. There is a likelihood, a functional form, and a set of priors. Being able to understand these three parts of a model and to choose them appropriately for your situation will be much more important in practice. For every model introduced, there is a detailed discussion in English about the considerations and choices that go into the model specification. Once we have gone through this discussion in English, we put the model in a compact notation that summarizes the model. This summary is useful for describing your model to others as well as for writing programming code in the *WinBUGS* software. The *WinBUGS* software will handle estimation of your model using MCMC.

Like many business processes, Bayesian statistics is modular. Once you become familiar with the different kinds of models (modules), it is very easy to (re)assemble them into a structure that is applicable to your data. The customized model can then be estimated using MCMC. This characteristic of modern Bayesian methods has been quite important in its evolution to becoming an empirical method of choice in current statistical practice where customization to data characteristics or theoretical needs is important. After getting some experience, you will be in a position to come up with your own models that help you address your own business data.

Using the "In Practice" Sections

The *In Practice* sections provide worked examples using real data and *WinBUGS* code listings. The practically oriented reader can focus primarily on these sections to build up a library of Bayesian models for working with business data. Prior to each *In Practice* section, there is a discussion of conceptual material. However, different learning styles may benefit from different approaches. Therefore, feel free to consider reading and working with the *In Practice* material first if that better suits your learning style. Once you are comfortable with the *In Practice* material, you can then return to an examination of the more conceptual material.

Not all people are the same, so I have tried to make the book have something to offer to a variety of readers. Different pathways through the book can be used for people with different interests and different backgrounds. For MBA students with a solid grounding in the regression model and its prerequisites, the pathway might include Chapters 1, 2, 5–7, with other chapters being optional. For Masters of Science in Analytics students, the earlier parts of Chapters 8 and 9 could be added. Readers who would prefer a less technical, applications-oriented approach could cover Chapters 1, 2, 5 and 6, then focus on *In Practice* sections in Chapters 7 through 11. A solid grounding in the foundations of Bayesian inference (Chapters 3 and 4) will be helpful for those who want to make the most of Bayesian inference. A traditional approach would place foundational concepts early in the process and the book follows this approach. However, these concepts can be postponed for readers who are more interested in immediate solutions.

This book makes use of the freely available software *WinBUGS* and *R*. Code listings appear throughout the book. The code listings are available at the author's Web site at `http://faculty.salisbury.edu/~edhahn`.

Using the "In Detail" Sections

The book also features *In Detail* sections where somewhat more advanced material is presented. These sections are for readers who would like to drill down more into a topic or who are looking for additional skill-building. Again, you (the reader) get to choose what you would like to learn about, and at what time. The *In Detail* sections can be read in their original sequence, or you can return to them as the need/interest arises.

1

INTRODUCTION TO BAYESIAN METHODS

1.1 BAYESIAN METHODS: AN AERIAL SURVEY

The modern business environment is awash in data. As a result, managers seek ways to summarize and simplify data to emphasize a select number of key aspects. They may also wish to examine whether certain kinds of structure and patterns are present in the data. Or, they may wish to use data to draw conclusions about other kinds of unobserved or latent phenomena they believe exist with respect to their businesses, customers, materials, and so on. These kinds of activities managers undertake are not mutually exclusive, but rather emphasize different aspects of the data discovery process.

Statistical methods are some of the most widely-used methods for data discovery. Many managers who have gone through an undergraduate or graduate business education will have encountered some of these methods. The methods that are typically taught to managers are called *classical statistical methods*. Classical methods are also known as *frequentist methods* because they derive from a frequency-based view of probability. Classical methods can be summarized as statistical methods that can be arrived at based on consideration of the likelihood function alone (Fisher, 1922). These include the familiar *t*-test, simple linear regression, and logit analysis by maximum likelihood. The likelihood function can be thought of as the "data function" since it quantifies the relative likelihood of param-

Bayesian Methods for Management and Business: Pragmatic Solutions for Real Problems, First Edition. Eugene D. Hahn.

eter values in the data we have observed. The likelihood function allows the manager to summarize or estimate unknown parameters such as the lifetime value of a customer or the average failure rate of an important component.

In addition to the likelihood function, Bayesian methods also incorporate a prior distribution for parameters. If the manager has preexisting beliefs about parameters, such as an intuition about the average failure rate of a component, he/she can often express this in terms of a probability distribution. Having done so, Bayesian methods can proceed and the final results (summarized by the posterior distribution) will contain a blend of the information arising from the prior (i.e., from personal beliefs) and from the likelihood function (i.e., from the data). This blend will reflect the relative weight of information arising from the two sources. For example, if the prior is extremely concentrated, very large quantities of discrepant data will be required in order to produce a substantive change in the parameter estimate. No doubt you have met someone at some point in your life whose prior beliefs about something were rather difficult to change even with substantial evidence, and this is conceivably possible with the Bayesian approach as well. More commonly, however, a manager or researcher will adopt what is called a *non-informative prior*. Such priors are designed to have little effect on the conclusions that would be drawn from the data, and hence reflect an "open-mindedness" about the data in the sense that a broad range of values would be considered reasonably possible. Given the practical emphasis of business and management, these diffuse priors are almost always employed when one is interested in understanding the data because there is little point in analyzing data if we already wish to retain our preconceived notions. However, there are situations where one needs to be more careful about the impact of the prior, and the most common of these is the situation where the sample size is small. Since the posterior distribution reflects the relative weight of the data and the prior, we must exercise care when the data influence is light due to its scarcity (of course, we would also want to draw conclusions cautiously if classical methods were used with small samples).

The use of the prior distribution is an important differentiating aspect between Bayesian and classical methods. Historically it has also been a major point of contention (Gelman and Robert, 2013), with proponents of opposing viewpoints trading critiques (e.g., see Edwards, 1972, ch. 4, for an example of a critique of the Bayesian approach). Another historical challenge for the Bayesian approach has been mathematical. As will be shown in Chapter 2, substantive Bayesian methods require the evaluation of integrals, and, for complex or nonstandard problems this can be difficult, tedious, or worse. Earlier Bayesian reference texts such as those by Zellner (1971) and Box and Tiao (1973) show that much can be accomplished in the Bayesian context when one had the requisite mathematical background. However, the twin barriers of skepticism regarding priors and mathematical difficulty made the application of Bayesian methods less common for many years. The popularization of Markov chain Monte Carlo (MCMC) methods (Gelfand et al., 1990, 1992) dramatically reduced the latter barrier. MCMC also provided researchers with a powerful tool that could be effectively wielded against complex and/or nonstandard problems. This raw power enabled individuals with a knowledge of Bayesian methods to examine important problems in new and revealing ways.

Businesses have been also able to take advantage of the benefits offered by Bayesian methods. For example, Medtronic was able to shorten the Federal Drug Administration (FDA) approval timeline for the development of a therapeutic strategy for a spinal-stabilizing device (Lipscomb et al., 2005). TransScan Medical was able to establish efficacy of its T-Scan 2000 device for mammography with a smaller sample size by incorporating prior information from previous studies (FDA, 1999). Enterprise software by Autonomy used Bayes' rule to uncover patterns in large corporate databases and has been deployed to unravel the events that occurred prior to the collapse of Enron as well as to detect terrorists (Fildes, 2010). The creators of the Web site `homeprice.com.hk` used Bayesian hierarchical models to provide consumers with pricing information on over 1 million residential real estate properties in Hong Kong and surrounding areas (Shamdasani, 2011). The energy industry has used Bayesian methods to understand petroleum reservoir parameters (Glinsky and Gunning, 2011) and update uncertainty regarding possible failures in underground pipelines (Francis, 2001). Finally, recent Bayesian work by Denrell et al. (2013) suggests that long-term superior corporate performance may depend considerably on early fortunate outcomes.

In a few management disciplines, particularly marketing (Rossi et al., 1996; Arora et al., 1998; Ansari et al., 2000; Rossi et al., 2005), Bayesian methods have been extensively and fruitfully applied. However, in many others, their full potential has yet to be realized. In part, this may be due to a lack of material showing the relevance of Bayesian statistics to a variety of business disciplines. This book aims to fill this gap.

1.1.1 Informal Example

We can begin with an informal example from a small business. Suppose you are a restaurant owner who wants to estimate how much a diner spends on average. Initially, based on a hunch, you estimate that the average is about \$25. Following up on your hunch, you pick a random day and obtain the data on how much each diner has spent. You calculate the average for the data and find the sample average is \$28.23. Based on this, you might intuitively update your initial estimate. You might revise your estimate to \$28 as a compromise between your hunch and the data.

Your hunch comes from your informal assessment over a long period, so you wouldn't want to completely discard it. However, the data seems to indicate that your hunch may have been a little on the low side.

We can try that process again in a slightly more sophisticated manner. Suppose your hunch was that the average amount spent was $25 and that you're fairly sure that the average will be within $5 of that. By "fairly sure" you mean that you think there is about a 95% chance that the average amount spent will be between $20 and $30. Suppose you think that the average amount spent approximately follows a normal distribution. You do a back-of-the-envelope calculation based on the normal distribution. You recall that the 95% central probability interval for the normal distribution uses the formula $25 \pm 1.96\sigma$ where σ is the standard deviation. A side calculation shows that, if $\sigma = 2.55$, then the 95% probability interval for the normal distribution is 25 ± 5. You decide that $\sigma = 2.55$ sounds reasonable here.

The Bayesian terminology for your hunch is the *prior*. More formally, this is called *prior distribution* since we were able to represent your beliefs with a statistical distribution. Your final estimate of $28 involves what is called the *posterior* in Bayesian terminology. You used empirical data to update your prior and came up with a revision, the posterior. Your intuitive update method is similar to what happens when we formally apply Bayes' theorem. A formal application of Bayes' theorem will give you a *posterior distribution*. The posterior distribution depends on both the prior and the data. The posterior distribution combines both sources of information in a sort of "weighted average" of the information available. If there is a lot of data and little prior information, the posterior distribution will be heavily influenced by the data. Conversely, if there is little data but the prior belief is strong, the posterior distribution will tend to look like the prior distribution.

1.2 BAYES' THEOREM

Bayesian methods utilize a formula obtained by an amateur mathematician in Bayes (1763). The Reverend Thomas Bayes, a Presbyterian minister, framed his mathematical development in terms of billiard balls, but the implications of his discovery were far wider. Indeed Bayes' theorem has often served as a model for how living creatures learn in an uncertain world. For example, athletes seem to follow Bayes' theorem intuitively (Wolpert, 2004) in order to hit a ball or to defend against a kick. This is because, in order to act, an athlete must incorporate both what she sees currently on the field and her past knowledge about what the ball or the opponent might do. In the context of business, one might have some beliefs about the price of a particular stock prior to the opening bell, then observe the stock's price over the course of the trading session, and use this to predict where it will go on the following day. Since every day we combine current information with past information to get an updated perspective, Bayesian methods have been described as instinctive for business (Hubbard, 2007).

Conceptually speaking, Bayes' theorem says the following:

$$\text{Prior beliefs} \Rightarrow \text{Data} \Rightarrow \text{Updated beliefs.}$$

More formally, we have

$$p(\theta) \times p(y|\theta) \propto p(\theta|y). \tag{1.1}$$

Here θ is the variable of interest, such as the price of a stock. Based on our prior beliefs, we assign a probability distribution to θ, which is denoted as $p(\theta)$. The likelihood function appears as $p(y|\theta)$. The likelihood function indicates the sampling distribution of the data y. The conditional relationship $p(y|\theta)$ indicates that the variable θ is believed to be relevant to the sampling distribution of y, i.e., y depends on θ. The symbol $\propto$ indicates the concept "is proportional to" and $p(\theta|y)$ is the distribution of our updated belief, the posterior distribution. Equation (1.1) is Bayes' theorem expressed up to a constant of proportionality; that is, on a relative basis, the posterior evidence is heavier where the product of the prior and the likelihood is large.

If θ is a discrete variable such as "buy" versus "sell" or bond rating ("AAA," "AA," ...), Bayes' theorem can be written as

$$p(\theta|y) = \frac{p(y|\theta)p(\theta)}{\sum_\theta p(y|\theta)p(\theta)}, \tag{1.2}$$

where we have rearranged the terms and a proportionality constant appears in the denominator. The numerator functions like the weighted-average formula. The information from the data, $p(y|\theta)$, is multiplied by its prior weight $p(\theta)$. The denominator ensures that all the posterior probabilities, $p(\theta|y)$, add up to 1.

In many business areas, we do not have a daily need to use calculus. Hence this book does not require you to perform calculus. An intuitive understanding of the ideas behind calculus is sufficient. We use calculus to handle continuous (or smooth) distributions and functions just like we handled discrete distributions and functions with the sum in (1.2). We can define a continuous function in terms of taking smaller and smaller discrete slices as in Figure 1.1. Calculus then allows us to work with the continuous function.

With this in mind, Bayes' theorem for continuous distributions is

$$p(\theta|y) = \frac{p(y|\theta)p(\theta)}{\int p(y|\theta)p(\theta)d\theta}. \tag{1.3}$$

Again, the proportionality constant appears in the denominator. It ensures the posterior distribution integrates (sums) to 1.

Note that in both Equations (1.2) and (1.3) the denominator term can be replaced with $p(y)$. This is because these denominators indicate that we have "averaged over" θ so it can be ignored. More formally we can say that θ has been either summed or integrated out of $p(y)$.

1.3 BAYES' THEOREM AND THE FOCUS GROUP

As an example of Equation (1.2), suppose that a company assembles a focus group. First, the participants are asked to read the background stories on a number of previously introduced products (suitably disguised). Next, they are asked to predict whether each product was later considered to be successful. By looking at the performance of the focus group, the company hopes to assess the potential of a new product in terms of whether it will be successful or not.

In order to do this, we begin by reviewing basic probability results relating marginal, conditional, and joint probabilities. These results are important in developing a knowledge of Bayesian statistics, and so they are repeated here. Hence, for a moment we will arrange a break for our focus group and consider a deck of cards.

The joint probability is the probability of separate events co-occurring. For example, it is possible to draw a card that has both the rank of Queen and the suit of Clubs. There is only one way to get an outcome such as this in the set of 52 cards. From set theory notation, one common way to write the intersection of two events is with the symbol $\cap$. So we could write a joint probability as $p(\text{Queen} \cap \text{Clubs}) = 1/52$. However, in Bayesian statistics another common way to denote joint probability is with the comma, as in $p(\text{Queen}, \text{Clubs}) = 1/52$. This is the convention that will be followed in this book.

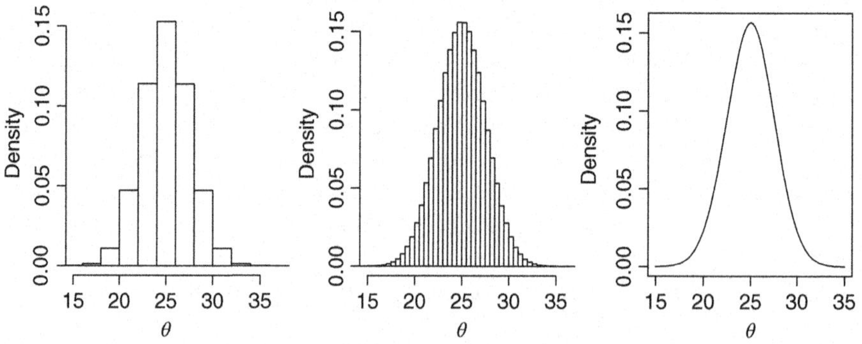

Figure 1.1 Moving from a Discrete Function to a Continuous Function

The marginal probability is the overall probability of an event for a given sample space. For example, the marginal probability of selecting a queen (the event) from a deck of cards (the sample space) is $p(\text{Queen}) = 4/52 = 1/13$, as can be seen in Table 1.1 by summing over the joint probabilities for each suit. The conditional probability $p(\text{Club}|\text{Queen})$ indicates the probability of an event given that another event is known to have occurred. Here, if we look only at line 3 of Table 1.1, we see that $p(\text{Club}|\text{Queen}) = 1/4$. Finally, we have the complement rule which indicates the probability of not obtaining the event. So the probability of not drawing a queen is $1 - p(\text{Queen}) = 12/13$. From the basic rules of probability, we may relate the joint, marginal, and conditional probabilities using the result $p(A, B) = p(A)p(B|A)$. We are free to condition on either variable and so, equivalently, $p(A, B) = p(B)p(A|B)$.

We now return to our focus group. For the products that were not considered successful, the participants were able to correctly identify them as unsuccessful 90% of the time. The successful products were correctly identified 95% of the time. The participants were next asked to read the background story on the company's new product that is being considered for development. The focus group responds that this product indeed seems to be successful. However, the industry is very competitive and only one new product out of 100 is considered a successful by the company's definition. Is the company's new product worth pursuing if the company wants to produce a successful product?

A successful product can be written B, while the complement $\overline{B}$ indicates an unsuccessful product. Similarly, A indicates the case where the focus group believes that the product will be successful, while $\overline{A}$ indicates

TABLE 1.1 Card Deck Contents

	Suit			
	Clubs	Diamonds	Hearts	Spades
Rank	Ace	Ace	Ace	Ace
	King	King	King	King
	Queen	Queen	Queen	Queen
	Jack	Jack	Jack	Jack
	10	10	10	10
	9	9	9	9
	8	8	8	8
	7	7	7	7
	6	6	6	6
	5	5	5	5
	4	4	4	4
	3	3	3	3
	2	2	2	2

it does not. Then we have $p(\overline{A}|\overline{B}) = 0.9$, $p(A|B) = 0.95$, and $p(B) = 0.01$. The company will then be interested in $p(B|A)$. We have from Equation (1.2)

$$\begin{aligned} p(B|A) &= \frac{p(A|B)p(B)}{p(A)} \\ &= \frac{p(A|B)p(B)}{p(A,B) + p(A,\overline{B})} \\ &= \frac{p(A|B)p(B)}{p(A|B)p(B) + p(A|\overline{B})p(\overline{B})} \\ &= \frac{0.95 \times 0.01}{0.95 \times 0.01 + 0.1 \times 0.99} \\ &= 0.0876. \end{aligned}$$

In reporting back to the company, the probability of their new product being a breakthrough innovation is 8.76%, which in absolute terms is still fairly low despite the promising reaction from the focus group. However, in relative terms $p(B|A)$ is over 8 times larger than $p(B)$, reflecting the positive assessment. Bayes' theorem gives a mechanism for weighting the prior evidence and the data evidence and combining the two to provide an updated result as discussed in Section 1.1. Additional independent focus groups could be used to provide yet more evidence about the product so as to sharpen the results for the company.

In terms of computational matters, we can see that the denominator of the equation required some treatment, and we will see in later chapters that this term (the normalizing constant) will require consideration. Here, to obtain $p(A)$ we expand it as a sum of joint probabilities in line 2 according to the basic rules of probability. We then expand these joint probabilities into a product of marginal and conditional probabilities in line 3. For line 4, we need $p(A|\overline{B})$, which is $1 - p(\overline{A}|\overline{B}) = 0.1$. We similarly find $p(\overline{B}) = 1 - p(B) = 0.99$.

1.4 THE FLAVORS OF PROBABILITY

In the previous section, both the company and our focus group were kind enough to provide us with information that we could use to assign probabilities. But what are probabilities? In this section, we will discuss additional properties of probabilities as well as different ways by which we can arrive at numbers that we can call probabilities.

1.4.1 Common Ground

Just about all notions of probability start with some commonly accepted premises and axioms (Kolmogorov, 1956). Probabilities are numbers that range from 0 to 1. An event with probability zero is an impossible event, whereas an event with probability 1 is a sure or certain event. Uncertain events are those that lie between these two extremes, as the complement rule indicates that the impossible event is certain not to happen. In terms of mathematical properties, events can have an important property known as *independence*. If A and B are independent, then $p(A)p(B) = p(A, B)$. Thus, for independent events $p(B|A) = p(B)$, and so whatever is the outcome of A, it has no effect on the chance of B occurring. Mutually exclusive (or disjoint) events cannot co-occur, i.e., $p(A, B) = 0$ if A and B are disjoint. For example, in our deck of cards one cannot draw a card that is both the rank of King and also Queen. If we have a complete set of mutually exclusive joint probabilities, they can be summed to produce a marginal probability. In the deck of cards, we obtained $p(\text{Queen}) = 4/52$ by using this rule because it is not possible to draw a card that is both a queen of spades and a queen of "not" spades at the same time.

1.4.2 Frequency-Based Probability

The most commonly encountered notion of probability involves frequencies. The counts appearing in Table 1.1 are an example of frequencies that were used to obtain probabilities in Section 1.3. In particular, it seems reasonable to believe that in the long run a Queen would be drawn once out of 13 times on average. More formally, this notion of probability requires one to be able to perform numerous independent replications of an experimental task with discrete outcomes under identical conditions using a predefined sampling scheme, such as sampling with replacement. The probability then arises as a long-run limit of our intuitions about the proportional occurrence of physical outcomes. In situations where this procedure can be undertaken, frequency-based probabilities are very compelling. In our cards example, with a well-shuffled deck of standard playing cards being used for each draw (and cards being replaced once drawn), it would be difficult to conceive of the probability of drawing the queen being other than 1/13. As a result, frequency-based probability (von Mises, 1928; Kolmogorov, 1956) has been applied fruitfully to countless important problems and can be considered a key building block of twentieth-century science.

However, the powerful and compelling nature of this kind of probability is offset by the fact that it is not always easy to apply in the real world. Rigged (or poorly constructed) roulette wheels and shaven (or otherwise misshapen) dice reveal that our intuitions about physical outcomes may be in error. In business, familiar concepts such as the first-mover advantage or changing customer expectations show that there is no way to turn back the clock to repeat an experiment on an industry or a customer base. There may even be instances where it is important to determine the probability of an event that has never occurred. For example, one could ask: what is the probability that a single voter's vote would cause one presidential candidate or the other to win a U.S. presidential election (Gelman et al., 1998), despite the fact that such an event has yet to happen? Other problems for the objective bedrock of frequency-based probabilities are discussed by Jeffreys (1961, ch. 5).

1.4.3 Subjective Probability

Subjective probability is the major competing notion of probability. Here, probability measures the degree of personal belief or personal confidence in an outcome. Thus two persons may have different assessed probabilities for the same outcome, such as which team will win a sports game. Savage (1954) showed that subjective probability can be tied closely to utility theory, where economic actors are hypothesized to try to maximize expected outcomes based on payoffs and personal probabilities.

Subjective probabilities can be elicited in a variety of ways. We may directly ask for a person's probability assessment and use the number supplied. A second way is to frame an event in terms of a betting scenario and see how much a person would be willing to wager on the outcome versus the non-outcome of an event. A third way can involve logical considerations. For example, if a person has no knowledge about the prevalence of outcomes, he or she may assume that all outcomes are equally likely as a default initial position.

Subjective probabilities can be obtained for a much wider class of events than can frequency-based probabilities. However, for some it is philosophically troubling to involve personal beliefs in an attempt to learn from data. Historically, this has caused many to question the Bayesian approach because it makes use of the (often subjectively determined) prior distribution. Still, if we want to learn about a complex, ever-evolving world, it is very difficult to escape from subjective probabilities and concepts. A preference for frequency-based probabilities involves a subjective utility assessment. The famous $p < 0.05$ significance rule from classical statistics arises partly from subjective considerations (Cowles and Davis, 1982), and early in the history of statistics different authors had somewhat different opinions about what the decision rule should be. A number of other subjective aspects of non-Bayesian statistics are reviewed by Zellner (1995).

In current practice, the equivalent of an audit trail is recommended for the use of subjective probability in Bayesian methods. First, we clearly indicate what the subjective probability distribution is. For example, we might believe that a normal distribution with a certain mean and variance is to be used. Next we describe why it is reasonable and defensible in light of the evidence available including any pertinent data (Gelman et al., 1998). Finally, we ideally present a sensitivity analysis that shows the impact of alternative probability distributions. In this step, we document whether conclusions change on the basis of our subjective probability assessments.

1.5 SUMMARY

Bayes' theorem succinctly describes how we should revise our beliefs with respect to evidence if we wish to be consistent with the laws of probability. Applying Bayes' theorem to simple cases is simple, as we have seen here. However, historically Bayes' theorem could be difficult to apply for more complex cases. Fortunately, modern computational techniques allow Bayesian methods to be applied with comparative ease to even extremely complex problems. The remainder of this book illustrates how we may do this with a particular emphasis on business and management contexts.

1.6 NOTATION INTRODUCED IN THIS CHAPTER

Notation	Meaning	Example	Section Where Introduced
θ	(Usually) generic symbol for parameter (or set of parameters)	θ	1.2
$\propto$	Proportional to	$A \propto B$	1.2
$p(\cdot)$	Probability distribution	$p(\theta)$	1.2
$p(\cdot, \cdot)$	Joint probability distribution	$p(\mu, \sigma)$	1.3
$p(\cdot \mid \cdot)$	Conditional probability distribution	$p(\mu \mid \sigma)$	1.3

2

A FIRST LOOK AT BAYESIAN COMPUTATION

2.1 GETTING STARTED

In order to put Bayes' theorem to use, we first have to make three decisions. First, we must select a sampling distribution or likelihood function for the outcome data. Next we must think about what factors might be influencing the outcome data so as to identify parameters and select a particular structure for them. Finally, we must assign a prior distribution for all the parameters we have identified in the previous decision.

The first two decisions are usually typically discussed as one decision, which is called *selecting the model* by other authors. In this book, we will discuss the first two decisions separately. The ordering in which the decisions is made is not critical as long as all three decisions are made.

Inevitably, these decisions involve considerations about your goals and your data. Investing the time with these considerations ultimately makes the Bayesian approach very powerful because of the flexibility it provides. As an analogy, it is possible to travel between two cities by train, and this may even be the most economical option in a number of circumstances. But learning the different tasks required to operate a car gives the driver many more options.

This chapter will focus on proportion data, an important kind of business data. Examples of this kind of data include the percentage of people who remember an advertisement, or the percentage of people who return

Bayesian Methods for Management and Business: Pragmatic Solutions for Real Problems,
First Edition. Eugene D. Hahn.

a particular item to the store. Proportions can arise in *discrete* data, which means the data takes on only certain values or can be mapped to the integers.

2.2 SELECTING THE LIKELIHOOD FUNCTION

Selecting the likelihood function for our data involves thinking about what kind of data we have and what distribution might be appropriate for representing and understanding our data. In some cases, we know how a random process behaves. For example, we might have a fair coin that has a 50–50% chance of coming up heads or tails. We might wonder how many heads we would observe or sample in five flips of the coin. A common distributional model for this process is the binomial distribution, written as

$$p(y|n, \pi) = \binom{n}{y} \pi^y (1 - \pi)^{n-y}. \tag{2.1}$$

Here, the number of heads y is a function of the total number of flips n and the probability π of observing a head. The conditioning arguments of (2.1) (after the vertical bar) indicate that we need to know n and π. Suppose we know $n = 5$ and $\pi = 0.5$, for example. Then we can find the probabilities of the possible values of y (see second column of Table 2.1). These probabilities sum to 1, as expected. The outcomes $y = 2$ and $y = 3$ are the most probable. Hence, if we were to repeatedly take samples from the binomial distribution with the given values of π and n, we would expect our the results of observations for y to converge to those given by the binomial sampling distribution.

The third through fifth column of Table 2.1 illustrate the separate quantities that were used to get the sampling distribution probabilities in the second column. The third column contains the numerator of the binomial coefficient, n. Since n is known, this column is a constant. The fourth column is the denominator of the binomial coefficient. These values vary

TABLE 2.1 Binomial Sampling Distribution, $n = 5$, $\pi = 0.5$

y	$p(y\|n, \pi)$	$n!$	$(y!(n-y)!)^{-1}$	$\pi^y(1-\pi)^{n-y}$	Product
0	0.03125	120	0.00833	0.03125	0.000260
1	0.15625	120	0.04167	0.03125	0.001302
2	0.31250	120	0.08333	0.03125	0.002604
3	0.31250	120	0.08333	0.03125	0.002604
4	0.15625	120	0.04167	0.03125	0.001302
5	0.03125	120	0.00833	0.03125	0.000260
	1				0.008333

with the unknown value of y. The fifth column contains terms that involve the parameter π. It just so happens that, in this example, the values in this column are constant across the dataset. This is not typical. It happens because here $\pi = (1 - \pi) = 0.5$, so, as a result, the formula collapses to 0.5^n. For any other value of π this would not occur. The final column, labeled Product, is the result of multiplying the values in the fourth and fifth columns. It is also the unnormalized probability distribution. We can see it is unnormalized because the sum of the numbers in the final column does not add up to 1. Instead, it equals 0.008333 (after rounding). However, the unnormalized probability distribution gives us the same relative evidence as the normalized one. In particular, $y = 2$ and $y = 3$ have the highest unnormalized probabilities with values of approximately 0.0026.

It is convenient that we know π in this situation, but in many others we will not know the value of parameter. Instead, we are interested to estimate and make inferences about a parameter's unknown values from known data. Put symbolically, we want $p(\pi|y,n)$ since the data y and n have been collected and are known, as opposed to the situation above where n and π are known for $p(y|n,\pi)$. However, $p(\pi|y,n)$ is the posterior distribution and, historically speaking, classical statistics has sought to avoid the posterior distribution because it requires the specification of a prior distribution.

Instead, Fisher (1922) advocated the idea of using the sampling distribution to assess the relative evidence for different values of the parameter. In this context, the sampling distribution is called the *likelihood function*. The symbol $\ell(\cdot)$ is used to denote the likelihood function. We write $\ell(\pi|y,n)$ to indicate that we are treating π as the unknown parameter in the current example. Larger values of the likelihood function indicate that the data provide greater evidence for a given value of the parameter. The point where the likelihood function is maximized is the parameter value that has the most evidence. This maximized value is called the *maximum likelihood estimate* (*MLE*) and it is an important quantity in classical inference.

Turning to the likelihood function, suppose we receive a new coin that may be biased. We flip it five times and get four heads. One immediate consequence of this situation is that a tabular display of sampling distribution results as in Table 2.1 is no longer relevant to our interests because we have observed $y = 4$ but we do not know π. Since y and n are known while π may have any value from 0 to 1, we can plot the likelihood function across the range of π. Figure 2.1 displays this plot. The bulk of the evidence for π places it in the vicinity of 0.4–1.0, with the maximum occurring at 0.8. We might say our best guess for π is 0.8, the MLE.

Table 2.1 and Figure 2.1 help to emphasize the distinctions between the sampling distribution and the likelihood function, even though the two functions are the same mathematically. The sampling distribution

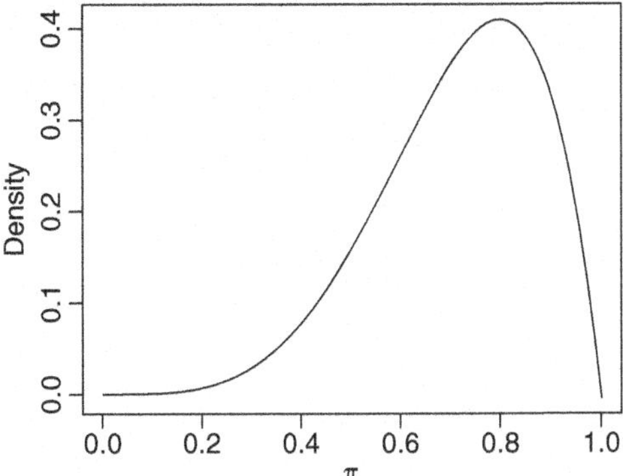

Figure 2.1 Density of Binomial Likelihood Function

for discrete data outcomes is best summarized as a table like Table 2.1, because y is considered unknown while π and n are considered known. In Figure 2.1, the likelihood function of the unknown but continuous proportion π can be better summarized as a plot.

The curve given by plotting the binomial likelihood function appears to be a probability distribution. However, it is not because the area under the curve (found by integration) does not equal to 1. We can see this in the plot by noting a few details. Recall that the standard uniform distribution (or rectangular distribution) ranges from 0 to 1 on the x-axis. Its height is 1 on the y-axis and hence there is a horizontal line at 1 extending across its range. Although the standard uniform distribution is sometimes depicted as a rectangle (because of unequal axis scaling), it is the same as the so-called unit square because it has the height, width, and area of 1. We can see that the plot in Figure 2.1 would fit well within the unit square, so it cannot be a probability distribution with area equal to 1. In general, a likelihood function will not be a probability distribution (despite appearances) because what is considered known and what is considered unknown will have changed.

Since the binomial coefficient with $y = 4$ and $n = 5$ is a constant (here $120/24 = 5$), we can plot the results from just the formula appearing at the top of column 5 in Table 2.1. This formula, which is $\pi^y(1-\pi)^{n-y}$, is called the *kernel* of the binomial likelihood function. It has this name because only the terms involving the unknown parameter π are kept, while the "husk" of the less important constant has been discarded. Figure 2.2 shows a plot of the kernel graphically. The same conclusion can be reached by the use of either Figure 2.1 or 2.2. Either way, $\pi = 0.8$ is the most promising estimate. Only the y-axis scaling has changed from one figure to the next. Since the normalizing constant adds nothing to our relative understanding, it can be omitted and inference (Bayesian and classical) can proceed using the kernel of the likelihood function.

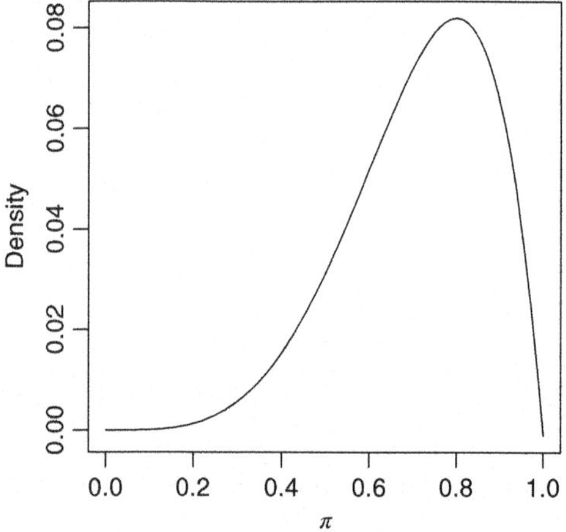

Figure 2.2 Density of Binomial Likelihood Kernel

An understanding of the properties of the data is important in selecting the likelihood function. A discrete-data likelihood function such as the binomial distribution could be very plausible for the number of coin flips, while the use of the continuous-data normal distribution is implausible for the number of coin flips. Another way to select the likelihood function is to consider what might be a good distribution for the model's errors or residuals. Assuming that we are doing a satisfactory (though not perfect) job of predicting the data, there will be deviations or errors from our predictions. A plausible distribution for these errors is an important criterion in the selection of the likelihood function.

2.3 SELECTING THE FUNCTIONAL FORM

The next decision we need to make is deciding on what structure we believe is appropriate for the relationship between the parameters and the data. Here, we call this selecting the functional form for the parameters. Using typical notation for functions, below are some examples of the functions one might select.

$$f(x) = x,$$

$$f(x) = 2 + x,$$

$$f(x) = x_1 + x_2 + x_1 x_2.$$

These functions will be populated with parameters relevant to the quantities of interest in the data we have. For example, we might make the following substitutions:

$$\pi = \pi,$$
$$\lambda = 2 + \gamma x,$$
$$\mu = \beta_1 x_1 + \beta_2 x_2 + \beta_3 x_1 x_2.$$

The last of these is a form often used in regression analysis where we wish to model the expected value μ of the dependent variable based on predictors x and coefficients β.

Selection of the functional form depends on a number of factors including what variables are available, what relationships are of interest, and what characteristics the data have. We might consider a large number of functional forms for a given dataset for exploratory reasons where little is known in advance. Alternatively, we might look at a number of functional forms to find support for specific beliefs about data relationships that we might have in advance. Selection of the functional form can be more formally called *specification of the deterministic portion of the model,* while selection of the likelihood function can be more formally called *specification of the stochastic portion of the model.* For our current example, we select the form $\pi = \pi$ because we do not have any other variables that influence π.

2.4 SELECTING THE PRIOR

The last decision we will need to make is the choice of the prior distribution. In most cases, priors are chosen so as to have little to no influence on the data. More formally, we seek a prior such that $\ell(\theta|y)p(\theta)$ is approximately proportional to $\ell(\theta|y)$. Priors that have these properties are called *vague, diffuse, weak,* or *noninformative priors.* Pause for a moment and consider the information in Figure 2.1. What kind of prior could we choose that would have little to no effect on this curve?

If we choose a prior that is equal to 1 for all possible values, then we obtain a noninformative prior because $\ell(\pi|y) = \ell(\pi|y) \times 1$. Hence, a noninformative prior in this instance is the standard uniform distribution. Figure 2.3 depicts these two densities together. While the standard uniform distribution is a noninformative prior for a proportion, we can see that it might be considered very informative about other kinds of parameters. The standard uniform distribution has a density of zero outside of its 0–1 range. Since the posterior is proportional to the prior times the likelihood, we see that the posterior distribution must also have zero density outside the 0–1 range. So for quantities such as a stock price or company profit, a standard uniform prior would be unsuitable for a noninformative

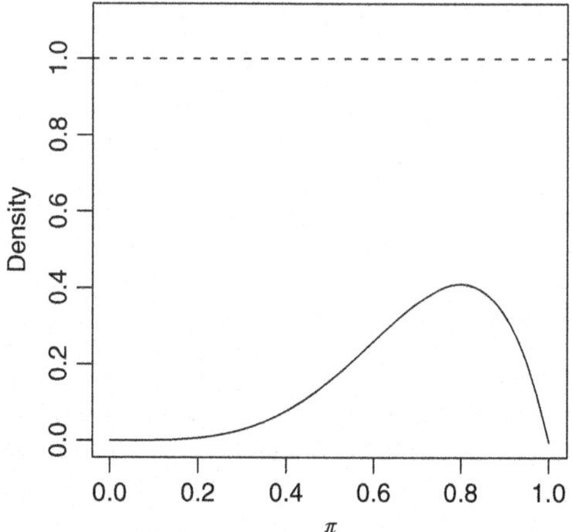

Figure 2.3 Noninformative Prior with Binomial Likelihood Function

prior because it would make values outside the range of 0–1 impossible. It is important to remember that the prior must be nonzero for any plausible values of the posterior.

For stock prices, we might consider using a uniform distribution from zero to positive infinity. Then, the posterior distribution could take on any possible positive price. The problem with this infinite-width prior is that it cannot be made to integrate to 1 and is therefore not a proper probability distribution. Such a prior is known as an *improper* prior. In some cases, improper priors can be used and a proper posterior will result; in other cases the improper prior will not produce usable results. Whenever an improper prior is considered, the propriety of the posterior must be checked to ensure the reportability of the results. This involves ensuring that the posterior distribution has a finite integral. The much simpler alternative is to always use a proper prior, i.e., a prior that is a valid probability distribution. This will ensure that the posterior distribution is proper.

2.5 FINDING THE NORMALIZING CONSTANT

We now have both numerator terms for Bayes' theorem (Equation 1.3), and for completeness we need the normalizing constant in the denominator. This is often the most difficult part of using Bayes' theorem. Fortunately, later chapters will show how we can bypass this step. This is in part due to the proportionality relationship in (1.1) and in part due to modern simulation-based computational methods, which have not been discussed yet. Modern Bayesian methods in essence get the computer to do the calculus for us, as we will see in the next chapter. It turns out that for this example we can find that the denominator is equal to $1/6$ when $y = 4$ and $n = 5$ (see chapter exercises for details).

2.6 OBTAINING THE POSTERIOR

To find the posterior distribution, we need a way of combining the prior and the likelihood. It turns out that the binomial likelihood and the uniform distribution are easily combined. This is because the uniform distribution is a member of the beta distribution family.

The beta distribution has the form

$$p(\pi|\alpha, \beta) = \frac{\Gamma(\alpha+\beta)}{\Gamma(\alpha)\Gamma(\beta)}\pi^{\alpha-1}(1-\pi)^{\beta-1}. \tag{2.2}$$

Here, $\Gamma(\cdot)$ indicates the gamma function and it generalizes the more familiar factorial function. You may recall the factorial function and formulas such as $3! = 3 \times 2 \times 1$. The gamma function extends the factorial function. The relation is $\Gamma(\alpha) = (\alpha - 1)!$ when α is a positive integer. Fortunately, the vast majority of the time we will easily be able to avoid calculating the gamma function by making sure α and β are integers. However, for the curious, it is possible to calculate it in, for example, Microsoft Excel using the formula `=EXP(GAMMALN(·))`. For example, setting $(\cdot)$ to 2 returns the value 1 in Excel. This is the same as $1! = 1$.

Some intuition into the beta distribution's parameters can be had by examining Figure 2.4. The distribution spans the range 0–1, as do proportions. As α increases, the distribution shifts to the right. This is because $\alpha - 1$ can be thought of as the number of successes in a given number of trials. Conversely, as β increases, the distribution shifts to the right. This is because $\beta - 1$ can be thought of as the number of failures in a given number of trials. When both α and β become larger, the distribution becomes concentrated around a particular point. This is analogous to having a large sample of successes and failures, enabling a fairly precise estimation of a proportion.

Returning to (2.2), the uniform prior we used was equivalent to using a beta distribution with $\alpha = \beta = 1$ as a prior. So, all terms in (2.2) equal 1 as desired. We can now assemble all of the components needed for Bayes' theorem. Keeping the kernels intact and setting constants to their numeric values (5 for the likelihood and 1 for the prior), we have

$$p(\pi|y, n, \alpha, \beta) = \frac{5\pi^{y}(1-\pi)^{n-y} \times \left(1 \times \pi^{\alpha-1}(1-\pi)^{\beta-1}\right)}{1/6}. \tag{2.3}$$

We can tidy this up by rearranging and combining terms, which gives

$$p(\pi|y, n, \alpha, \beta) = 30\pi^{\alpha+y-1}(1-\pi)^{\beta+n-y-1}. \tag{2.4}$$

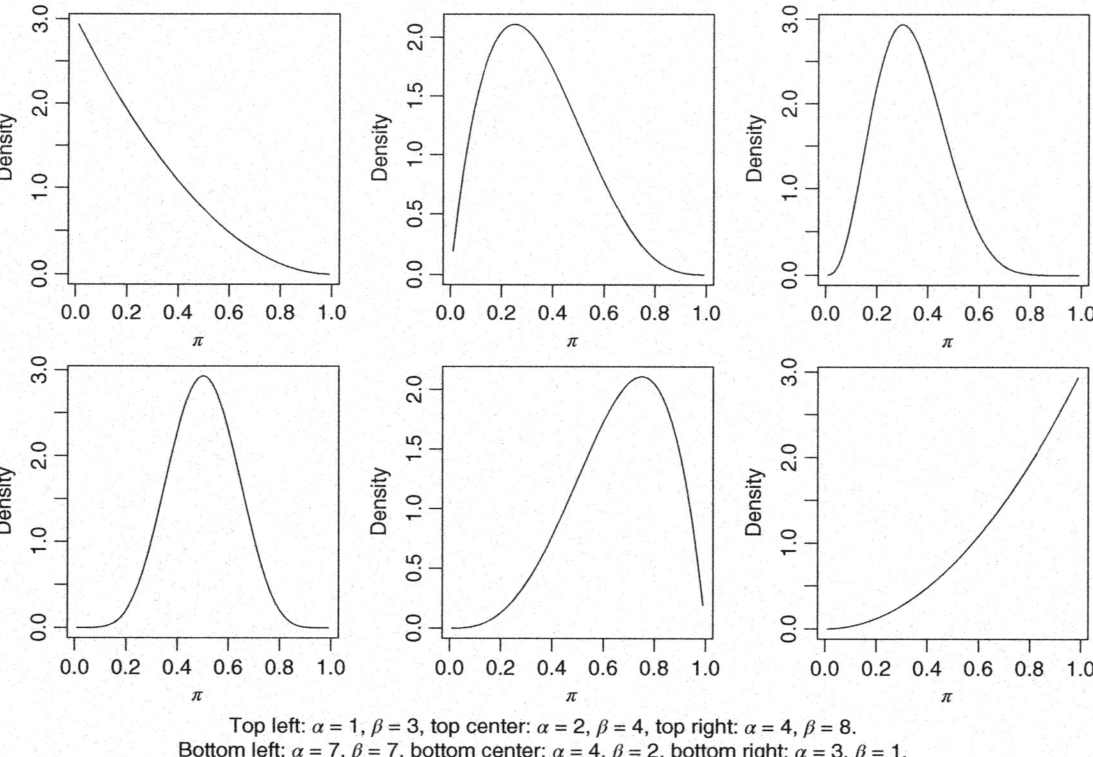

Top left: $\alpha = 1, \beta = 3$, top center: $\alpha = 2, \beta = 4$, top right: $\alpha = 4, \beta = 8$.
Bottom left: $\alpha = 7, \beta = 7$, bottom center: $\alpha = 4, \beta = 2$, bottom right: $\alpha = 3, \beta = 1$.

Figure 2.4 Examples of Beta Distributions

We see that the kernel remains intact except for changes in the exponents. Many times, this is referred to as *updating*, since the information from the likelihood is causing us to revise our prior beliefs. Here, our prior kernel started out with parameters α and β, then we collected some data, and the data caused the kernel to be updated to parameters $\alpha + y$ and $\beta + n - y$.

In addition, a little consideration reveals that the constant term 30 can also be expressed in terms of $\alpha + y$ and $\beta + n - y$. Hence, it is equivalent to the normalizing constant of a beta distribution. Putting that insight together with the beta distribution kernel means that the posterior distribution is a beta distribution. This is an important outcome. If the prior and posterior distributions are in the same family, we call the prior a *conjugate* prior. A conjugate prior can speed up Bayesian analysis considerably, since the posterior distribution will be in a known form and the posterior distribution's parameters will be simple combinations of the data and prior parameters. The general form for the conjugate posterior using the beta prior and the binomial likelihood is

$$p(\pi|y,n,\alpha,\beta) = \frac{\Gamma(\alpha+\beta+n)}{\Gamma(\alpha+y)\Gamma(\beta+n-y)}\pi^{\alpha+y-1}(1-\pi)^{\beta+n-y-1}. \tag{2.5}$$

As a practical matter this is very useful. We can now bypass all the work we undertook in this section and the previous one whenever we encounter a binomial likelihood and whenever our prior beliefs can be represented by any member of the beta distribution family. We can instead proceed directly to (2.5). Also, we are now in a position to verify that the constant "30" that appeared in (2.4) corresponds to the constant term in (2.5). Performing the substitutions shows that $6!/(4!1!) = 30$, as required.

Going forward, we can continue to collect data and update the posterior distribution as the data arrives. Suppose we were considering flipping the coin another five times. Our current posterior distribution captures all the information we have before performing these additional flips so it would be a reasonable prior for the next five flips. To obtain the posterior, we would simply add the new values of y and n to the old ones and we would obtain a valid updated posterior. This contrasts strongly with classical inference where such "peeking" at the data complicates statistical inference assuming that one wants to control Type I error over the course of the research.

To summarize, in classical inference, data that could have been obtained, but were not, are relevant for inference. In order to maintain a predefined Type I error rate and look at the data multiple times, it is necessary to make adjustments to the critical value such as Bonferroni's adjustment (Bayarri and Berger, 2004; Schulz and Grimes, 2005). In Bayesian inference, the posterior is exact at all times, so there is no need to perform these kinds of adjustments to make correct probability statements. There are many situations in business and management contexts where it is useful to be able to sequentially examine the data as it arrives. For example, media companies may wish to see how a recent movie, music album, or book release is performing, while goods-producing firms may want to track the adoption of a new product. In marketing research, advertising awareness studies tracking ad performance over time are of great interest to sponsoring firms, and the urge to review the data as it arrives is strong. There are no complications for doing so in Bayesian inference.

A final implication of conjugate analysis is that our original prior parameters α and β can be interpreted in terms of a prior "sample." Noting the forms in (2.2), we see $\alpha - 1$ corresponds to a prior y. Similarly, $\beta - 1$ corresponds to a prior $n - y$. So our prior of $\alpha = \beta = 1$ corresponds to zero "prior successes" out of zero "prior tries." We can explore other prior beliefs as well.

Having found the posterior distribution, we can plot it (Figure 2.5) but a numerical summary of the posterior distribution would be attractive. There are several ways we might summarize the posterior distribution. First, we might adopt the approach used in classical statistics with the MLE and take the value where posterior density is maximized. This is called the *maximum a posteriori* (*MAP*) estimate. This is a conceptually attractive estimator; however, it requires us to either know the expression for the mode or to numerically search for it. Here, it turns out that the posterior distribution is a beta distribution and so we can use a known expression for its mode. The formula for the beta distribution's mode

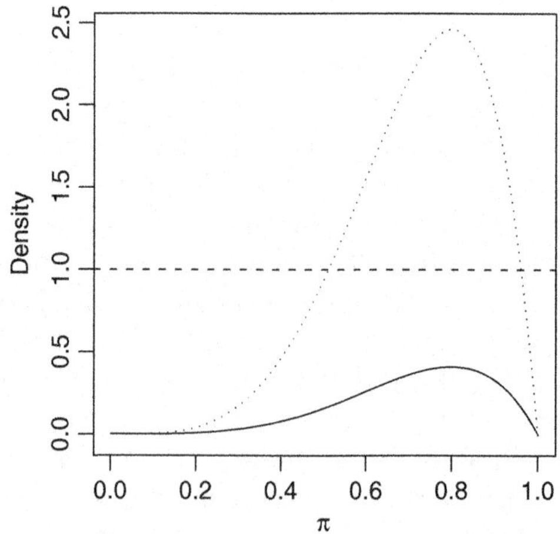

Solid line: likelihood, dashed line: prior, dotted line: posterior.

Figure 2.5 Binomial Likelihood Function and Conjugate Beta Prior and Posterior

is $(\alpha + y - 1)/(\alpha + \beta + n - 2)$, placing the mode at 0.8. This coincides with y/n.

Instead of the mode, we could use the mean as a summary statistic of the posterior. The expression for the mean is $E(\pi) = (\alpha + y)/(\alpha + \beta + n)$ or 0.7143. Another commonly reported summary statistic is the variance. The variance of the beta distribution posterior is

$$\mathrm{Var}(\pi) = \frac{(\alpha + y)(\beta + n - y)}{(\alpha + \beta + n)^2(\alpha + \beta + n + 1)}. \tag{2.6}$$

For the current example, we can report the posterior variance as 0.0255. We might also want to report an interval estimate for the parameter similar in spirit to the classical confidence interval. For a simple model such as this one, we may be able to use information about the distribution's quantile function. The quantile function for the beta distribution does not have a *closed-form solution*, i.e., it lacks a simple analytic expression such as (2.6). Instead, the solution has to be found by numerical methods. For this and other reasons, we will increasingly come to rely on the powerful computational tools found in the freely available *WinBUGS* and *R* software. For now, the common Microsoft Excel software suffices since it can be used to compute the beta distribution's quantile function. Since our posterior beta distribution has parameters $\alpha + y$ and $\beta + n - y$, inserting the formulas `=BETAINV(0.025,5,2)` and `=BETAINV(0.975,5,2)` into Excel produces the 95% posterior credible interval of 0.3588–0.9567.

2.7 COMMUNICATING FINDINGS

We have conducted a Bayesian analysis on our data, and now we would like to communicate our findings to others. A convenient way to summarize the work we did in Sections 2.2–2.4 is to present them in the following format:

$$\begin{aligned} &y \sim \text{Binomial}(n, \pi) && \text{(likelihood specification)},\\ &\pi = \pi && \text{(functional form specification)},\\ &\pi \sim \text{Beta}(\alpha, \beta) && \text{(prior specification)}. \end{aligned}$$

In the first line, note the use of the symbol $\sim$, which has the meaning "is distributed as." Thus the first line can be translated literally as "y is distributed as binomial with parameters n and π." In general, we provide the likelihood specification on this line and indicate how the outcomes y are a function of parameters.

The second line describes the functional form. The functional form specifies how data and unknown parameters are related to one another.

This portion of the specification would often be called the *model* in other statistical contexts. However, in Bayesian inference the statistical model also includes the priors and the likelihood, so we use the term *functional form* instead. In this particular example, the second line stating the functional form is not entirely necessary because it is self-evident. However, if π were a function of other parameters, then we would indicate what those parameters were and how they were related to π. For example, we might have a regression-like specification, as discussed in Section 2.3.

The final line describes the distribution we have selected for the prior on our unobserved parameter. Notice that on the left-hand side of the last line we could have written $p(\pi)$ instead of π if we wanted to accentuate the fact that we are discussing the prior *distribution* of π (and we also could have written the first line similarly). However, this added emphasis is not really necessary in our summary notation since it is implied by $\sim$ and also it is understood that we are discussing distributions when we assign priors, specify likelihoods, and obtain posterior distributions in Bayesian analysis. Throughout this book, we will drop the $p(\cdot)$ notation when we are using this summary notation.

For more complex models, we would likely need more than three lines of specification information, whereas for simple ones such as this two lines of specification would be sufficient. Of course, the information above would need to be accompanied by some discussion of the data and the reasoning by our choices for the functional form and the priors. With a little thought, we may be able to resolve questions that might arise in the reader's mind and thereby streamline our accompanying discussion. For example, restating the above specification in the following equivalent way

$$y \sim \text{Binomial}(n, \pi)$$
$$\pi \sim \text{Uniform}(0, 1)$$

is likely to eliminate some questions from the reader's mind about the prior since he or she will understand that the specific uniform distribution has been used instead of a general member of the beta distributions.

In communicating findings, it is typical for *prior sensitivity analysis* to be conducted and discussed. Prior sensitivity analysis involves the consideration of reasonable alternative priors and examination of whether their use leads to substantially different findings. For our example above, one could make the case that the uniform distribution is the most reasonable prior for expressing a lack of prior information. Another often-used candidate for a noninformative prior is the Beta$\left(\frac{1}{2}, \frac{1}{2}\right)$ distribution (Figure 2.6). This prior is called a *Jeffreys' prior* (1946) and this family of priors has the feature of being invariant under parameter transformation. Using the Jeffreys' prior for the binomial distribution,

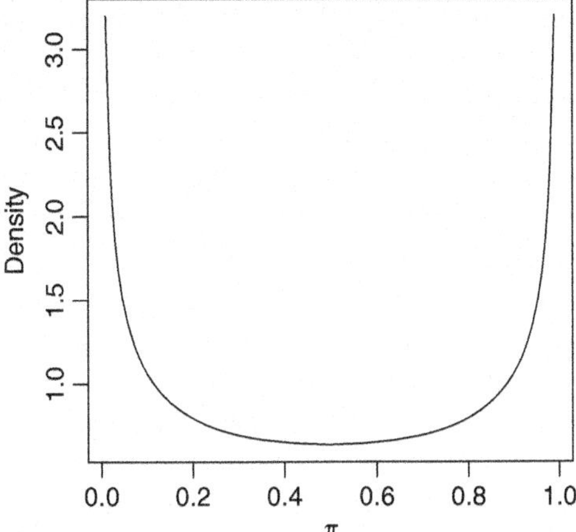

Figure 2.6 Beta $\left(\frac{1}{2}, \frac{1}{2}\right)$ Distribution

we find little difference in our results. The mean is slightly higher at 0.75, the variance is slightly higher at 0.02678, and the 95% credible interval is 0.3714–0.9775. We see that the Jeffreys' prior concentrates weight at both endpoints. The information arriving from the likelihood will counteract this and concentrate weight in a single area. We could devise other priors that have a similar property as the Jeffreys' prior. For example, a Beta(0.01, 0.01) prior density would be very concentrated at the ends and so the data would have to "work harder" to counteract this. Taking this idea to the limit, a final prior we might consider in terms of a sensitivity analysis is the Beta(0, 0) prior. This prior is an improper prior and, as mentioned before, must be used with some caution. As long as $y \geq 1$ and $n \geq 1$, the resulting posterior will be proper. If not, we will have an improper posterior and inference will be meaningless.

The results could be quite different from what we obtained above if we were to use a highly informative prior. A prior that would concentrate most of its weight toward one extreme, such as Beta(10, 100) prior, would have a major effect on the results. This is because such a prior can be considered equivalent to having a large quantity of prior data that concentrates π to be principally in the range 0.05–0.15. Results based on this prior would probably not be considered reasonable by most readers unless there was a compelling justification for such a strong prior.

Prior sensitivity analysis is always advisable, but it may be more critical in certain cases than others. When the sample size is small, the choice of prior can be influential. In our above example, we found that using two different kinds of noninformative priors did not have much impact, but if the sample size were smaller still, such as if $y = 0$ and $n = 1$, we would see more of a difference in the posterior distributions arising under the uniform and Jeffreys' priors. A multiparameter variation of the small-sample size case is the small-sample size per parameter case. As the ratio of the sample size to the number of parameters approaches 1, the ability of the data to provide meaningful information about the parameters becomes curtailed. In some cases, this problem can be reduced by respecifying the model as a hierarchical model, which will be discussed in Chapter 8.

Some parameters may be difficult to estimate well because they are correlated or have some kind of data-related overlap. For example, the Student-*t* distribution can be parameterized in terms of a location parameter μ, a scale parameter σ, and a degrees of freedom parameter ν. The degrees of freedom parameter controls the heaviness of the tails (kurtosis), assuming $\nu > 4$. However, in practice it may be difficult to statistically separate the standard deviation (σ) from the heaviness of the tails (ν). In this situation, a more informative prior on ν may be needed. For example, instead of trying a flat prior on ν over the range 4–1000, we may wish to redefine the prior so that it only allows ν to take on integer values in a smaller range. We will revisit this issue in Section 10.1.13.

Another type of variable that may be sensitive to the prior are latent variables. These variables, representing constructs that are only indirectly inferred from the data, may be substantially affected depending on prior specification, so careful thought and sensitivity analysis is typically required. We discuss latent variables in Chapter 11.

2.8 PREDICTING FUTURE OUTCOMES

Many times, we are interested in predicting some future business outcome. This could be the probability that a given consumer will buy a product, or the value of a particular equity's share price. In order to do this, we can use existing data to learn about unknown parameters. Learning about the parameter will reduce our prior uncertainty and help sharpen our predictions. After we have used the data to learn about the parameters, we will examine what kinds of data outcomes are likely given what we know about the parameters. Let $\tilde{y}$ represent future values of y that we have not observed yet. Then the formula for the posterior predictive distribution for a discrete variable is

$$p(\tilde{y}|y) = \sum_{\theta} p(\tilde{y}|\theta)p(\theta|y). \tag{2.7}$$

Here, we see the rightmost term on the right in (2.7) is the posterior distribution for θ given the existing data y. Next to this we have the distribution of future data $\tilde{y}$ that would arise if we had a particular value of θ given to us. On the left-hand side we see that the result is the distribution of the future data given the current data. Thus by summing over all the values of unknown variable θ, we "average out" θ given the current data to form the future data's distribution. What happens is that the more likely values of θ have greater influence on the future data's distribution (and the less likely values of θ have less influence). Phrased differently, we use the current data to refine our knowledge about θ and then use our knowledge of θ to gain insight into what we might expect from the future data. In the case of a continuous variable, we would use an integration in (2.7) instead of a summation.

It is worth pointing out that the right-hand side of (2.7) looks a bit like the right-hand side of Bayes' theorem. Here, our current state of knowledge (the posterior from a previous analysis) is the prior, and we want to see what the "likelihood" of various future *data* values (as opposed to parameter values) might be. For example, $p(\theta|y)$ might contain our current information about parameters for the value of homes in a particular part of the country, but this parameter information would also possess some uncertainty. Given what we do know as well as that uncertainty, we can predict what particular homes might be valued at in the near future.

Now that we have the formula for the posterior predictive distribution, how can we use it? The easiest way to use it is through computer simulation. We have not introduced the topic of computer simulations for Bayesian methods yet. Once this topic is introduced in Chapter 5, we will be better able to make use of this formula.

As you might guess, it is also possible to predict our current data when we have not yet used it to sharpen our knowledge of θ. In this situation, we have only our prior assessment of θ instead of the posterior. Inserting the prior where the posterior was gives

$$p(y) = \sum_{\theta} p(y|\theta)p(\theta). \tag{2.8}$$

Here, we average over the prior distribution instead of the posterior distribution. This distribution is known as the *prior predictive distribution*. It might seem unusual to try to predict our current data without bothering to learn about it first by obtaining the posterior. However, the prior predictive distribution has other uses. For example, it has already made an appearance under the name of normalizing constant. This is the quantity appearing in the denominator of Bayes' theorem. A third name for this quantity is the *marginal likelihood*. This name arises from the fact that we are averaging the data likelihood over the prior.

We again require computer simulation to make the most out of the prior predictive distribution. It is not as practical as the posterior predictive distribution. In Chapter 7, we will also see that the marginal likelihood can be used for model comparison. Just as with the posterior predictive distribution, we would replace the summation with an integration in (2.8) when we have a continuous variable.

2.9 SUMMARY

In this chapter, we have seen in greater detail how the likelihood function and the prior are combined to update our beliefs. We have seen an example of conjugate updating, which is a particularly convenient form of updating our beliefs as all it requires are simple changes in the exponents of the likelihood function. We have also introduced the idea of predicting future outcomes using Bayes' theorem. Predicting future outcomes involves very few new ideas because Bayes' theorem is in essence a belief revision algorithm. We can use Bayes' theorem to update our present beliefs into what we should predict about the future just as we can use Bayes' theorem to update our past beliefs to what we should believe about the present.

The most important thing we have discussed in this chapter is the broader topic of Bayesian modeling. A Bayesian model has three parts: the likelihood, the functional form, and the prior. The power of the Bayesian approach comes from the flexible way in which you can create models for your own business needs. For the time being, we will focus on simple models. In later chapters, you will see your modeling skills advance as we consider different scenarios and different kinds of data.

2.10 EXERCISES

1. What is considered the unknown quantity in the sampling distribution? In contrast, what is considered the unknown quantity in the likelihood function?

2. Small-business clients of a particular accounting firm have a 90% chance on average of showing up on time for a tax preparation appointment. Using the binomial sampling distribution, what is the probability that only two of the next three clients will show up on time?

3. Suppose that in your retail business out of 842 purchases of an item, 46 purchasers returned the item. Suppose we use a binomial likelihood with a uniform prior. What is the posterior mean of the probability of returns, π? What is the maximum *a posteriori* estimate of π?

4. Consider the retail business data in the previous exercise. Use Microsoft Excel to find the 95% posterior credible interval for π. Use the formula `=BETAINV (quantile,` $\alpha + y$, $\beta + n - y$`)` in Excel to find these values.

5. Consider the retail business data mentioned above. Use a binomial likelihood with the Jeffreys' beta prior. What is the posterior mean for π when the Jeffreys' prior is used?

6. Suppose that in a group of people, 14 are human resources professionals and 20 are not. Under a uniform prior and a binomial likelihood, what is the posterior mean for π, the probability of a group member being a human resources professional? What is the posterior variance of π?

7. Describe in your own words what the kernel of a statistical distribution refers to.

8. Referring to Section 2.7, what are the three components of a Bayesian model?

9. (If you are familiar with calculus) Considering Section 2.5, use (1.3) to show that $\int_0^1 p(y|\theta)p(\theta)d\theta$ is $1/6$ as mentioned.

2.11 NOTATION INTRODUCED IN THIS CHAPTER

Notation	Meaning	Example	Section Where Introduced
$\ell(\cdot)$	Likelihood function	$\ell(y\|\theta)$	2.2
$\Gamma(\cdot)$	Gamma function	$\Gamma(2) = 1!$	2.6
$\sim$	"Is distributed as"	$\pi \sim$ Uniform(0, 1)	2.7
$\tilde{y}$	Future unobserved data	$p(\tilde{y}\|y)$	2.8

3

COMPUTER-ASSISTED BAYESIAN COMPUTATION

3.1 GETTING STARTED

In Chapter 2, we were able to obtain all of our results using an analytic approach to Bayesian inference. The analytic approach is attractive in smaller problems. This is because we can use formulas and formula manipulations to find exact values for the parameters of interest.

The analytic approach can be contrasted with the numeric approach, where we use a computer to provide approximate answers. Why would anyone be interested in an approximate answer when exact answers are available? There are at least three reasons. First, as the number of parameters increases, the complexity of the math required to produce an exact result also tends to increase. Hence, for more advanced models that we might find useful in business, the analytic approach can become very burdensome. Second, we can improve the accuracy of our approximations by instructing the computer to run for a longer period. With ever-increasing computing power, we may be able to get a result that is exact to many significant digits in a few minutes. Hence, the result could be considered exact for practical business purposes if the approximation error is below the threshold of having a substantial impact on business undertakings. Third and finally, there is freely available software that is very useful for performing the kinds of analyses we will be considering. These software packages contain a variety of capabilities that further speeds up the

Bayesian Methods for Management and Business: Pragmatic Solutions for Real Problems, First Edition. Eugene D. Hahn.

process of performing analyses. As a result, it will be worthwhile to incorporate these tools at an early stage, so we begin doing so in this chapter.

The numerical methods we will use fall under the category of *Monte Carlo methods*. This term comes from Metropolis and Ulam (1949), who were referencing Monaco's palatial casino. The evocative name could be considered a marketing triumph for a down-to-earth idea familiar from a basic statistics course. This idea is that a carefully selected sample can be very informative about a population while saving us the labor of carrying out a completely exact census.

In the current context, we will draw samples from the terms appearing in Bayes' theorem (1.3) with an eye toward understanding the posterior distribution. This approach has two benefits. First, when we are using Bayesian methods in practice, we will almost always be able to avoid doing calculus. This is significant because the calculus in complex problems can be extremely burdensome without Monte Carlo methods. Second, we can now estimate a parameter by taking the average from our sample. This is just like estimating the average height of a sample of males and females from basic statistics, except that now our sample is made up of simulated values from the posterior distribution.

The first software package we will use is *R*. The installation files for *R* can be found at `http://www.r-project.org/`. As of this writing, accessing the URL will provide a welcome screen and recommend that you select a convenient mirror site for download. Once you have selected a mirror site, you can download a precompiled binary distribution that runs on your operating system. Select the *base* distribution and follow the accompanying download instructions. After running the installer program, you should have a working copy of *R* on your computer. The graphical user interface (GUI) version of the software is the easiest to work with for most users. This should be located in the `bin` folder of the installation and is called `rgui.exe` on Windows systems. Launch the program and you should see a console window. This window is where you enter commands for *R* to execute. The symbol > (greater than) is the prompt at which you can enter commands. Above the > symbol you should see information about the version of *R* you have installed such as its release date. As a test command, enter `license()` at the prompt. You should see informational text about the licensing conditions for *R*.

3.2 RANDOM NUMBER SEQUENCES

When undertaking sampling from a population, it is important to draw a truly random sample. Without a random sample, we may be inadvertently exaggerating some characteristics and ignoring others. This leads to biased results which are of questionable usefulness in solving problems.

Truly random numbers can be found in the real world. The drifting blobs in lava lamps, for example, have been used to generate them (Whitfield, 2004). In Monte Carlo methods, our source of randomness is a computer. Computers follow their programming exactly and so cannot be sources of true randomness (late-night disk failures notwithstanding). Hence it is essential for computer programs to be able to produce numbers that are not easily distinguishable from truly random numbers. If in a sequence of numbers it is difficult to predict a subsequent number (or set of numbers) based on past numbers, then we say the sequence is pseudo-random. Conversely, patterns and predictability in the numbers indicate that the sequence may not be suitable for numerical work because of the potential for introduction of bias. Several routines for producing random numbers have been found to be predictable such that simple plots can reveal patterned shapes that seem almost artistic as opposed to random (Ripley, 1987, ch. 3). Historically, the randomness of the routines of major statistical software packages such as SAS and SPSS has also been questioned (McCullough, 1998, 1999), so having access to pseudo-random numbers is not necessarily a trivial matter.

In addition to the consideration of predictability in the short to medium term, users of Monte Carlo methods must also be aware of the *periodicity* of the random number generating routine over the long term. At some point, the cycle of random numbers for any given routine will begin to repeat. When this happens, no new information is being obtained by further continuation of Monte Carlo estimation. Thus it is important to have an awareness of the period of the algorithm being used for random numbers. At the time of writing, the default random number generating algorithm in *R* is the Mersenne twister (Matsumoto and Nishimura, 1998). This algorithm has been found to have desirable properties in terms of producing random numbers. Moreover, its period is enormous since the Mersenne twister generates $2^{19937} - 1$ random numbers before repeating. Typing `?RNGkind` in *R* will cause a listing of the different kinds of random number generating routines available to be displayed.

A second software package that will be used later in this book is *WinBUGS*. This software has been designed specifically for Bayesian computation unlike the general-purpose *R*. As we shall see, *WinBUGS* has a number of attractive features and can dramatically speed up the practice of Bayesian inference. For example, it contains an expert system to decide how to best sample from posterior distributions. This means the user will not have to write her or his own code to do so, but rather can focus on design of models and examination of results. As for random numbers, *WinBUGS* uses a type of algorithm called a *linear congruential generator*. The period of the specific algorithm that is used is 2^{31} (Altman et al., 2004, p. 31). This number is in excess of 2 billion and so this should be sufficient for many kinds of problems. However, if the number of parameters is large and divisible by 2, the effective period may be shorter than it appears. Suppose the number of parameters is 1000. The random samples will be assigned to parameters in a fixed order over time. Since 1000 is divisible by 2, after $2^{31}/1000$ samples (i.e., after slightly over 2 million samples for the entire parameter set), the random numbers will begin to repeat. In all fairness, such a situation would be a very uncommon occurrence because there is typically no need to run the analysis this long. Moreover, one could easily add another ignorable parameter so as to make the number of parameters not divisible by 2. But being aware of technical limits (even if they are scarcely encountered) is worthwhile.

3.3 MONTE CARLO INTEGRATION

Our first foray into computer-assisted Bayesian computation begins with basic Monte Carlo integration. In Section 2.5, we were interested in finding the normalizing constant of Bayes' Rule. We did this analytically via calculus, but we can also find this value numerically. Figure 2.3 contained a plot of the unnormalized posterior density $p(y|\pi)p(\pi)$. Suppose this picture was enlarged and pasted on a wall. If random darts were to be thrown at it, the number of darts landing under the unnormalized density could be a numerator, and the total number of darts landing in the unit square could be the denominator, yielding an estimate of the area under the curve. More formally, suppose we specify a multidimensional volume V in which we draw S randomly sampled points, $\theta^{(1)}, \ldots, \theta^{(S)}$. Then we can write a formula

$$V\frac{1}{S}\sum_{i=1}^{S} f(\theta^{(s)}) \tag{3.1}$$

which can produce an approximate estimate of the area under the density.

To guide intuition, we will randomly sample a point along the x-axis where x corresponds to π. Then we will randomly sample a point along the y-axis, obtaining our dart's landing position. Finally, $f(\theta)$ will be the indicator function that takes the value 1 if the random value of x and y lies below the unnormalized posterior, and 0 otherwise. Our strategy can be called a *rejection* or an *accept–reject* approach. We will accept values that are in the region of interest by setting them to 1. Values outside the region will be rejected or set to zero here. We then can replace the integration by a sum of all the values of 1 divided by S. It also turns out we can ignore V here because $V = 1$ for the unit square. Since $V = 1$ is also the total area, we do not need to explicitly divide our average of $f(\theta)$ by V.

Let us try an example of this formula using our likelihood and prior from Chapter 2. We have already found the normalizing constant to be $1/6$, so we can check this result by using Monte Carlo. The likelihood was the binomial distribution with $y = 4$ and $n = 5$. Substituting $y = 4$ and

$n = 5$ into Equation (2.1) gives $5\pi^4(1-\pi)$. The prior was the uniform distribution, which was equal to 1 for all values of π. So the unnormalized posterior is $5\pi^4(1-\pi) \times 1 = 5\pi^4(1-\pi)$. Typing the following commands into *R* produces an estimate of the normalizing constant (the text after the # sign is a comment and does not need to be typed into *R*). *R* already has an internal definition for `pi` so we will use `x` instead of `pi` in our *R* code below.

```
set.seed(123)           # initialize random number routine
n.iter <- 1000          # set n.iter ('S') to 1000 for iterations
x <- runif(n.iter)      # store 1000 uniform random numbers in x
y <- runif(n.iter)      # same as above for y
indic <- ifelse(y < 5*x^4*(1-x),1,0)    #compute f(theta)
mean(indic)             # return the mean of indicator function
```

That's it—we just evaluated an integral by taking the mean. The value returned is 0.175. This is already somewhat close to 1/6 (0.1667) but perhaps not as close as we would like. If we did not know any better, perhaps we would accept that as the estimate. It is always worthwhile doing some checking with Monte Carlo methods, as there are any number of reasons why an estimate may be "too approximate" (as is the case here) or have other more serious limitations.

A first thing to examine is the running mean of our results. The running mean at iteration *i* is defined as the mean of the values sampled up to and including the value at iteration *i*. We would like to see the running mean stabilizing toward one value. If this were not occurring, it would indicate that we are not sharpening our estimate as the number of Monte Carlo iterations increases. The following code creates the graph on the left side of Figure 3.1.

```
iter <- seq(1:n.iter)   # create vector 1, 2, ..., 1000
running <- iter         # set running = iter
for (i in 1:n.iter) {   # for loop, calculates running mean
  running[i] <- mean(indic[1:i])}
plot(iter,running)      # plot iter on x-axis, running on y-axis
```

As shown on the left of Figure 3.1, we can see that the estimate is converging toward its true value, but the graphics in the plot obscure the behavior somewhat. Replacing the last two commands with

```
plot(iter,running,pch=NA)   # make plot with no data in it
lines(running)              # add running in line format to plot
```

changes the circles appearing in the left plot to lines. We see in the plot on the right in Figure 3.1 that the lines are easier to follow.

Changing `n.iter` to 10,000 and rerunning the code changes our estimate of the normalizing constant to 0.1705. Changing `n.iter` to

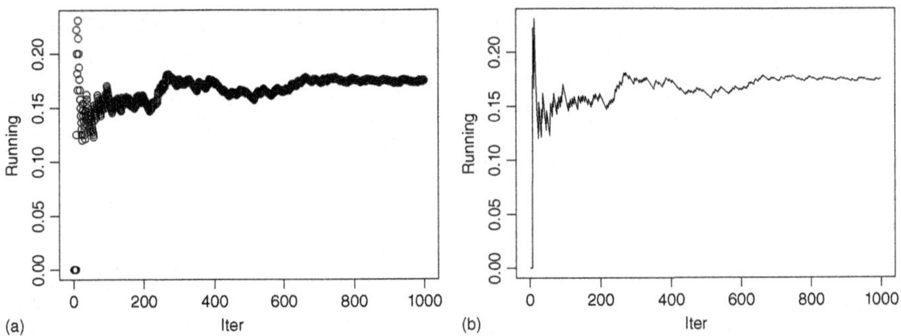

Figure 3.1 (a,b) Running Average of Normalizing Constant Estimate

100,000 and rerunning the code gives an estimate of 0.16692. The estimate is clearly improving. You can continue adding zeros to `n.iter`, and you will see that it gradually stabilizes closer and closer to its true value. You can also recreate the figure with this greater numerical precision but be forewarned. *R* may slow down quite a bit when using `for` loops such as we have done here when `n.iter` is large. *R* is much faster when it is given a large column of numbers than it is when using loops. For example, our `ifelse` statement performs an `if` comparison on the columns of `y` and `5*x^4*(1-x)`, and then places the result (1 if the condition is true, 0 if the condition is false) in a new column `indic`. It is better for beginners to use a slower but familiar method for accuracy's sake. Space prevents us from getting deep into the subtleties of *R* in this book, but Venables and Ripley (2002) has more on the art and science of working with *R*.

Another way to do Monte Carlo integration in one dimension is as follows: to approximately integrate from location a to location b with respect to a function $f(\theta)$, we calculate

$$(b-a)\frac{1}{S}\sum_{i=1}^{S} f(\theta^{(s)}). \tag{3.2}$$

This follows from the mean value theorem of calculus. In this formulation, we omit the y variable and just average over the sampled values of $f(\theta)$. Code to perform this might look like this:

```
set.seed(123)
n.iter <- 100
x <- runif(n.iter)
f.x <- 5*x^4*(1-x)      #calculate function of interest
mean(f.x)
```

The result is 0.1672 and so this method gives two digits of accuracy with `n.iter` of only 100. With `n.iter` of 1000, the estimate is 0.1658. At 10,000 iterations, it is 0.16645 and we have three digits of accuracy. The previous approach required 100,000 iterations to achieve three digits of accuracy. Intuitively, we can see that code based on (3.2) is getting a certain level of accuracy faster than code based on (3.1). This makes sense when we note that the approach of (3.1) involves uncertainty in `x` and `y` and many simulations are rejected. In (3.2), we use all simulations to produce the average of the exact function of interest.

We see there are a number of trade-offs we have to consider when using Monte Carlo methods. Some methods may be more intuitive but require more computer time. Others may be less intuitive but simpler to program, reducing time spent on debugging efforts. Yet other methods may be more complex but may provide gains in terms of decreased computer time and lower usage of computer resources, such as memory. A final point is that to gain an additional digit of accuracy (10 times more accurate), we are very likely to find that we must run the sampler more than 10 times longer. For example, the approach in (3.2) increases in accuracy with the square root of S, so it is necessary to increase `n.iter` by a factor of 100 to gain another digit of accuracy (excluding Monte Carlo error). Lastly, it is always worthwhile to check our results for problems. Graphical tools and comparison with known benchmarks, as illustrated here, are simple but useful ways for laying a solid groundwork in advance of moving on to new data or more complex models.

3.4 MONTE CARLO SIMULATION FOR INFERENCE

As we have seen in the previous section, Monte Carlo integration is a useful tool for performing the integration tasks needed for inference. Monte Carlo simulation from distributions is another useful tool, as it greatly expands the kinds of questions we can answer. This is because Monte Carlo simulation from distributions allows us to easily draw inferences based on functions of distributions. In particular, the expectation function is an important function because we are almost always interested in the expected value of parameters. With Monte Carlo methods providing a large number of *independent* samples from posterior distributions, the law of large numbers assures us that, as the sample size of the Monte Carlo run goes to infinity, the expected value of the Monte Carlo sample converges to its true value. In terms of a formula, we have

$$E\left(f(y)\right) \approx \frac{1}{S}\sum_{s=1}^{S} f(y^{(s)}) \tag{3.3}$$

And this approximation improves as the length of the Monte Carlo run, S, increases. Intuitively, we can see this happening in Figure 3.1 for example.

Since we specify S, we are free to make our results as numerically precise as we wish. Moreover, virtually any kind of transformation or function that is tailored to be relevant for our interests can be obtained with a relative minimum of human effort. Once we have the function or distribution of interest, we conduct inference using the expectation function as usual. We might ordinarily think of the expected value as being relevant to the mean, but in our Monte Carlo sample we can also find the expected value of the variance parameter, a proportion parameter (as in the next section), or any other parameter. The law of large numbers ensures that these expected values approach their true values with large sample sizes.

3.4.1 Testing for a Difference in Proportions

Problem: A local business decides to allocate shelf space to a new product category. For future marketing purposes, they would like to understand whether men or women are more likely to purchase items from this category. Over the first day, 14 out of 37 female shoppers purchase the item, whereas 3 out of 22 male shoppers purchase the item.

Solution: We confirm with the manager that a uniform prior represents her beliefs for each gender's purchase propensity. It seems reasonable to assume that men's and women's behavior are independent and exchangeable. The binomial distribution is a natural choice for this data. Data for women is denoted y_w and for men y_m. Then

$$\begin{aligned}
y_w &\sim \text{Binomial}(n_w, \pi_w) && \text{(likelihood for women)},\\
y_m &\sim \text{Binomial}(n_m, \pi_m) && \text{(likelihood for men)},\\
\pi_w &\sim \text{Uniform}(0,1) && \text{(prior for women's propensity)},\\
\pi_m &\sim \text{Uniform}(0,1) && \text{(prior for men's propensity)}.
\end{aligned}$$

Since women's and men's data and parameters are independent above, we can compute the posterior distribution for each independently. The prior distributions are in the beta family, so we can use conjugate updating to obtain the posterior distributions. From Equation (2.5), the posterior beta distribution for women is Beta(15, 24). For men the posterior is Beta(4, 20). To obtain our numeric results, the following commands are entered in *R*.

```
set.seed(123)
n.iter <- 10000
pi.w <- rbeta(n.iter,15,24)
pi.m <- rbeta(n.iter,4,20)
diff <- pi.w-pi.m                         # function of interest

mean(pi.w)                                # summary statistics
```

```
sd(pi.w)                                  # for pi's
mean(pi.m)
sd(pi.m)

hist(diff,50)                             # summary statistics
mean(diff)                                # for diff
sd(diff)
quantile(diff, probs=c(0.025, .975))  #95% credible interval
quantile(diff, probs=c(0.05,   .95))  #90% credible interval
```

Our main interest centers on the difference in purchase propensity, `diff`. Taking the posterior mean of `diff` as our estimate, the difference in the proportions is estimated as 21.8%. The 95% credible interval just barely includes the value of zero, while the 90% interval excludes it (3.61%, 39.1%). Women appear more likely to purchase the item at the 90% level.

As a sensitivity analysis, the results are reconsidered under the Jeffreys' prior. The posterior beta distribution for women is Beta(14.5, 23.5), and for men it is Beta(3.5, 19.5). Hence the above code for `pi.f` and `pi.m` is slightly modified so as to use the new lower values for α and β. Given the modest sample size, the change in the prior does have a modest effect on the results. The difference in the proportions is 22.9%. Also, the 95% credible interval now excludes 0 (1.16%, 42.8%). The more conservative solution from the uniform prior is used, although the client is briefed that the evidence is on the borderline and under slightly different assumptions, a 95% credible interval would exclude zero.

Comment: Contrast the approach here with a classical approach to testing differences in proportions. The classical test for differences in proportions is usually performed using a normal approximation based on the central limit theorem. If sample sizes are large and if π tends toward 0.5, the normal approximation works well. In some situations, however, the normal approximation is less accurate. For example, Figure 3.2 shows the posterior distribution of `diff` under uniform priors. We see that the posterior distribution of `diff` is negatively skewed since the tail on the left is longer than that on the right. This indicates that employing the normal approximation would indeed give us only an approximation of the 95% confidence interval in this data. By contrast, the Bayesian approach used here is exact at all sample sizes. Alternatively, a classical test could be devised using the exact distribution of the difference of two proportions (Pham-Gia and Turkkan, 1993), but this distribution is very complicated.

3.4.2 Predicting Customer Behavior

Problem: With the previously obtained information, the local business turns to the more crucial problem of predicting consumer behavior. The decision to allocate shelf space would be better informed by having an understanding of the forecasted purchase quantities. The values of

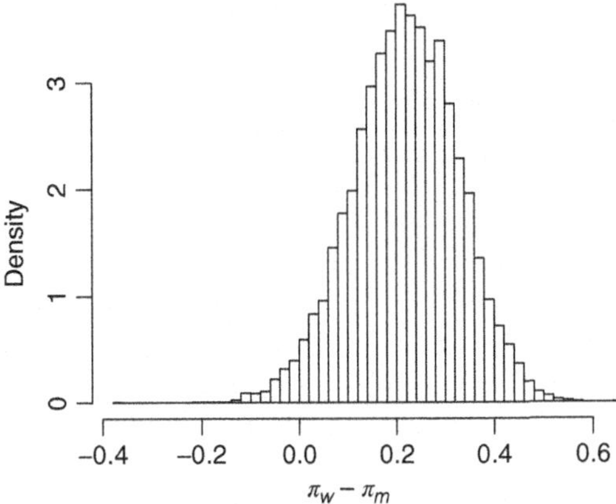

Figure 3.2 Posterior Distribution of $\pi_w - \pi_m$ under Uniform Priors

π by contrast are considered slightly abstracted away from immediate business concerns. Assuming that 500 female customers would enter the store in the next month, what can we learn about the forecasted sales among women for this time period?

Solution: We use the posterior predictive distribution from (2.7). We insert into (2.7) the binomial likelihood with $n = 500$ and $\pi = \pi_w$. The posterior for π_w is also needed. To accomplish this in *R*, first we need to make sure that the results under the Jeffreys' prior are not active by rerunning the original code from Section 3.4.1 if needed. The following code is then entered.

```
sales.w <- rbinom(n.iter,500,pi.w)   #500 female customers
hist(sales.w,100)
mean(sales.w)
sd(sales.w)
quantile(sales.w, probs=c(0.025, .975))
```

The command `rbinom` instructs *R* to randomly draw `n.iter` samples from a binomial distribution where n is equal to 500 and π is equal to `pi.w`. These samples are then stored in a variable `sales.w`. We then ask *R* to provide summary statistics for `sales.w`.

The predicted number of purchases by women among 500 female shoppers is 192.2. There is a fair amount of uncertainty in this estimate, as the posterior predictive's standard deviation is 40.0. The 95% credible interval for the number of sales is (117, 272). Since we have the predictive distribution of sales, it is a straightforward matter to find the predictive distribution of sales revenues. We have already seen that we may obtain distributions of functions of interest by applying functions to distributions. So, for sales revenue this would be obtained in *R* by a line of code such as `sales.rev.w <- revenue.per.item * sales.w`. We can then summarize `sales.rev.w` by calculating its summary statistics as we have done previously.

3.4.3 Predicting Customer Behavior, Part 2

Problem: The local business is intrigued by the findings but a new question is raised. Instead of using 500 shoppers as a known quantity, they would like to incorporate shopper traffic forecasts for the next month into the analysis. What would the forecasted sales be if next month's traffic forecasts were used instead of exactly 500 shoppers?

Solution: There is little conceptually new here, as Bayes' rule allows for combining information both certain and uncertain in extremely flexible ways. To summarize, we need to find the joint posterior distribution where uncertainty about n is accounted for. However, it would be better to use more advanced tools for more advanced problems. We politely tell the local business that we will look into the matter for them. The tools for this scenario will be discussed in Section 5.5.2.

3.5 THE CONJUGATE NORMAL MODEL

Perhaps the most well-known and most used distribution in business is the normal distribution. The normal distribution is often applied when *continuous* data such as profit/loss or time-to-completion are examined. The normal distribution is especially applicable when we think that many factors may have combined additively to produce the outcome. Profit/loss, for example, is the overall result of accumulations from sales and reductions from purchases and expenses. The central limit theorem indicates that sums of random quantities tend toward the normal distribution with sufficient sample size, so the normal distribution may be plausible. Since the mean of the data is just the sum divided by the sample size, the central limit theorem also applies to means. Another reason for the normal distribution's popularity is its ease of use. This property among other things makes Bayesian inference using the normal distribution straightforward.

3.5.1 The Conjugate Normal Model: Mean with Variance Known

We first consider the situation where the variance σ^2 is known. Although this situation is somewhat uncommon, it is an essential starting point for subsequent work. It can simplify matters to use the *precision* τ where $\tau = 1/\sigma^2$ (see Section 3.11). We can consider either the precision to be known or the variance to be known—either assumption will give us the same result. We can work out the posterior distribution analytically using some algebra and basic probability concepts. A detailed discussion of this process appears in Section 3.11. Here we only present the key outcomes from the process of Section 3.11.

We will need to supply our prior estimate of the mean, μ_0. We will also need to have a prior estimate of the variance, σ_0^2. We can then obtain a prior estimate of the precision τ_0 by taking the reciprocal of σ_0^2. Then we will need the mean of our dataset, $\overline{y}$, and the sample size of our dataset, n. Lastly, we will need the known variance of our dataset, σ^2, which gives us the known precision of the dataset, τ. In Section 3.11, we find the posterior kernel. From it we can conclude that the posterior mean and posterior variance of μ are

$$E(\mu|y) = \frac{\overline{y}n\tau + \mu_0\tau_0}{n\tau + \tau_0}, \tag{3.4}$$

$$\mathrm{Var}(\mu|y) = (n\tau + \tau_0)^{-1}. \tag{3.5}$$

Here, $E(\cdot)$ indicates the posterior mean of μ. We see that the posterior mean of μ is a weighted average of the prior mean μ_0 and the data mean $\overline{y}$. As the sample size increases, the posterior mean gets closer and closer to the data mean $\overline{y}$. Also, if the prior precision is made very small (i.e., the prior variance is very large), the posterior mean of μ will also become increasingly closer to $\overline{y}$.

We can also write the posterior precision of μ as the reciprocal of the posterior variance of μ. So the posterior precision of μ is $\mathrm{Prec}(\mu|y) = n\tau + \tau_0$. We see that the posterior precision of μ is a weighted sum. As either n gets larger or τ_0 gets smaller, the posterior precision of μ gets closer to $n\tau$. Similarly, as either n gets larger or τ_0 gets smaller, the posterior variance of μ approaches σ^2/n. This resembles the quantity from classical statistics called the *standard error of the mean*.

To finish up our work here, we take our results and put them in a compact form for communication to others. Suppose we use $N(\cdot,\cdot)$ as an abbreviation for "the normal distribution with a given mean and a given variance." Then we can write the posterior distribution in the following compact form:

$$\mu|y \sim N\left(\mathrm{Mean} = \frac{\overline{y}n\tau + \mu_0\tau_0}{n\tau + \tau_0}, \mathrm{Variance} = (n\tau + \tau_0)^{-1}\right). \tag{3.6}$$

In words, the posterior distribution of μ is normally distributed with the mean and variance indicated.

We can also use the more convenient precision to summarize our findings. So we could also write (3.6) as follows:

$$\mu|y \sim N\left(\mathrm{Mean} = \frac{\overline{y}n\tau + \mu_0\tau_0}{n\tau + \tau_0}, \mathrm{Precision} = n\tau + \tau_0\right).$$

Yet another possibility would be to summarize the posterior using the standard deviation instead of the variance or the precision. Since you may encounter the posterior distribution being expressed in different parameterizations, it is important to check what parameterization is being used.

Also, notice that the conditioning arguments after the | on the left-hand side of (3.6), μ_0 and τ_0, have been omitted as opposed to results such as (2.5) where the full list has been retained. This is because, as the number of parameters grows, providing a full list of all of the conditioning arguments can become cumbersome. Throughout this book, it should be clear from the formula which parameters are conditioned for even if they do not appear on the left-hand side. This is because, if a parameter appears on the right-hand side of a formula, then the left-hand side must depend on that parameter. Retaining y as a conditioning argument will always indicate that we have a posterior distribution where the parameter has been updated by the data.

3.5.2 The Conjugate Normal Model: Variance with Mean Known

In this section, we will examine the posterior distribution of the variance when the mean is known. It is very rare that we would somehow know the mean with certainty but not the variance. However, understanding this situation also provides an important building block for the future.

We have seen that it is easier to work with the precision instead of the variance, so in this section we will use the precision. Also, we will need to know about a distribution called the *gamma distribution*. The gamma distribution is related to the familiar chi-squared distribution. You may recall that the chi-squared distribution is used to model the sums of squares of normal variables. The chi-squared distribution has one parameter which is called the *degrees of freedom parameter*. The gamma distribution has two parameters and can be thought of as a scalable chi-squared distribution. A certain way of writing the gamma distribution gives us the following density function:

$$p(y|\alpha, \beta) = \frac{\beta^\alpha}{\Gamma(\alpha)} y^{\alpha-1} \exp(-y\beta). \tag{3.7}$$

The gamma distribution takes on various shapes depending on the values of α and β, as seen in Figure 3.3. At the top center of Figure 3.3 we see the Gamma(1/2, 1/2) distribution, which is identical to the chi-squared distribution with one degree of freedom. At the top right left of the figure, we see that the gamma distribution is mostly flat with an almost vertical spike near zero when α and β are both set to small values.

The mean of the gamma distribution is α/β, and the variance of the gamma distribution is α/β^2. Setting $\alpha = \nu_0/2$ and $\beta = 1/2$ gives the chi-squared distribution. We will use the closely related substitution $\beta = \nu_0 s_0^2/2$. Here we can think of ν_0 and s_0^2 as expressing information about

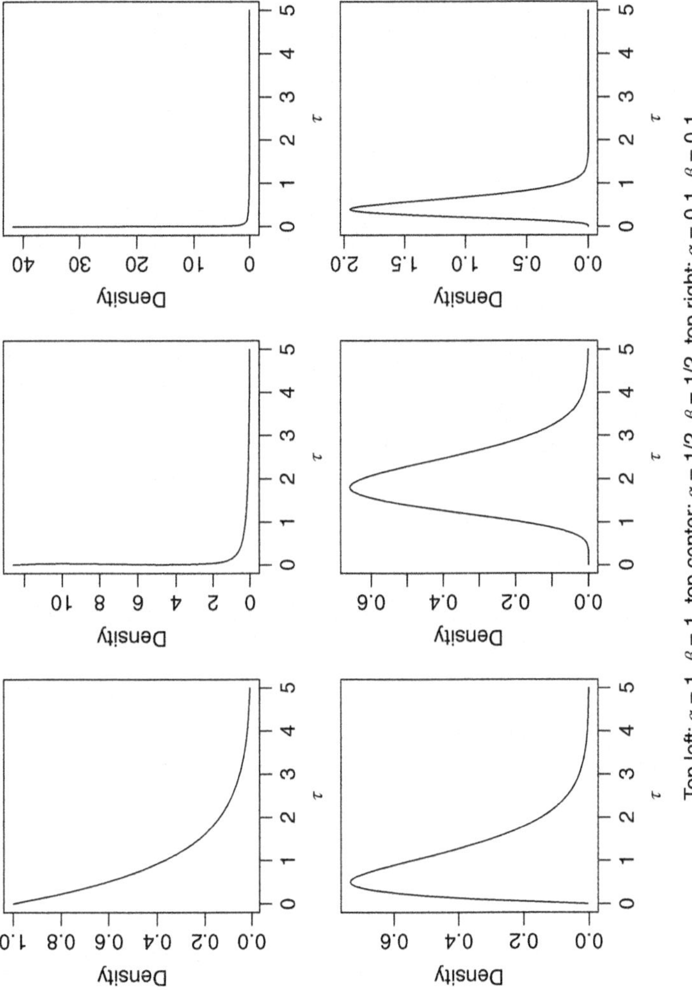

Top left: $\alpha = 1$, $\beta = 1$, top center: $\alpha = 1/2$, $\beta = 1/2$, top right: $\alpha = 0.1$, $\beta = 0.1$.
Bottom left: $\alpha = 2$, $\beta = 2$, bottom center : $\alpha = 10$, $\beta = 5$, bottom right: $\alpha = 5$, $\beta = 10$.

Figure 3.3 Examples of Gamma Distributions

our prior degrees of freedom and variance, respectively. We will also need to have the sample variance of our data, s^2, and the sample degrees of freedom, $v = n - 1$.

In Section 3.11.2 we analytically find the posterior for the precision, τ. As before, we will summarize our result for the posterior distribution of τ in a compact form. This gives

$$\tau|y \sim \text{Gamma}\left((v_0 + n)/2, \frac{1}{2}[v_0 s_0^2 + vs^2 + n(\mu - \overline{y})^2]\right). \tag{3.8}$$

In written language, this means the posterior distribution of τ has a gamma distribution. The first parameter of the gamma distribution is $\alpha = (v_0 + n)/2$, and the second parameter is $\beta = \frac{1}{2}[v_0 s_0^2 + vs^2 + n(\mu - \overline{y})^2]$. We can now use our knowledge of the gamma distribution to shed light on τ. For example, since the mean of the gamma distribution is α/β, we can now find the posterior mean of τ by calculating α/β using the values in (3.8).

It is also possible to specify a distribution for the variance directly. If the *inverse gamma* distribution is used as a prior for the variance, then we can obtain an inverse gamma posterior distribution for the variance. Other names for this prior include the scaled inverse chi-squared distribution (Gelman et al., 2003). However, in this book we will avoid the use of the inverse distributions, in part to concentrate on a fewer number of workhorse distributions and in part because the inverse distributions have parameter restrictions which have to be observed in order for key quantities such as the mean and variance to exist.

3.5.3 The Conjugate Normal Model with Mean and Variance Both Unknown

The more usual case is that both the mean and the variance are unknown. In this situation, we can again use the likelihood in (3.20). Suppose we have little prior information about μ and τ. We use the following flat priors:

$$p(\mu) \propto 1/c, \quad -\infty < \mu < \infty \tag{3.9}$$

$$p(\tau) \propto 1/\tau, \quad 0 < \tau < \infty. \tag{3.10}$$

The prior for μ indicates that it is proportional to some constant and is therefore flat from negative to positive infinity. This is similar to the uniform prior we saw for the Beta distribution; however, there is an important difference. The Beta(1, 1) prior was uniform over the range (0, 1) and was a valid probability distribution. The prior for μ has an infinite range and so there is no way to make this prior integrate to 1 as would be needed for a proper probability distribution. Hence, this is an improper prior. It also happens to be the Jeffreys' prior for the normal mean. The prior for τ is also flat in the sense that it corresponds to a uniform prior on $\log \tau$. It is also an improper prior. Despite this, we can obtain a proper posterior for our parameters.

After finding the posterior distribution (Section 3.11.3), we obtain the following results:

$$\mu|y, \tau \sim N(\overline{y}, \text{Variance} = (n\tau)^{-1}). \tag{3.11}$$

This says that, conditional on τ, μ has a normal distribution with mean $\overline{y}$ and the indicated variance. We can also write the conditional distribution of μ in terms of the more familiar σ^2. This gives

$$\mu|y, \sigma^2 \sim N(\overline{y}, \text{Variance} = \sigma^2/n). \tag{3.12}$$

We might also want to know the marginal distribution of μ. It can be shown that the marginal distribution of μ has a location-scale t-distribution. This kind of t-distribution allows the mean to be any value and the standard deviation to be any positive value. The location-scale t-distribution is slightly different from the central t-distribution found in many textbooks. The central t-distribution relies on a data standardization. Because the data has been standardized for the central t-distribution, the central t-distribution can be easily summarized in a textbook table using only the degrees of freedom parameter. However, we cannot assume the data has been standardized here. So, instead, we have to use the location-scale t-distribution. With this distribution we may write

$$\mu|y \sim t(\overline{y}, \sigma^2/n, \nu). \tag{3.13}$$

In words, the posterior distribution of μ has a location-scale t-distribution with mean μ, variance σ^2/n, and degrees of freedom ν. While the full formula for this distribution is somewhat complex, with Monte Carlo it is easy to simulate from this distribution. First, we simulate a large sample of values from a central t-distribution, then we multiply them by $\sigma/\sqrt{n}$, and finally add $\overline{y}$ to the result.

The marginal posterior distribution of τ has a gamma form. In particular, we have

$$p(\tau|y) \sim \text{Gamma}\left(\nu/2, \nu s^2/2\right). \tag{3.14}$$

Remembering our results from the gamma distribution, we see that the posterior mean of this distribution is $1/s^2$.

3.6 IN PRACTICE: INFERENCE FOR THE CONJUGATE NORMAL MODEL

We have obtained expressions for the posterior distributions of the mean and the precision in terms of conjugate normal models. Here we will examine the use of these posterior distributions to learn about data.

Lenz and Engledow (1986) studied firms' responses to volatile organizational environments such as changing market conditions, increased competition, and new technologies. They performed a field study on the environmental analysis units of leading-edge firms. Sales data (in billions) reported for the firms appears in Table 3.1. We will examine this data using the results of Section 3.5.

3.6.1 Conjugate Normal Mean with Variance Known

In order to use the posterior distribution given in (3.6), we will need to assess our priors for μ_0 and τ_0. Ideally, this would occur before reviewing the data of Table 3.1. Since the data is available though, it might occur to you to use the sample mean and the sample precision for priors. This is not consistent with fully Bayesian practice for two reasons. First, the prior conceptually represents your prior beliefs before seeing the data, not the exact values from the data. Second and perhaps more importantly, using specific quantities from the sample data in both the prior and the likelihood in essence double-weights the data as opposed to the scenario where the prior is assessed independently from the data.

While using the data to inform the prior is discouraged in fully Bayesian practice, there is a school of thought called *empirical Bayesian methods* where prior values are routinely based on the data. The empirical Bayesian school points to the ease of prior specification and a decrease in subjectivity of the priors as motivations for this approach. However in this book we will use a fully Bayesian approach.

If we have little prior knowledge, then we would say our prior precision is very low. Suppose we entertain a prior standard deviation of 100 (in billions) for our data. This implies a prior variance of 10,000 (in billions) and therefore $\tau_0 = 0.0001$. As for μ_0, it is common to set priors for means equal to zero when little is known in advance. While this prior value could be utilized here, we will suppose that our prior mean for leading-edge companies is larger than zero because leading-edge companies rarely have zero dollars in gross sales on average. We select \$1 (billion) for μ_0 as a modest but more plausible value.

Calculations show $\overline{y} = 15.8$ and the sample variance $s^2 = 405.47$ for the data in Table 3.1. Hence $\tau = 0.02466$. Substituting into (3.4), we find the

TABLE 3.1 Gross Sales Data

Company	Gross Sales	Company	Gross Sales
Pipeco	\$4.5B	Consco	\$2.1B
Retco	\$30.0B	Telco	\$65.8B
Foodco	\$7.8B	Diverco	\$26.5B
Servco	\$7.5B	Enerco	\$5.2B
Chemco	\$6.3B	Appco	\$2.3B

posterior mean of μ is $E(\mu) = 15.74$. The posterior mean is slightly lower than $\bar{y}$ because of our conservative prior mean, $\mu_0 = 1$. The posterior variance of μ is $\sigma_\mu^2 = 40.83$. Standard normal theory results ($E(\mu) \pm 1.96\sigma_\mu$) can be used to provide a 95% posterior credible interval of ($3.28B, $28.20B). We can therefore conclude that there is a 95% chance that the true mean lies in the interval $3.28B to $28.20B.

3.6.2 Conjugate Normal Variance with Mean Known

To perform inference for the variance assuming a known mean, we will need to assess a value for the prior degrees of freedom, ν_0, and the prior variance, s_0^2. Before we do this, we must bear in mind that the gamma distribution introduced in (3.7) is defined only when α and β are greater than 0. So ν_0 and s_0^2 must be greater than 0 for our prior to be proper. It may be worthwhile to think in terms of a prior sample size and use this to guide our intuition about ν_0. If we assume a prior sample size of $n_0 = 2$, then a prior value for $\nu_0 = n_0 - 1$ could be 1. Alternatively, we could take a prior sample size of a value very close to, but just larger than, 1. This will make ν_0 close to zero and yield results that are similar to the flat Jeffreys' prior for $\log \tau$ if we take s_0^2 to be similarly small. We will take $\nu_0 = 1$ here, and also $s_0^2 = 1$. For our known mean, here we suppose that μ is known to be 20.

The posterior distribution is given in (3.8). The first parameter of the distribution, $\alpha = (\nu_0 + n)/2$, is 5.5. For the second parameter, term-by-term addition gives $\beta = (1 + 3649.26 + 176.4)/2 = 1919.33$. To summarize the resulting posterior, there are two things we need to know about the gamma distribution. First, we recall that the mean of the gamma distribution is α/β. Focusing on the mean, we can now obtain the expected precision as $E(\tau) = 0.0028746$. We can take the reciprocal of this value as an estimate of the posterior variance. Doing so gives 347.88 as the estimate for σ^2.

The second thing we need to know about the gamma distribution is that forming a 95% credible interval will be somewhat more difficult since the gamma distribution is highly skewed and has no simple expression for its quantiles. However, with Monte Carlo simulation and the values of α and β, we can obtain these quantities and more with relative ease. The following code in *R* will give us the information we want.

```
set.seed(123)
  n.iter <- 100000
  alpha  <- 5.5
  beta   <- 1919.33

tau <- rgamma(n.iter, alpha, beta)
mean(tau)
sd(tau)
```

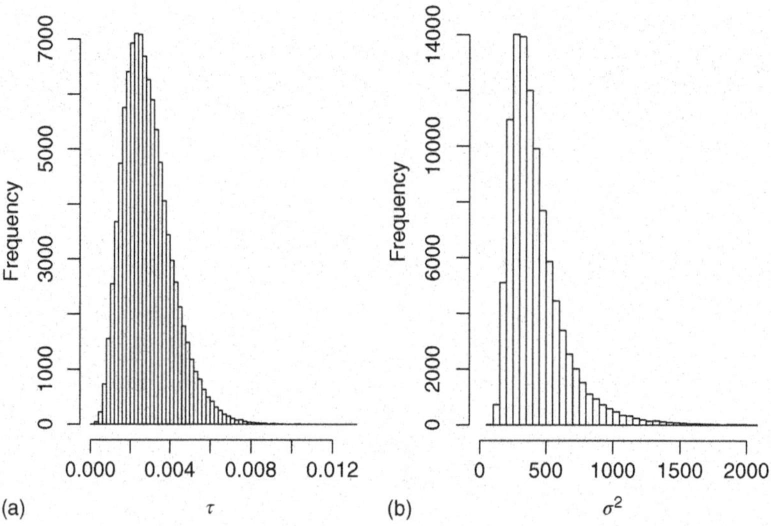

Figure 3.4 Posterior Distributions for τ (a) and σ^2 (b)

```
quantile(tau, probs=c(.025, .975))
hist(tau,100)
```

The command `set.seed` makes the random number generator in *R* start at a fixed point. The command `rgamma` instructs *R* to randomly draw `n.iter` samples from a gamma distribution with the given values of `alpha` and `beta`. These samples are then stored in a variable `tau`. We then ask *R* to provide summary statistics for `tau`.

R gives the posterior mean of τ as 0.0028688, which is fairly close to what we obtained analytically. The mean of the reciprocal of τ, given by `mean(1/tau)`, is 425.8. This value is our Monte Carlo-based posterior estimate of σ^2. To obtain the 95% credible interval for σ^2, we make the appropriate substitution and enter

```
quantile(1/tau, probs=c(.025, .975))
```

which gives (175.07, 1004.09) as the result. The skewness of the posterior distributions for both τ and σ^2 can be seen in Figure 3.4. If, instead, we preferred to summarize the posterior distribution for σ, it would be a simple matter of making the substitution `sqrt(1/tau)` for `tau` and calculating the relevant quantities.

3.6.3 Conjugate Normal Mean and Variance Both Unknown

When the mean and variance are both unknown and prior information is very minimal, a direct approach to obtaining the posterior distribution for μ is to use (Equation 3.13). The sample mean and variance remain at $\bar{y} = 15.8$ and $s^2 = 405.47$, respectively. Simulating from the central t-distribution in *R* is done using the `rt()` command. Once we have the central t-values, we make the location-scale transformation by multiplying them with the standard error and adding $\bar{y}$ to the result. The following code illustrates this:

```
set.seed(123)
  n.iter <- 100000
  y.bar  <- 15.8
  std.error <- sqrt(405.47/10)
  nu <- 10-1

central.t <- rt(n.iter, nu)
mu <- central.t*std.error + y.bar
mean(mu)
quantile(mu, probs=c(.025, .975))
```

Here, the *R* command `rt` generates random samples from the central t-distribution. We multiply these random samples by `std.error` and then add `y.bar` to the result. This gives the location-scale t-distribution. We see that the posterior mean of μ is 15.79 and the 95% credible interval is $(1.320, 30.17)$.

Suppose we wanted to do a single-sample Bayesian t-test of the hypothesis that $\mu > 10$. We can calculate the probability that this occurs in our sample as our estimate.

```
prob <- ifelse(mu>=10,1,0)
mean(prob)
1-mean(prob)
```

We find all the instances where a sampled value of μ is >10, and set `prob` equal to 1 in those instances. The remaining values of `prob` are set to zero, then its mean gives our estimate. Here, the posterior probability that $\mu > 10$ is 0.806. The posterior probability that $\mu \leq 10$ is therefore 0.194.

We can contrast the Bayesian t-test with a classical one. The alternative hypothesis is that $\mu < 10$. We can perform this test in *R* as follows:

```
data <- c(4.5, 30, 7.8, 7.5, 6.3,
          2.1, 65.8, 26.5, 5.2, 2.3)
t.test(data,
   alternative = c("less"),
   mu = 10)
```

Here we use the collection operator `c(...)` to place a collection of data into the variable named `data`. We then use *R*'s `t.test` command with the relevant options. *R* returns the value of the central t-statistic as $t = 0.911$ and the p-value as 0.807. We see that the two approaches provide essentially the same numerical result. Note that the posterior probability is easy to understand. However the p-value's interpretation is somewhat less straightforward. The classical p-value does not let us make any probability statements about the current data except for the following relatively limited statement: the probability that $\mu \geq 10$ in our data is either 0 or 1 and we do not know which of these applies here. Instead, the p-value is a statement about how likely it would be to obtain a value of the test statistic this large or larger in a long series of trials if the null hypothesis were true.

The direct approach using (3.13) is one method to obtain the posterior distribution of μ. This involved obtaining the joint distribution of μ and σ^2, and then integrating out σ^2 using calculus. However, we can obtain it another way by using the relationship between the joint, marginal, and conditional distributions from basic probability. Thinking in terms of the more convenient τ, note that $p(\mu, \tau|y) = p(\mu|y, \tau)p(\tau|y)$. So we can simulate from the marginal $p(\tau|y)$ and the conditional $p(\mu|y, \tau)$ to obtain the joint distribution $p(\mu, \tau|y)$. To then get the marginal distribution of μ, we "integrate out" τ by the simple matter of ignoring the computer's results for τ and looking at the computer's results for μ. Compared to a direct integration, this is a bit more relaxing. The *R* code is as follows:

```
set.seed(123)
n.iter <- 1000000
  y.bar <- mean(data)
  nu <- length(data)-1
  s2 <- var(data)
  n  <- 10

#simulate tau and the marginal standard error
tau <- rgamma(n.iter, nu/2, nu*s2/2)
mgl.std.error <- sqrt((1/tau)/n)

#given the standard error, simulate mu
mgl.mu <- rnorm(n.iter, y.bar, mgl.std.error)

#summarize the distribution of mu
mean(mgl.mu)
quantile(mgl.mu, probs=c(.025, .975))
```

The command `rnorm` instructs *R* to randomly draw `n.iter` samples from a normal distribution. In our code above, the random samples will come from a normal distribution with a mean of `y.bar` and a standard deviation of `mgl.std.error`. Because `mgl.std.error` itself is randomly generated, each time *R* simulates a normal sample, the standard deviation will be somewhat different. The normal samples are stored in a variable `mgl.mu`. We can then find summary statistics for `mgl.mu`. The posterior mean is 15.80, and the 95% posterior interval is (1.366, 30.24). Additional calculations show that the posterior probability that $\mu \geq 10$ is 0.805. The posterior distribution of μ appears in Figure 3.5.

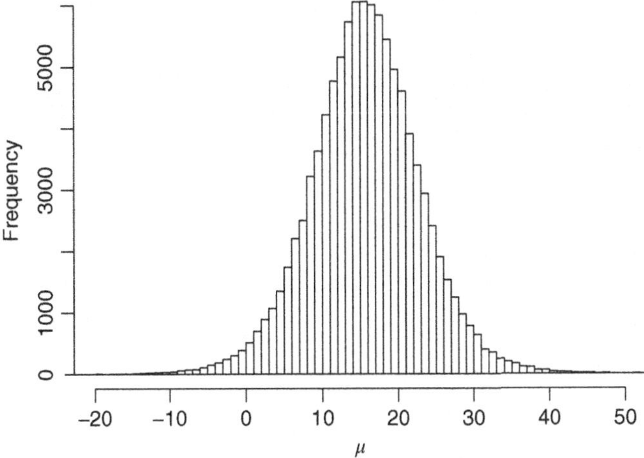

Figure 3.5 Posterior Distribution of μ when Variance is Unknown

It may seem strange that we simulated from the conditional distribution of $p(\mu|y, \sigma^2)$ but are now calling `mg1.mu` the marginal distribution. The reason we have obtained the marginal distribution of μ is that σ^2 is no longer constant but has its own variability arising from the gamma distribution of τ. When we simulate the conditional distribution of μ in `mg1.mu`, the argument to the standard deviation of `mg1.mu` is `mg1.std.error`. This variable has the distribution of simulated values of $(\sigma^2/n)^{1/2}$, and the extra variability from σ^2 carries over into the distribution of μ. As a result, μ has a marginal distribution that is more spread out than it would be if we had placed a constant value in the standard deviation argument to `rnorm` (Section 3.9). It also has heavier tails. This is known as *overdispersion*, and is a useful concept for business and management data. We may often find that a simple statistical model does not do a good job of accounting for outlying observations. In particular, we may find the outlying observations occur more often than our model suggests. If this is the case, our results will be excessively precise and fail to model the data realistically. This may lead to poor business decisions. A straightforward way to capture overdispersion is to allow the uncertainty regarding one parameter to propagate over to the uncertainty in another. Here our uncertainty about σ^2 is carried over into μ. As a result, the t-distribution is overdispersed compared to the normal distribution. Allowing for overdispersion is easy in a Bayesian framework—we structure the model so that parameter uncertainty can propagate. We will see more examples of this in the coming chapters.

Above, we have considered testing a one-sided alternative hypothesis such as $\mu < 10$. Additional considerations arise if we have a two-sided alternative hypothesis such as would result if the null hypothesis is $\mu = 4$. If a parameter such as μ has a continuous distribution, then the probability that we would simulate μ exactly equal to a given value becomes infinitely small. So we cannot directly use an approach like the one we used for prob above. Instead, one strategy is to look at the (two-sided) credible interval and see whether or not the null value falls into it (Box and Tiao, 1973, ch. 2.8; Zellner, 1971, ch. 10.2). This strategy is most appropriate when prior information is weak or diffuse as we have here. With the current data, $\mu = 4$ falls into the 95% credible interval so the plausibility of this value cannot be ruled out at the 95% level. Running the code

```
quantile(mgl.mu, probs=c(.05, .95))
```

shows that $\mu = 4$ does not fall inside the 90% credible interval. Hence we could conclude that $\mu = 4$ is not credible at the 90% level.

3.7 COUNT DATA AND THE CONJUGATE POISSON MODEL

Many important business and management contexts generate count data, that is, data which consists of values from the nonnegative integers. In this section, we take a look at some of these contexts and then explore basic models for this type of data.

Corporate Innovation: Firms regularly undertake innovation efforts to provide new products and new revenue streams. Some of these efforts may be patentable, especially in science/technology-related fields such as biotech. Hence innovation studies may look at patent counts to understand what factors influence corporate innovation (Stuart, 2000). Diffusion of innovative organizational practices may also be considered. For example, Guler et al. (2002) examined the diffusion of ISO 9000 certification among firms using count data methods.

Start-Up Creation/Job Creation: Within a country, certain regions or certain local policies may encourage the creation of firms and/or jobs, while others may discourage them. Internationally, certain countries and parts of the world have positive job growth prospects or may favor job creation or facilitate start-up firms. Hence it is natural to look at the count of new start-ups or jobs created to understand differences in international environments (Capelleras et al., 2008) or local/regional environments. The opposite point of view can also be taken. For example, factors leading to managers exiting a firm can also be examined from a count data perspective (Broschak, 2004).

Strategic Business Actions: Usually, there are numerous opportunities for firms to improve their market position or respond to external threats. Corporate acquisitions and divestures take the form of count data (Sanders, 2001), as do defensive efforts designed to repel takeover initiatives (Field and Karpoff, 2002). Firms must also provide justifications for their actions. Thus the counts of different types of verbal justifications for chief executive officer (CEO) compensation have been investigated by Wade et al. (1997). Finally, life cycle theories such as those for industry life cycles, product life cycles, or technology life cycles, often involve countable events or outcomes. This motivates the application of count data models to the examination of business life cycles (Hahn and Bunyaratavej, 2011).

3.7.1 In Detail: Conjugate Poisson Model Development

Because count data involve nonnegative integers, statistical models must accommodate these characteristics of the data. The starting point for many count data models is the Poisson distribution. The Poisson distribution has the following formula:

$$p(y|\lambda) = \frac{\lambda^y}{y!}\exp(-\lambda). \tag{3.15}$$

Here, λ is the rate parameter, where the rate is defined as the count per unit of measure such as per unit of time. It so happens that both the mean and the variance of the Poisson distribution are equal to λ. If we have a collection of n data points that we believe arose independently from a given Poisson distribution, then we again multiply all of the likelihood values for our likelihood function. Like the familiar summation symbol $\sum$, there is a useful symbol for multiplication which is $\prod$. For example, the product $x_1 \cdot x_2 \cdot x_3$ can be written as $\prod_{i=1}^{3} x_i$ using this notation. So, the likelihood function is

$$\begin{aligned} \ell(y|\lambda) &= \prod_{i=1}^{n} \frac{\lambda^{y_i}}{y_i!}\exp(-\lambda) \\ &\propto \lambda^{\sum y_i}\exp(-n\lambda). \end{aligned} \tag{3.16}$$

Having identified the likelihood function, we then consider what might be a relevant prior distribution. Referring back to (3.7), we notice that a good candidate for a conjugate prior for the Poisson distribution is the gamma distribution because the gamma distribution (omitting normalizing constants) has the same form as the likelihood (3.16). The gamma prior and the Poisson likelihood are then combined to obtain the posterior distribution

$$\begin{aligned} p(\lambda|y) &\propto \lambda^{\sum y_i}\exp(-n\lambda) \times \lambda^{\alpha-1}\exp(-\lambda\beta) \\ &\propto \lambda^{\sum y_i+\alpha-1}\exp(-(n+\beta)\lambda). \end{aligned} \tag{3.17}$$

Therefore, we see that the posterior distribution of λ is

$$\lambda|y \sim \text{Gamma}\left(\sum y_i + \alpha, n + \beta\right). \tag{3.18}$$

The posterior mean of λ is then $(\sum y_i + \alpha)/(n + \beta)$, and the posterior variance of λ is $(\sum y_i + \alpha)/(n + \beta)^2$. Bearing (3.18) in mind, since n is the number of observations or number of trials in the data, the prior value for β can be thought of as the number of prior observations or prior trials. Similarly, the correspondence between $\sum y_i$ and α indicates that α can be thought of as the total of the prior countable events. Taking α and β to be close to zero will give a weak prior for λ because these terms will then have minimal impact on the posterior relative to the data.

3.7.2 In Practice: Inference for the Conjugate Poisson Model

Birley (1986) studied the births and deaths of firms in St. Joseph County, Indiana, which at the time of study was close to U.S. income and population norms. Data was collected over a 5-year period from 1978 to 1982 and was subcategorized by sector. The data appears in Table 3.2.

Here we examine the data for the manufacturing sector. We separately examine both the birth rate and the death rate over the study period. To do so, the following code can be entered in *R*.

```
set.seed(123)
 n.iter <- 100000
 births <- c(24, 36, 20, 26, 30)
 deaths <- c(17, 14, 24, 20, 27)
 alpha  <- beta  <- .1
 n.b    <- n.d   <- length(births)
 sum.b  <- sum(births)
 sum.d  <- sum(deaths)

lambda.b <- rgamma(n.iter, sum.b+alpha, n.b+beta)
mean(lambda.b)
quantile(lambda.b, probs=c(.025, .975))
lambda.d <- rgamma(n.iter, sum.d+alpha, n.d+beta)
mean(lambda.d)
quantile(lambda.d, probs=c(.025, .975))
```

TABLE 3.2 Firms Created and Dissolved by Sector

Year		Agriculture	Manufacturing	Transport	Service
1978	Dissolved	4	17	12	52
1978	Created	13	24	20	78
1979	Dissolved	2	14	12	45
1979	Created	8	36	16	102
1980	Dissolved	7	24	24	61
1980	Created	3	20	20	77
1981	Dissolved	2	20	21	66
1981	Created	4	26	24	82
1982	Dissolved	2	27	19	50
1982	Created	5	30	19	89

The results show that the birth rate λ_b has a posterior mean of 26.69 manufacturing firms being created per year. The 95% credible interval for λ_b is $(22.40, 31.37)$. For the death rate λ_d, the posterior mean is 20.02 and the 95% credible interval is $(16.32, 24.10)$. We might wonder whether there is evidence for net manufacturing firm growth during the study time period. As has been described previously, one way to look at this is to examine the posterior distribution of $\lambda_b - \lambda_d$. Another way is to form a new dataset such as

```
net <- births - deaths
```

and then to conduct inference on `net`. These analyses appear in the chapter exercises.

A basic graphical depiction of the posteriors appears in Figure 3.6. This figure can be created with the following code.

```
dists <- cbind(lambda.b, lambda.d)
boxplot(as.data.frame(dists), boxwex=.3,
   range=0.85, outline=FALSE)
```

It is important to note that Figure 3.6 is a simple *R* box-and-whisker plot, and thus the length of each plot's whiskers do not exactly correspond to the 95% quantiles. The length of the whiskers can be changed by using the `range` option to `boxplot`. Experimentation with values of `range` here created whiskers that approximated the 95% credible interval. *R* has many advanced graphical tools which permit the exact display of

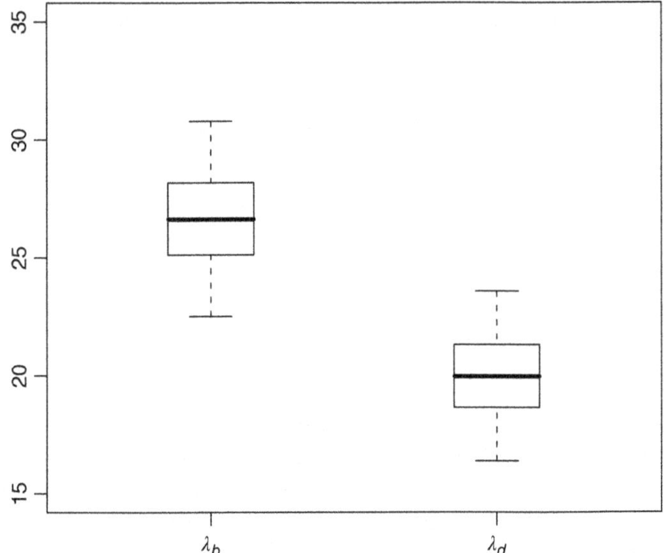

Figure 3.6 Boxplots of Posterior Distributions for λ_b and λ_d

values. Graphics programming for *R* is a book topic in itself (Murrell, 2011; Wickham, 2009) and would take us too far afield to cover in great detail. Fortunately, exact plots of the type in Figure 3.6 can be produced with no programming effort in a software package called *WinBUGS*. *WinBUGS* will be introduced in Chapter 5.

3.8 SUMMARY

This chapter has illustrated the steps that are undertaken in a conjugate Bayesian analysis. The likelihood function is obtained first after consideration of the type of data to be analyzed. Next, a prior distribution is selected that matches the functional form of the likelihood. Bayes' theorem then indicates that the posterior is proportional to the prior times the likelihood. We multiply the two and examine the result to find the posterior distribution.

Conjugate Bayesian analysis is attractive because it allows us to obtain the exact posterior distribution in terms of a known family of distributions. Key statistics such as the mean and the variance of the posterior distribution are then straightforward functions of distributional parameters. However, conjugate analysis does place restrictions on what the distribution of prior belief looks like. If prior beliefs do not resemble a member of the conjugate prior family, then conjugate analysis cannot be used.

In practice, however, many times we wish to de-emphasize the prior through the use of weak or vague priors. In these circumstances, we can typically find conjugate priors which have these weak properties. For this reason, conjugate techniques are still very widely used in Bayesian analyses. Monte Carlo simulation from the posterior allows the straightforward calculation of summary statistics and other functionals even when closed-form expressions for them are not available.

3.9 EXERCISES

1. In Firm A, 27 employees out of 92 work in sales. In Firm B, 82 out of 273 employees work in sales. Test whether the proportion of employees who work in sales is different across the two firms at the 95% level. Use a uniform prior for π for both firms.

2. Suppose Firm B above plans to hire 10 more employees across the entire business. The firm tends to hire in proportion to current staffing levels, but there is some variability depending on the quality and fit of the applicants. Simulate from the posterior predictive distribution for the number of new sales employees out of the 10 total hires, assuming

that the proportion of sales hires to total hires is 82/273. What is the 95% credible interval for the number of new sales hires?

3. A supplier has a fabrication process with a known variance of 1/10 of an inch. Your prior mean is 0 inch and your prior precision is 0.001. You take a random sample of 10 products from the supplier. The sample mean is 6.2 inches. What is the posterior mean and the posterior 95% credible interval?

4. An internet Web site contains advertising which generates revenue when clicked. As the site is new, the revenue's mean and variance are not known. Your prior mean for daily revenue is $3 and your prior precision for daily revenue is 0.00001. Fourteen days of daily revenue data are collected. The sample mean of the data is $5.48, and the sample standard deviation is $0.29. What is the 95% credible interval for daily revenue?

5. Reexamine the data of Table 3.2. Simulate from the difference of the birth rates in agriculture, transport, and service as compared to the birth rate for manufacturing. Namely simulate from $\lambda_{\text{Agriculture}} - \lambda_{\text{Manufacturing}}$, $\lambda_{\text{Agriculture}} - \lambda_{\text{Manufacturing}}$, and $\lambda_{\text{Service}} - \lambda_{\text{Manufacturing}}$. What can you conclude about the relative birth rates of these industries compared to manufacturing during this time period based on the 95% credible intervals?

6. Using the data of Table 3.2 and code of Section 3.7.2 simulate from `lambda.b` and `lambda.d`. Then form a difference variable using the code `diff <- lambda.b - lambda.d`. Obtain the 95% credible interval for `diff`. Next form a variable `net` as described in Section 3.7.2. Find the posterior distribution for `net` using a prior α and a prior β of 0.1. Then find the 95% credible interval for `net`. Are the two credible intervals substantially different and, if so, how?

7. Toward the end of Section 3.6.3 we discussed that the marginal distribution of μ, `mgl.mu`, had extra variability that carried over from the variability in τ. Investigate this using *R*. Enter the data into *R* using the command `data <- c(4.5, 30, 7.8, 7.5, 6.3, 2.1, 65.8, 26.5, 5.2, 2.3)`. Calculate `y.bar` and `s2` as before. Add a line of code which reads `fixed.std.error <- sqrt(s2/n)`. Then add a line of code which reads `mu.fixed <- rnorm(n.iter, y.bar, fixed.std.error)`. Next find the 95% `quantiles` of `mu.fixed` and compare them with those in the text. How do the new quantiles differ?

8. In Detail: Rederive (3.24) using the precision. First reexpress (3.20) and (3.22) in terms of τ and τ_0. Then find the posterior distribution.

Comment on which approach to finding the posterior is easier, this one or the approach leading up to (3.24).

3.10 NOTATION INTRODUCED IN THIS CHAPTER

Notation	Meaning	Example	Section Where Introduced
ν	Degrees of freedom	$\nu = n - 1$	3.5
τ	Precision	$\tau = 1/\sigma^2$	3.5
$\prod$	Product	$x_1 x_2 x_3 = \prod_{i=1}^{3} x_i$	3.5
$E(\cdot)$	Expected value	$E(\tau)$	3.6

3.11 APPENDIX—IN DETAIL: FINDING POSTERIOR DISTRIBUTIONS FOR THE NORMAL MODEL

This appendix goes into detail working out how the posterior distribution is found for the normal model. It is up to you whether you would like to see these details on a first reading. We will do this using algebra and some concepts from probability. The algebra is not important in itself, but the concepts from probability (joint, conditional, and marginal distributions) are worth knowing when we discuss Markov chain Monte Carlo methods in Chapter 4. If you would instead prefer to get started with modeling and just take the posterior distributions as given, feel free to focus on Section 3.6.

The normal distribution has the following probability density function:

$$p\left(y|\mu,\sigma^2\right) = \frac{1}{\sqrt{2\pi\sigma^2}} e^{-\frac{1}{2\sigma^2}(y_i-\mu)^2}. \tag{3.19}$$

Instead of the variance, we can also think of the normal distribution as being parameterized in terms of the standard deviation σ. The normal distribution is symmetric around μ and so the median and the mode coincide with μ.

Typically, we will have a number of data points to examine. Suppose we have n data points. We assume they all independently arise from a common normal distribution with parameters μ and σ^2. Phrased differently, they are *conditionally independent* given μ and σ^2. We recall that independent probabilities can be multiplied to give the overall joint probability. We use the symbol $\prod$ to indicate multiplication. For example, the product $x_1 \cdot x_2 \cdot x_3$ is the same as $\prod_{i=1}^{3} x_i$. Given our collection of data y, the normal likelihood function is then

$$\ell(y|\mu,\sigma^2) = \prod_{i=1}^{n} \frac{1}{\sqrt{2\pi\sigma^2}} e^{-\frac{1}{2\sigma^2}(y-\mu)^2}$$

$$= (2\pi\sigma^2)^{-n/2} \exp\left(-\frac{1}{2\sigma^2}\sum_{i=1}^{n}(y_i-\mu)^2\right). \tag{3.20}$$

In the second line above, the initial fraction has been converted to a power, while in the second term the product can be rewritten as a sum in the exponential function's exponent using the relationship $e^a e^b = e^{a+b}$. In order to begin working with (3.20), we need to recall two things. The first is a mathematical "trick" called *completing the square*. Recall that $(a-b)^2 = a^2 - 2ab + b^2$. Suppose we only have $a^2 - 2ab$. Then to complete the square, we perform the following steps:

$$a^2 - 2ab,$$
$$a^2 - 2ab + b^2 - b^2,$$
$$(a-b)^2 - b^2.$$

Hence we add in the term we need to complete the square while also subtracting the same term to cancel out the addition. The second thing to recall is the definition of the sample variance s^2, and then show some equivalencies regarding it. Recall that

$$s^2 = \frac{1}{n-1}\sum_{i=1}^{n}(y_i-\bar{y})^2,$$

$$(n-1)s^2 = \sum_{i=1}^{n} y_i^2 - 2\sum_{i=1}^{n} y_i\bar{y} + n\bar{y}^2,$$

$$(n-1)s^2 = \sum_{i=1}^{n} y_i^2 - 2n\bar{y}\,\bar{y} + n\bar{y}^2,$$

$$(n-1)s^2 = \sum_{i=1}^{n} y_i^2 - n\bar{y}^2. \tag{3.21}$$

In the third line we have used the relationship between the sample mean and the sample sum to transform the sum into n times the mean.

3.11.1 Analysis of the Normal Mean with Variance Known

We first consider the situation where the variance is known. Although this situation is somewhat uncommon, it is an essential starting point for subsequent extensions. The normal distribution in (3.20) will be our likelihood function, but we still need a prior for unknown parameters. A conjugate normal prior for μ is

$$p(\mu) = \frac{1}{\sqrt{2\pi\sigma_0^2}} e^{-\frac{1}{2\sigma_0^2}(\mu-\mu_0)^2}, \tag{3.22}$$

where μ_0 and σ_0^2 are the prior mean and prior variance for μ, respectively. For a conjugate approach, we need the likelihood to have the same form as the prior. Since the prior has the term $(\mu - \mu_0)^2$, we need to rearrange (3.20) to have the same form. We also can use a helpful strategy, which is to suppose that the sample mean $\bar{y}$ and the sample variance s^2 are going to be important here. These two quantities are key quantities for the normal distribution because, once we know $\bar{y}$ and s^2, we can characterize a given normal distribution in a given sample. Focusing on the kernel, we have

$$\begin{aligned}
\ell(y|\mu,\sigma^2) &\propto \exp\left(-\frac{1}{2\sigma^2}\sum_{i=1}^{n}(y_i-\mu)^2\right) \\
&\propto \exp\left(-\frac{1}{2\sigma^2}\left(\sum_{i=1}^{n} y_i^2 - 2\mu n\bar{y} + n\mu^2\right)\right) \\
&\propto \exp\left(-\frac{1}{2\sigma^2}\left(\sum_{i=1}^{n} y_i^2 + n(\mu^2 - 2\mu\bar{y})\right)\right) \\
&\propto \exp\left(-\frac{1}{2\sigma^2}\left(\sum_{i=1}^{n} y_i^2 - n\bar{y}^2 + n(\mu^2 - 2\mu\bar{y} + \bar{y}^2)\right)\right) \\
&\propto \exp\left(-\frac{1}{2\sigma^2}\left((n-1)s^2 + n(\mu-\bar{y})^2\right)\right) \\
&\propto \exp\left(-\frac{1}{2\sigma^2}n(\mu-\bar{y})^2\right).
\end{aligned} \tag{3.23}$$

We expanded and rearranged the original squared term in the first three lines above. Then we completed the new square in line 4. In doing so, we also produced $(n-1)s^2$, which is a constant for any given data we might have. Constants can be dropped when working with the kernel, and we can again use $e^{a+b} = e^a e^b$ to push the constant term out of the exponent's sum into a multiplier. Once the constant multiplier is out, it can be dropped via proportionality, giving (3.23). We can now use Bayes' rule to obtain the posterior

$$p(\mu|y) \propto \exp\left(-\frac{1}{2\sigma^2}n(\mu-\bar{y})^2\right)\exp\left(-\frac{1}{2\sigma_0^2}(\mu-\mu_0)^2\right)$$

$$\propto \exp\left(-\frac{1}{2}\left[\frac{1}{\sigma^2/n}(\mu^2 - 2\mu\bar{y} + \bar{y}^2) + \frac{1}{\sigma_0^2}(\mu^2 - 2\mu\mu_0 + \mu_0^2)\right]\right)$$

$$\propto \exp\left(-\frac{1}{2}\left[(\frac{1}{\sigma^2/n} + \frac{1}{\sigma_0^2})\mu^2 - 2\mu\left(\frac{\bar{y}}{\sigma^2/n} + \frac{\mu_0}{\sigma_0^2}\right) + \cdots\right]\right),$$

where $\cdots$ indicates an expression that is entirely made up of constants which can be dropped. We see that completing the square would be worthwhile here. After factoring out the constant in front of μ^2 and continuing on, we find

$$p(\mu|y) \propto \exp\left(-\frac{1}{2}\left(\frac{1}{\sigma^2/n} + \frac{1}{\sigma_0^2}\right)\left[\mu^2 - 2\mu\frac{\frac{\bar{y}}{\sigma^2/n} + \frac{\mu_0}{\sigma_0^2}}{\frac{1}{\sigma^2/n} + \frac{1}{\sigma_0^2}}\right]\right)$$

$$\propto \exp\left(-\frac{1}{2}\left(\frac{1}{\sigma^2/n} + \frac{1}{\sigma_0^2}\right)\left[\left(\mu - \frac{\frac{\bar{y}}{\sigma^2/n} + \frac{\mu_0}{\sigma_0^2}}{\frac{1}{\sigma^2/n} + \frac{1}{\sigma_0^2}}\right)^2 - \cdots\right]\right). \tag{3.24}$$

The posterior kernel indicates that the mean of the posterior distribution is $\left(\frac{\bar{y}}{\sigma^2/n} + \frac{\mu_0}{\sigma_0^2}\right)\Big/\left(\frac{1}{\sigma^2/n} + \frac{1}{\sigma_0^2}\right)$. The posterior variance is $\left(\frac{1}{\sigma^2/n} + \frac{1}{\sigma_0^2}\right)^{-1}$. This gives the results of (3.4) and (3.5).

3.11.2 Analysis of the Normal Variance with Mean Known

In Section 3.5.2, we assessed our gamma prior using ν_0 and s_0^2. We said that these quantities express information about our prior degrees of freedom and variance, respectively. Making these substitutions into the gamma distribution gives us our prior distribution for the precision as follows:

$$p(\tau|\nu_0 s_0^2) \propto \tau^{(\nu_0-2)/2}\exp\left(-\frac{\tau}{2}\nu_0 s_0^2\right). \tag{3.25}$$

To find the posterior distribution, we reexpress the likelihood function from (3.20) using the second-to-last line of (3.23). It will be necessary to convert our variances to precisions. We will also use the notation $\nu = n - 1$. With these things in mind, we have

$$\ell(y|\tau) \propto \tau^{(n/2)}\exp\left(-\frac{\tau}{2}\left(\nu s^2 + n(\mu - \bar{y})^2\right)\right) \tag{3.26}$$

and multiplying (3.25) and (3.26) gives a gamma posterior distribution

$$p(\tau|y) \propto \tau^{(\nu_0-2)/2}\exp\left(-\frac{\tau}{2}\nu_0 s_0^2\right)\tau^{(n/2)}\exp\left(-\frac{\tau}{2}\left(\nu s^2 + n(\mu - \bar{y})^2\right)\right)$$

$$\propto \tau^{(\nu_0+n-2)/2} \exp\left(-\frac{\tau}{2}\left(\nu_0 s_0^2 + \nu s^2 + n(\mu - \bar{y})^2\right)\right). \tag{3.27}$$

The above may seem a little esoteric if it is the first time you are seeing this, but all that is involved is recognizing a distribution in our likelihood function (here the gamma distribution) and then performing some arithmetic inside and outside of the exponents. Note also that we can again see conjugate updating in action. We started with a prior degrees of freedom ν_0, and then this was updated to $\nu_0 + n$. A similar update occurred for s_0^2. We can now summarize the posterior distribution for τ as in (3.8).

3.11.3 Analysis of the Conjugate Normal Model with Mean and Variance Both Unknown

The priors for this model are given in (3.10). Multiplying (3.20) by these priors leads to the joint posterior

$$\begin{aligned} p(\mu, \tau|y) &\propto \tau^{-1}\tau^{n/2} \exp\left(-\frac{\tau}{2}\sum_{i=1}^{n}(y_i - \mu)^2\right) \\ &\propto \tau^{(n-2)/2} \exp\left(-\frac{\tau}{2}\sum_{i=1}^{n}(y_i - \mu)^2\right) \\ &\propto \tau^{(n-2)/2} \exp\left(-\frac{\tau}{2}\left[\nu s^2 + n(\mu - \bar{y})^2\right]\right), \end{aligned} \tag{3.28}$$

where the last line reuses the form of (3.23). The conditional posterior for μ is easy to work out. Writing it out symbolically gives away the clue—$p(\mu|y, \tau)$ is the case where both y and τ are treated as known. We have already gone through this development in Section 3.5.1. We could of course rederive the result, but we will take a more intuitive approach that yields the same outcome. Since $p(\mu)$ has an infinite range, suppose the prior variance is therefore "infinite" and so we take the limiting value of τ_0 as it goes to zero. Then inserting $\tau_0 = 0$ in (3.4) and (3.5) gives (3.11) and (3.12).

In contrast to the conditional posterior, the marginal posterior of μ does not yield so easily. Now τ must be integrated out of (3.28) instead of being held constant. It can be shown that the marginal posterior distribution has a Student t-distribution. In particular, $t = \dfrac{\mu - \bar{y}}{\sigma/\sqrt{n}}$ is (central) t-distributed with ν degrees of freedom. The central t-distribution is the one found often in textbook appendices. We see that, when we use flat priors in this situation, we find that both Bayesian and classical results lead to the same t-distribution.

We could also integrate to get the marginal distribution of τ. However, a conceptually simpler way is to use the relationship between the joint, marginal, and conditional distributions from basic probability. Using this approach, we observe that $p(\tau|y) = p(\mu, \tau|y)/p(\mu|y, \tau)$. Then, if we take (3.28) and (3.11), we have

$$
\begin{aligned}
p(\tau|y) &\propto \frac{\tau^{(n-2)/2} \exp\left(-\frac{\tau}{2}[\nu s^2 + n(\mu - \bar{y})^2]\right)}{\tau^{1/2} \exp\left(-\frac{n\tau}{2}(\mu - \bar{y})^2\right)} \\
&\propto \tau^{(n-3)/2} \exp\left(-\frac{\tau}{2}\nu s^2\right) \\
&\propto \tau^{\nu/2-1} \exp\left(-\frac{\tau}{2}\nu s^2\right). \qquad (3.29)
\end{aligned}
$$

The marginal distribution of τ therefore has a gamma distribution.

4

MARKOV CHAIN MONTE CARLO AND REGRESSION MODELS

In the previous chapter, we used Monte Carlo simulation to estimate quantities of interest. The values arising from Monte Carlo simulations were completely independent from sample to sample across the simulation run. In this chapter, we introduce the idea of Markov chains. Markov chains are sequences in which each sample is generated independently, but overall the samples may be correlated or dependent across the run. Perhaps the most famous kind of Markov chain is the random walk (Figure 4.1). The dependence across time is visible in the figure even though each move has been generated independently. We can use the behavior of a Markov chain inside of a Monte Carlo simulation, and it turns out this is very useful for Bayesian statistics.

4.1 INTRODUCTION TO MARKOV CHAIN MONTE CARLO

In Chapter 3, we obtained the posterior distributions of model parameters through conjugate analysis. To obtain numerical summary statistics, we performed Monte Carlo simulations from these distributions. These simulation runs had the property of independence across iterations since each individual simulated value was an independent random draw from the parameter's posterior distribution. The Monte Carlo methods we have used so far have at least three attractive properties. First, they are extremely fast in *R*. *R* can obtain hundreds of thousands of simulations

Bayesian Methods for Management and Business: Pragmatic Solutions for Real Problems, First Edition. Eugene D. Hahn.

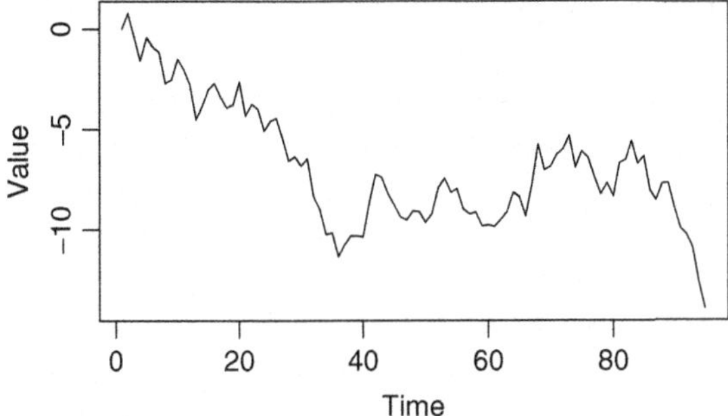

Figure 4.1 Example of a Random Walk

from distributions that it knows about in less than a second when we supply *R* with arguments that are either scalars or vectors. Second, we can be assured by the law of large numbers that our simulation-based numerical estimates are converging to their true values. Third, they are easy to program as we have seen, at least for smaller and simpler models.

However, there are a few limitations to the methods we used in Chapter 3. First, as mentioned in Section 3.8, we were confined to certain types of priors in order to obtain conjugacy. It would certainly be possible to hold prior beliefs that did not have the necessary conjugate form, but if this occurred, further analysis would have been impossible. Hence we will want methods that can accommodate a greater variety of prior distributions. Second, as we have more and more parameters, it requires greater and greater sophistication to derive the marginal distributions of parameters. We saw in Section 3.5.3 that integration was needed to obtain the location-scale t-distribution as the marginal distribution of μ when τ was unknown. As the number of parameters increases, this becomes more and more difficult. Moreover, part of the attraction of Monte Carlo simulation was in avoiding these integration difficulties.

The idea behind Markov chain Monte Carlo (MCMC) methods is that we can use *dependent* simulations from the posterior distribution. The benefit of this approach is that we now have an extremely powerful tool for simulating from virtually any kind of posterior. The cost is that more attention needs to be paid to ensure that we have indeed simulated from the posterior. Because the benefit is so much greater than the cost, virtually all contemporary Bayesian methods employ MCMC for estimation. Indeed, the large upsurge in the usage of Bayesian methods in recent years is directly attributable to the development of MCMC.

In this chapter, we will examine the two most important MCMC methods: the Gibbs sampler, and the Metropolis–Hastings algorithm. We will discuss theoretical considerations regarding Markov chains and investigate practical examples of implementation.

4.2 FUNDAMENTALS OF MCMC

In this section, we cover essentials of Markov chain theory from a conceptual perspective. A detailed treatment of the mathematics justifying the use of Markov chains to sample from posterior distributions is outside of the scope of this book (Tierney, 1994, 1996; Norris, 1997, for more details). A Markov chain is a simple kind of mathematical process which includes the familiar random walk of Figure 4.1. The process has no memory of its past trajectory, but instead makes a random move according to some predefined rule or distribution as it proceeds. As it proceeds, the Markov chain can be thought of as wandering or exploring through some mathematical space. For this to be useful in Bayesian statistics, we would like the Markov chain to explore the posterior distributions of parameters. In particular, our goal is to have the Markov chain explore the posterior distribution in sufficient detail so that we can understand the distribution's features. We can draw an analogy with geographical maps. A crude map that is based on limited exploration of an area may have omissions, distortions, and other incorrect features. Hence, conclusions that one might draw about an area from this kind of map could easily be wrong. We do not want this to happen in our analyses. Instead, we would like the Markov chain's exploration to be reasonably complete and thorough in the sense that the "mapping" process gives us the information we need at the level of detail we require. Just as any modern map is unlikely to be precise to the millimeter, we do not need our results from the Markov chain to be exact as long as the results are quite close at the relevant level of detail. We will be interested in how fast the Markov chain explores the posterior, and whether it has become trapped or is otherwise blocked from accessing certain parts of the posterior.

There are three formal conditions that must be satisfied for the Markov chain to sample satisfactorily from the posterior. If it is possible for a Markov chain to travel to all areas of the distribution in a finite number of iterations, then we say the Markov chain is *irreducible*. The Markov chain should not become stuck in a loop where it visits the same exact areas repeatedly. If it does not exhibit this looping behavior, then it is *aperiodic*. Finally, the Markov chain is *positive Harris recurrent* if it will eventually return to all areas and that this behavior does not depend on the starting point (Chan and Geyer, 1994; Tierney, 1994). If these three conditions apply, then the Markov chain is called *ergodic*.

If we have properly constructed our ergodic Markov chain for our analysis, after a certain number of iterations it should reach its *stationary distribution* (Robert and Casella, 2004, ch. 6). The stationary distribution is the final target distribution for the parameter we are trying to estimate. The stationary distribution should be invariant: that is, it should not fundamentally drift, change, or evolve as the iterations continue. The stationary distribution also should not depend on the starting value of the Markov chain.

Once the chain has reached its stationary distribution, the values produced by the stationary distribution are likely to be dependent but we can still take averages of quantities of interest as consistent estimators of their true values. This is because it can be shown that results as in (3.3), which justified regular Monte Carlo methods, apply to ergodic Markov chains as a result of the ergodic theorem (Roberts, 1996, p. 47). Namely, we can use sample averages from the Markov chain involving our functions of interest as approximations to the true values of our functions of interest. Moreover, these approximations get better as the number of iterations increases. Again, we are able to transform the difficult task of evaluating integrals into the much simpler task of calculating sample averages.

The trade-off is that our results will have approximation error, but this can be controlled by running the Markov chain until we are satisfied with the approximation. It can also be shown that the central limit theorem applies to ergodic Markov chains. In words, the difference between the Monte Carlo sample average and the corresponding true value has a normal distribution. This fact can be used to quantify the Monte Carlo error associated with a given Markov chain run. We can therefore calculate and review the error for a given run and thereby make a decision as to whether we are satisfied with the current accuracy or whether longer runs are needed.

To summarize, there are a number of conditions that have to be satisfied in order for our Markov chains to give us the insights we need. There are practical tools for checking whether these conditions are satisfied. We will begin by introducing some of these tools in the next section.

4.3 GIBBS SAMPLING

The first MCMC method we will consider is the Gibbs sampler. The Gibbs sampler relies heavily on a concept called the *full conditional distribution*. The word "full" in full conditional distribution emphasizes that we hold constant (i.e., condition on) every parameter except the parameter (or group of parameters) of interest. Phrased differently, in the full conditional distribution, we have temporarily made all variables into constants except for the one parameter (or group of parameters) we are focusing on. Then the idea behind the Gibbs sampler is simple: cycle through all of the full conditional distributions of the parameters and simulate once from each. So to simulate from a parameter's marginal distribution, hold all other parameters at their most recently sampled values and take a random draw from the parameter's conditional distribution. Then repeat this process for each of the remaining parameters, and cycle as long as needed. At the end of the run, we will have the marginal distribution of all parameters. Gibbs sampling allows us to simplify the problem greatly by allowing us to focus on only one variable at a time at a given point in the run.

With a little creativity, we can put this idea into a business context. Suppose you are forecasting both sales volume and raw material costs for a product. To begin, you use some initial information and forecast sales for the upcoming month of January. Next, you realize that raw material costs might be lower with higher sales because of quantity discount pricing. You decide to base your February raw material cost forecast in part on the sales forecast you just made. In thinking about your February sales forecast, you realize that, if raw material costs were lower, you could pass that on to the consumer through rebates. This would in turn impact sales. As a result, you make your February sales forecast based on your February raw material cost forecast. Continuing forward to later months, you would again cycle through the different forecasts using the data and the latest forecast information available. If we switch the idea of "forecasting X" with "simulating from the distribution of X," we have the concept behind Gibbs sampling.

To put it more specifically using notation, suppose the parameter vector is $\boldsymbol{\theta} = (\theta_1, \cdots, \theta_P)$ and that we are currently on iteration i of the Gibbs sampler run. Let $\theta_p^{(i)}$ indicate the ith simulated value for θ_p. Then, sample as follows:

$$
\begin{aligned}
\theta_1^{(i)} &\sim p\left(\theta_1 | y, \theta_2^{(i-1)}, \cdots, \theta_P^{(i-1)}\right) \\
\theta_2^{(i)} &\sim p\left(\theta_2 | y, \theta_1^{(i)}, \theta_3^{(i-1)}, \cdots, \theta_P^{(i-1)}\right) \\
&\vdots \\
\theta_P^{(i)} &\sim p\left(\theta_P | y, \theta_1^{(i)}, \cdots, \theta_{(P-1)}^{(i)}\right) \\
\theta_1^{(i+1)} &\sim p\left(\theta_1 | y, \theta_2^{(i)}, \cdots, \theta_P^{(i)}\right) \\
&\vdots
\end{aligned}
$$

and proceed until the desired number of sampled values is obtained for all parameters. Typically, each component of $\boldsymbol{\theta}$ is a single parameter (such as the mean or the regression coefficient). However, this is not a requirement. We could combine, for example, θ_2 and θ_3 into a group or block. The full conditional distribution for this block can then be inserted into the Gibbs sampling process. This strategy may add a small amount of complexity but can improve the performance of the Gibbs sampler in certain situations such as when a block of parameters is highly correlated. In practice, though, the simpler one-parameter-at-a-time approach is much more common. We illustrate this approach with an example.

4.3.1 Gibbs Sampling for the Normal Mean

In Section 3.5.3, we examined how to estimate μ and τ for normally distributed data. To do this, we had to find the conditional distribution of μ as given by Equation (3.11), which was $\mu|y, \tau \sim N\left(\overline{y}, (n\tau)^{-1}\right)$. In our previous analysis, we did not need the conditional distribution of τ, so it was not examined. However, to implement the Gibbs sampler we will need to find the full conditional distribution of all parameters including τ. We will do that now.

The joint posterior distribution for the normal model is given by (3.28)

$$p(\mu, \tau|y) \propto \tau^{(n-2)/2} \exp\left(-\frac{\tau}{2}\left[\nu s^2 + n(\mu - \overline{y})^2\right] \right).$$

To find the conditional distribution for τ, we treat all instances of μ as fixed. This gives

$$p(\tau|y, \mu) \propto \tau^{(n-2)/2} \exp\left(-\frac{\tau}{2}\left[\nu s^2 + n(\mu - \overline{y})^2\right] \right) \tag{4.1}$$

which happens to be identical on the right-hand side to the joint posterior, although in most cases some manipulation will be required. With a little examination, this can be seen to be proportional to a gamma distribution. Hence we find that

$$\tau|y, \mu \sim \text{Gamma}\left(n/2, (\nu s^2 + n(\mu - \overline{y})^2)/2\right). \tag{4.2}$$

With both full conditional distributions available, we can now proceed to write some *R* code to perform Gibbs sampling for the normal mean.

However, one thing to consider is that Gibbs sampling requires us to condition on existing values of parameters. At the beginning of the Gibbs sampling process, the sampler has just begun and there are no such values. What can the sampler condition on? The solution is to supply *initial values* to the program for it to use in the first iteration. It helps matters to provide reasonable initial values (i.e., values that are likely to occur in the posterior), and the more complex the model, the more important reasonable initial values become. For expository purposes, we will give a reasonable initial value to μ below but not for τ. Example code is as follows:

```
#Gibbs sampling for normal parameters
set.seed(123)

n.iter <- 100000
data  <- c(4.5, 30, 7.8, 7.5, 6.3,
  2.1, 65.8, 26.5, 5.2, 2.3)
y.bar <- mean(data)
```

```
s2    <- var(data)
n    <- length(data)
nu    <- n - 1

#storage for parameters
tau <- mu <- rep(0,n.iter)
#initial values
tau[1] <- 1
mu[1] <- 1

for (i in 2:n.iter) {
 if(i%%10000==0) cat(i, " iterations complete","\n")
 mu[i] <- rnorm(1, y.bar, sqrt(1/(n*tau[i-1])))
 tau[i] <- rgamma(1, n/2, (nu*s2 + n*(mu[i]-y.bar)^2)/2)
 }
 sd(mu[501:100000])
```

The majority of the program involves preparatory tasks. These include calculating necessary quantities such as ν and s^2. We also need to provide a storage location for `tau` and `mu` and we do so using the `rep` command which creates a vector of zeros for subsequent usage. We can then set the first element of each vector to its initial value. The final `for` loop implements Gibbs sampling. The first line in the `for` loop is a user convenience. This line outputs the iteration count to the *R* code window every 10,000 iterations. By clicking on the code window or pressing the space bar, this counter will refresh in real time. This allows the user to gauge how much more time is left before the run is complete. The next two lines perform the sampling for μ and τ. Here, `mu` is sampled from its conditional distribution (3.11) given the most current value of `tau`. Then `tau` is sampled from its conditional distribution (4.2) given the most current value of `mu`.

Figure 4.2 shows the movement process of the Gibbs sampler. To better guide intuition, the y-axis of Figure 4.2 displays the more familiar σ^2 instead of τ. The square to the right of the center of the figure shows the position of the Gibbs sampler at iteration 501. The point is ($\mu = 17.4, \sigma^2 = 648.7$). The Gibbs sampler then randomly moves to a larger value of μ ($\mu = 27.1$). Next it randomly moves to a smaller value of σ^2 (517.7). The location of the Gibbs sampler at iteration 502 (and subsequent iterations) is marked with a circle. Then the cycle begins again. We can observe that, for values toward the bottom of the plot where σ^2 is smaller, the values of μ tend to be more concentrated around the overall vicinity of 14–16. Conversely, when the sampled value of σ^2 is larger, μ is more likely to move farther away from its central vicinity of 14–16.

4.3.2 Output Analysis

We mentioned that MCMC methods are very powerful, but the cost is that we need to spend more time verifying that we are sampling from the posterior distribution. Hence, before taking our results seriously, we need

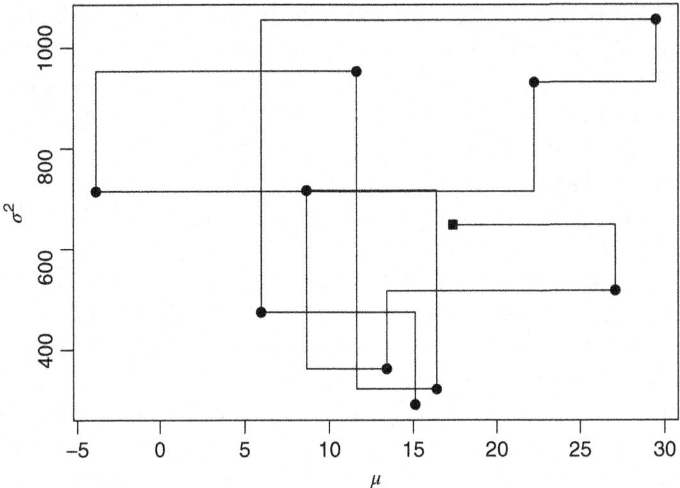

Figure 4.2 Movement of Gibbs Sampler for μ and σ^2 for Iterations 501–510

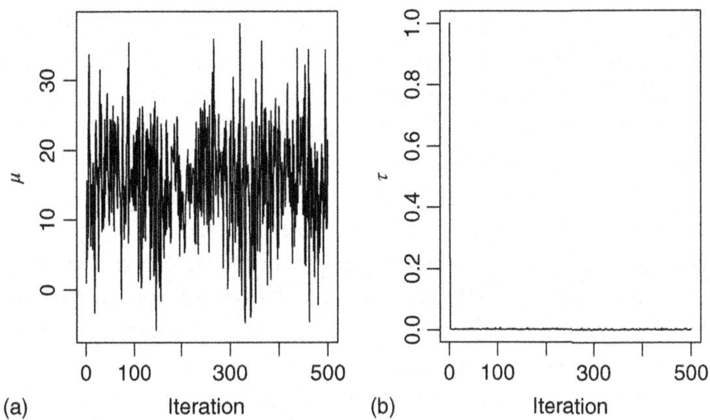

Figure 4.3 Posterior Traces for μ (a) and τ (b)

to perform checks on the output. The most basic and yet perhaps most worthwhile check is to look at the path that the sampler has taken across time. This is sometimes called the *trace* of the distribution, and plots of the current traces appear in Figure 4.3. The traces in Figure 4.3 show only the first 500 iterations for expository purposes. Many times, we would like to see the trace for the entire run, but "zooming in" as we have done here can also be instructive during analyses. The *R* command to produce plots such as this is `plot(ts(mu[1:500]))`, where `ts` stands for time series.

The trace for μ looks typical for acceptable MCMC output. We see the Gibbs sampler moving through the distribution with a fair amount of speed and some meandering. It does not appear to get stuck at any location. However, the trace for τ does not currently look as promising. We can see that the initial value we provided is very far away from the posterior distribution. We say that, at least for the early part of the chain, the chain has not *converged*, i.e., it is not yet producing values from the true posterior distribution of the parameter. If we were to estimate τ with this initial value and some of its successors included, it would bias our estimate. Therefore, typical practice is to omit a first portion of the Gibbs sampling run to allow the influence of initial values to be diluted. This first portion of the Gibbs sampling run is called the *burn-in*.

In our current example, we see that the Gibbs sampler for τ moves very quickly away from the poorly chosen initial value. Discarding 500 iterations is perhaps overly conservative, as the trace suggests that it would be sufficient to discard only the first 10 or so iterations. However, the cost of conservatism here is a fraction of a second of computer time, so virtually nothing is lost by being overly cautious. In more complex models, the influence of a poorly chosen starting value could take much longer to diffuse away. Run times can also be much longer, making the choice more consequential. The challenge is that a good choice for an initial value is not usually known in advance. Making the burn-in period longer is one option. Another is to consider several possibilities for initial values and perform multiple runs. The results from each run can be compared with each other to ensure that the effects of the choice of the initial values and the burn-in length are negligible.

We see that after 500 iterations the Markov chain appears to have settled in its final range and is producing samples reasonably consistently from there forward. This is one piece of evidence we could use to conclude that the Markov chain has converged (other recommended diagnostics are discussed in Chapter 6). Note that the available diagnostics can only detect nonconvergence, i.e., they can only raise a red flag about the Markov chain's output. There is no diagnostic that gives a 100% convergence "seal of approval," and so the final decision about whether a chain has converged always has to involve a case-by-case review of the evidence.

Figure 4.4 displays the traces after the first 500 iterations have been discarded as burn-in. A total of 5000 iterations are displayed. This is still a zoomed-in view since the total run contains 100,000 iterations. However, the zoomed-in view is more useful for seeing whether there is any local misbehavior in the Gibbs sampler. These traces look typical of a Gibbs sampler that is satisfactorily producing samples from the posterior distribution. To gain greater confidence in our output, we would typically want to take a zoomed-in look at other portions of the chain as well as a zoomed-out look of the entire sampling run to see whether evidence for convergence is consistent across the chain.

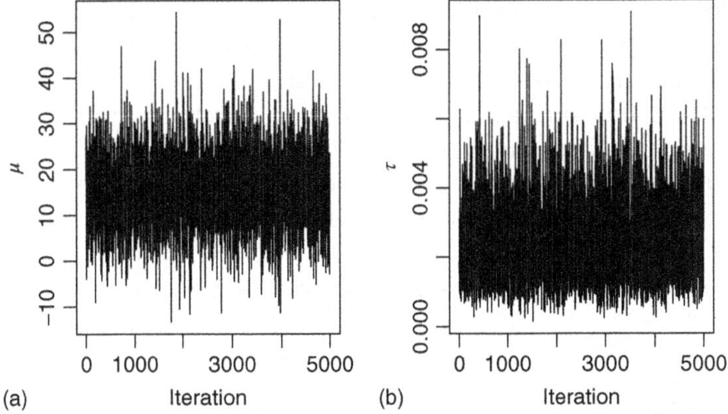

Figure 4.4 Posterior Traces for μ (a) and τ (b) after a 500-Iteration Burn-In

Although there are other MCMC diagnostics which we will consider in Chapter 6, we assume they have all produce satisfactory results. Then we can estimate our parameters with the following code:

```
mean(mu[501:100000])
mean(tau[501:100000])
mean(1/tau[501:100000])
```

In the code above, observe that we calculate the means after omitting the first 500 burn-in iterations of the Markov chain. The resulting estimates ($\mu = 15.8, \tau = 0.00247, \sigma^2 = 520.0$) are consistent with the results obtained from the code in Section 3.6.3. The results also shed light on the path of the Gibbs sampler in Figure 4.2. In the figure, we could visualize that when the variance σ^2 was small (i.e., when the precision τ is large), the Gibbs sampler tended to move μ to the vicinity of 14–16. Here we see that $E(\mu) = 15.8$ is in this range as we informally observed.

4.4 GIBBS SAMPLING AND THE SIMPLE LINEAR REGRESSION MODEL

The linear regression model is an important statistical model and one of the most widely used ones in business and management. In this model, the relationships between a set of predictor (or independent) variables $x_1, x_2, \ldots, x_{j-1}, x_j$ and an outcome (or dependent) variable y are estimated. The relationships are taken to be linear. An intercept β_0 and a series of slope terms $\beta_1, \beta_2, \ldots, \beta_{j-1}, \beta_j$ are used to assess the linear relationships of the predictor variables with y. These slope terms are an important part of the enduring popularity of the regression model. This is because these slope terms also express proportionality relationships. We often encounter proportionality relationships in business. For example, if a company has more assembly lines, it can produce more products. A larger advertising budget allows a firm to purchase more prime-time advertisements. These kinds of common relationships, where we believe that the more (or less) we have of x, the more (or less) we have of y, can be expressed in the regression model. The simple linear regression model is the special case where we have only one predictor variable. Thus the functional form of the simple regression model is

$$y = \beta_0 + \beta_1 x_1 + \epsilon, \tag{4.3}$$

where ϵ is the error term. The error term arises because, typically, prediction is not entirely perfect across all observations. Errors may occur because a relevant predictor variable has been omitted, or because of random fluctuations in the environment causing y to have some intrinsic unpredictability. The error term quantifies the difference between the observed value of y and its predicted value from the regression, $\beta_0 + \beta_1 x_1$. The predicted value can be more compactly written as $\hat{y}$ or as μ: the latter symbol emphasizes that the predicted value is the conditional mean of y given x.

With the functional form specified, we now consider the distribution of the data with the goal of selecting a likelihood function. The most common choice in classical statistics is to take y as having a normal distribution. While our predictions may not be perfect, it would be desirable that our predictions be accurate on average. This leads to the criterion that $E(\epsilon) = 0$ be used for the error terms. Under this criterion, the predicted values are called *unbiased*. The condition that the error terms are *homoscedastic* is also used for the simple linear regression model. This means that the variance of the errors ϵ, σ^2, is assumed to be constant across the different values of x_1. It will simplify the formulas below if we work with $(x - \bar{x})$ instead of x, and also with the precision τ of ϵ, $\tau = 1/\sigma^2$. Then the likelihood function for the simple linear regression model is

$$\begin{aligned} p(y|\beta_0, \beta_1, \tau) &\propto \prod_{i=1}^{n} \tau^{1/2} \exp\left[-\frac{\tau}{2}\left(y_i - \beta_0 - \beta_1(x_{1i} - \bar{x})\right)^2\right] \\ &\propto \tau^{(n/2)} \exp\left[-\frac{\tau}{2}\sum_{i=1}^{n}\left(y_i - \beta_0 - \beta_1(x_{1i} - \bar{x})\right)^2\right]. \end{aligned} \tag{4.4}$$

One possibility for a prior is a flat prior specification similar to that of (3.10), giving

$$\begin{aligned} p(\beta_0, \beta_1) &\propto 1/c, \quad -\infty < \beta_0, \beta_1 < \infty \\ p(\tau) &\propto 1/\tau, \qquad 0 < \tau < \infty. \end{aligned} \tag{4.5}$$

The product of the likelihood and the prior yields the joint posterior distribution

$$p(y|\beta_0, \beta_1, \tau) \propto \tau^{(n/2)-1} \exp\left[-\frac{\tau}{2}\sum_{i=1}^{n}\left(y_i - \beta_0 - \beta_1(x_{1i} - \overline{x})\right)^2\right]. \tag{4.6}$$

The full conditional posterior distributions can be found by making a few substitutions that can be also found in classical ordinary least-squares (OLS) simple linear regression. The sum of squared errors (SSE) is a quantity that contributes to the classical point estimate of the error variance. In OLS regression, the error variance is calculated as $SSE/(n-2)$. In Bayesian simple linear regression, the SSE also appears on the right-hand side of the joint posterior (4.6), namely

$$\sum_{i=1}^{n}\left(y_i - \beta_0 - \beta_1(x_{1i} - \overline{x})\right)^2 = SSE.$$

For Bayesian statistics, SSE is not a point quantity because we have some uncertainty about β_0 and β_1. However, for the full conditional of τ, we treat all of the variables in SSE as known. So the full conditional for τ is

$$p(\tau|y, \beta_0, \beta_1) \propto \tau^{(n/2)-1} \exp\left(-\frac{\tau}{2}SSE\right). \tag{4.7}$$

By now, this kind of formula may look familiar. Referring back to the results in (4.1) should make it appear even more familiar. We recognize (4.7) as a gamma distribution. Thus we write the full conditional as

$$\tau|y, \beta_0, \beta_1 \sim \text{Gamma}\,(n/2, SSE/2). \tag{4.8}$$

Next, for the mathematically curious, we can expand the trinomial in SSE, giving

$$\begin{aligned}
SSE &= \sum_{i=1}^{n}\left(y_i - \beta_0 - \beta_1(x_{1i} - \overline{x})\right)^2 \\
&= \sum_{i=1}^{n}\left(y_i^2 - 2\beta_0 y_i + \beta_0^2 - 2\beta_1(x_{1i} - \overline{x})y_i + \beta_1^2(x_{1i} - \overline{x})^2 + 2\beta_0\beta_1(x_{1i} - \overline{x})\right) \\
&= \sum y^2 - 2\beta_0 \sum y + n\beta_0^2 - 2\beta_1 \sum\left(y(x_1 - \overline{x})\right) + \beta_1^2 \sum(x_1 - \overline{x})^2. \qquad (4.9)
\end{aligned}$$

The last term of the second line sums to zero, causing this term to drop out. The full conditional for β_0 involves only the terms in (4.9) that have

β_0 in them. Factoring out the n and completing the square (Section 3.11) gives

$$\begin{aligned} p(\beta_0|y,\beta_1,\tau) &\propto \exp\left[-\frac{\tau}{2}(n\beta_0^2 - 2\beta_0 \textstyle\sum y)\right] \\ &\propto \exp\left[-\frac{n\tau}{2}(\beta_0^2 - 2\beta_0\overline{y})\right] \\ &\propto \exp\left[-\frac{n\tau}{2}(\beta_0 - \overline{y})^2\right]. \end{aligned} \tag{4.10}$$

This is a normal distribution with mean $\overline{y}$, so we have

$$\beta_0|y,\beta_1,\tau \sim N\left(\overline{y},(n\tau)^{-1}\right) \tag{4.11}$$

which happens to be very similar to (3.11). A similar process (see exercises) shows that

$$\beta_1|y,\beta_0,\tau \sim N\left(\frac{\sum(y(x_1-\overline{x}))}{\sum(x-\overline{x})^2},\left(\tau\textstyle\sum(x-\overline{x})^2\right)^{-1}\right). \tag{4.12}$$

It turns out that the posterior means for β_0 and β_1 are equal to the respective point values of the OLS estimates of β_0 and β_1. Also, it is possible with slightly more work to express the joint conditional posterior for β_0 and β_1 in terms of a bivariate normal distribution. This approach generalizes well to performing Gibbs sampling for multiple regression at the small cost of computing a covariance matrix. An alternative approach to multiple regression involves the Metropolis–Hastings algorithm, which is examined in Section 4.6.

4.5 IN PRACTICE: THE SIMPLE LINEAR REGRESSION MODEL

Bromiley and Marcus (1989) studied the impact of negative business information being made public by looking at major automobile recall announcements. In some cases, negative information gradually becomes disclosed over time, allowing market participants to anticipate negative information. In other cases, the information comes suddenly, creating unanticipated negative information. Bromiley and Marcus (1989) focused on the latter. They report the number of unanticipated major automobile recalls in four periods: 1967–1968, 1972–1973, 1977–1978, and 1982–1983. Here we analyze whether the number of major recalls increased, decreased, or stayed the same over time. We fit a simple linear regression model to their data where x takes on the value 1, 2, 3, or 4 corresponding to the four periods. The dependent variable y is the number of unanticipated major recalls for the two largest manufacturers

Ford (6, 7, 8, and 12 recalls, respectively) and GM (3, 6, 9, and 8 recalls, respectively). We could also consider a count regression for this dataset as in Section 9.2, but for now we will use simple linear regression. The following code is entered in *R*.

```
set.seed(123)
 n.iter <- 100000
 y <- c(6,7,8,12, 3,6,9,8)
 x <- rep(c(1,2,3,4),2)
 xdif    <- x-mean(x); sumx2 <- sum(x^2)
 n     <- length(y); y.bar <- mean(y)
 sumxdify <- sum(xdif*y)
 sumx2dif <- sum(xdif^2)

#storage for parameters
 tau <- b0 <- b1 <- rep(0,n.iter)
#initial values
 tau[1] <- 1; b0[1] <- b1[1] <- 0

#Gibbs sampling
 for (i in 2:n.iter) {
  if(i%%10000==0) cat(i, " iterations complete","\n")
 #sample tau
  error <- (y - b0[i-1] - b1[i-1]*xdif)
  tau[i] <- rgamma(1, n/2, sum(error^2)/2)
 #sample beta0
  b0[i] <- rnorm(1, y.bar, sqrt(1/(n*tau[i])))
 #sample beta1
  b1[i] <- rnorm(1, sumxdify/sumx2dif, sqrt(1/(sumx2dif*tau[i])))
 }
```

The first part of the code involves preparatory data tasks. Of particular importance here is the specification of `xdif`, where $\overline{x}$ is subtracted from x. Other quantities used in the Gibbs sampler are precalculated so that they do not need to be reevaluated `n.iter` $-$ 1 times inside the loop. Storage for the samples is allocated, and initial values are given. Then Gibbs sampling is undertaken using the full conditionals found in Section 4.4. When the run is complete, the behavior of the Markov chain is reviewed with commands such as `plot(ts(tau[1:500]))`, `plot(ts(tau[501:n.iter]))`, and related commands using the other parameters. In all cases, the Markov chain appeared to have moved to its stationary distribution within 10 iterations. We use the first 500 iterations as burn-in, to be conservative. Results can be calculated using `mean`, `sd`, and `quantile` as in Section 4.3. We also report the OLS regression results obtained from the *R* command `summary(lm(y~xdif))`, and contrast the Bayesian results with the OLS results. Results appear in Table 4.1.

TABLE 4.1 Regression Results: Automobile Recall Data

	Bayesian Estimates			OLS Estimates		
	Posterior Mean	Posterior Std. Dev.	95% Credible Interval	Point Estimate	*t*-statistic	*p*-value
β_0	7.372	0.652	(6.059, 8.680)	7.375	13.83	<0.0001
β_1	1.849	0.585	(0.681, 3.016)	1.850	3.88	0.008
σ	1.737	0.625	(0.971, 3.312)			

Interpretation of β_0 and β_1 is as follows: Because the x variable has been centered around its mean, β_0 refers to the predicted value of y at the *average* value of x. This interpretational difference due to the transformation of x is common to both the Bayesian estimates and the OLS ones, and it varies from the usual interpretation of β_0 as the predicted value of y when x is zero. The average value of x is 2.5 in the data here, corresponding to the midpoint between 1972–1973 and 1977–1978. So β_0 indicates that the expected number of unanticipated major recalls is 7.375 during the midpoint between 1972–1973 and 1977–1978.

Our main interest in the analysis is the findings for β_1. The interpretation of β_1 is the usual interpretation for a regression slope coefficient, the estimated change in y for a 1-unit change in x. β_1 is 1.737, indicating that the number of GM/Ford major automobile recalls was increasing over time in the study period. The Bayesian 95% credible interval excludes the value of 0 and the classical p-value also rejects the value of 0. However, the interpretation of the Bayesian and the classical finding has differences. The Bayesian interpretation is straightforward: there is a 95% probability that the true value of β_1 is in the range (0.971, 3.312). The classical two-sided p-value against the null hypothesis does not allow a direct probability conclusion about the data at hand, but rather indicates that, if we were to repeat the procedure a large number of times using samples from the appropriate population, then we would observe values of the β_1 test statistic as large or larger than the one we have observed here about 0.8% of the time if the null hypothesis were true.

Returning to the Bayesian approach, we can calculate the posterior probability that $\beta_1 \leq 0$ by entering

```
sum(ifelse(b1[501:100000]<0,1,0))/99500.
```

The result is that $p(\beta_1 \leq 0) = 0.004$. Because this is a probability of a particular business environment displaying certain characteristics, we could potentially generate income with this information. For example, assuming the situation to be unchanged, financial analysts or risk managers could note the probability above and hedge financial instruments or risk accordingly. Or an expected value could be calculated as the product of a monetary payoff times the probability, and this could be used to incentivize corporate change. However, these kinds of activities are not possible with the classical OLS results because classical inference does not produce posterior probability distributions.

4.6 THE METROPOLIS ALGORITHM

Along with the Gibbs sampler, the Metropolis–Hastings algorithm is one of the most important and widely used MCMC sampling tools. This section will focus on Metropolis' original description of the algorithm (Metropolis et al., 1953), and the next section will discuss Hastings' important extension of it. To summarize the idea behind the Metropolis algorithm, we will make a proposal move (known as a *jump*) to a random numeric value. Then, some of these moves will be accepted and others rejected in such a way as to obtain simulations from the distribution of interest. One way to implement the Metropolis algorithm is the random walk Metropolis algorithm. In the random walk Metropolis algorithm, the chain makes a random proposed jump from its current position. The Metropolis ratio then determines whether that proposed jump is accepted.

Intuition for the idea behind the random walk Metropolis algorithm can be described with reference to Figure 4.5. Suppose the simulation is currently at $x = 1$, indicated by the dashed vertical line, and our target distribution is the standard normal distribution as shown. At this current state, we propose to make a jump to a new location. Suppose for now that the only two places we might jump to are the end points of the horizontal line: that is, we might either jump to $x = 0.5$ or $x = 1.5$, and that a coin toss

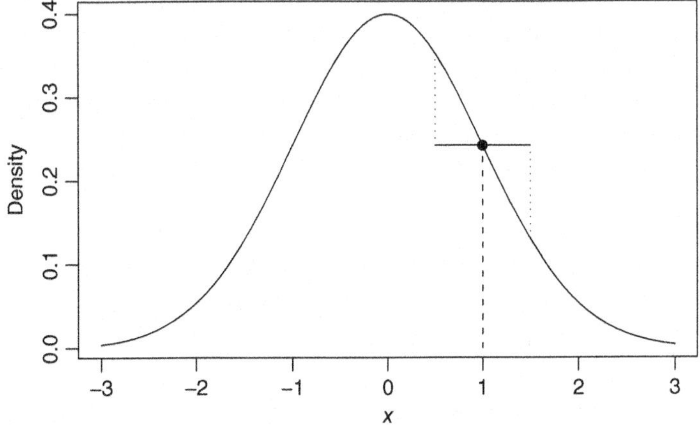

Figure 4.5 Metropolis Algorithm with Two Example Proposed Jumps

be used to make this decision. If the coin comes up heads, we propose to jump to $x = 0.5$, whereas if is tails, we propose to jump to $x = 1.5$.

Next we determine whether to accept the proposal. Metropolis et al. (1953) showed that by accepting some (but not all) proposed jumps according to certain rules, we can simulate from any target distribution. The first rule is that "uphill" jumps are always accepted. In the current example, if the coin comes up heads, we propose to move to $x = 0.5$ and then accept this upward move proposal, thus giving our new simulated value of x as 0.5. The second rule is that "downhill" proposals are accepted according to the ratio of the probability densities of the current location and proposed location. The probability density (or height) of the normal distribution at the circle in the figure is about 0.2419. The probability density of the normal distribution at $x = 1.5$, as shown by the descending dotted line, is about 0.1295. We would therefore accept the downward move 53.5% of the time (0.1295/0.2419). In the remaining cases, we would reject the proposal and remain at $x = 1$. Once the proposed move is either accepted or rejected, the cycle starts over again until the desired number of iterations is completed. Now, it is important to note that the current description of the proposed moves occurring only at the ends of the horizontal line is too restrictive. The reason why only two moves have been discussed is to permit a clear graphical representation. Instead, the Metropolis algorithm entails a symmetric proposal distribution (with infinite possible moves that are more difficult to show graphically). In particular, the Metropolis algorithm indicates that we could randomly propose to move to any location throughout our jumping proposal horizontal line (instead of just the ends), then accept the uphill proposals and accept downhill proposals according to the corresponding probability ratios. Once we have moved (or stayed in the same location), we repeat the process, and continue to do so a large number of times. Surprisingly enough, this simple procedure will allow us to simulate from the normal distribution and almost any other continuous distribution we might think of. A random proposal to move anywhere along the horizontal line is the same as a random draw from the uniform distribution, which is a symmetric distribution for a jumping proposal, and so the rules are applicable. We can summarize the Metropolis algorithm as follows:

1. Select an initial value for the simulation run. As in Gibbs sampling, selecting an initial value that is consistent with the distribution of interest helps speed convergence of the algorithm.
2. For the remaining iterations, i,
   (a) Propose a move to new location $\hat{\theta}$ from current location $\theta^{(i-1)}$ by simulating from a symmetric jumping proposal distribution.

(b) Compare the ratio of the probability densities for the move versus the current location to an acceptance probability, $a^{(i)}$. If

$$\frac{p(\hat{\theta})}{p(\theta^{(i-1)})} > a^{(i)}, \tag{4.13}$$

accept the move by setting $\theta^{(i)} = \hat{\theta}$. Otherwise, reject it and remain at the current location such that $\theta^{(i)} = \theta^{(i-1)}$. Since this ratio will be greater than 1 for all uphill moves, uphill moves will always be accepted.

The attractiveness of the Metropolis algorithm stems from the fact that it does not rely on the full conditional distribution. So when full conditional distributions are unavailable or unwieldy, we can still simulate from the posterior distribution using the Metropolis algorithm. It is perfectly acceptable to use Gibbs sampling for one parameter and the Metropolis algorithm for another in a given problem. This is because both methods produce simulations from the posterior even though they do it in different ways.

4.6.1 In Practice: Simulating from a Standard Normal Distribution Using the Metropolis Algorithm

A test case for putting the Metropolis algorithm to use is to try to simulate from the standard normal distribution (shown in Figure 4.5). The idea is to put the approach to test in a situation where the answers are known in advance. If there are substantial discrepancies from the known answer, we should check our implementation more closely. Here we know that the mean of the standard normal is zero and the standard deviation is 1; hence our results should approximate the known values. We enter the following code in *R* to implement the random walk Metropolis algorithm:

```
n.iter <- 100000
x <- acc <- rep(0, n.iter)
x[1] <- 1

accprob <- runif(n.iter,0,1) #acceptance probability
linelength <- 1 #length of proposal line in Fig 4.4
propmove <- runif(n.iter, 0, linelength) #uniform proposals
propmove <- propmove - 0.5 #allows both + and - moves

for (i in 2:n.iter)
 { prop <- x[i-1] + propmove[i]
  propratio <- dnorm(prop,0,1)/dnorm(x[i-1],0,1)
```

```
  if (propratio > accprob[i])
   {x[i]  <- prop  #accept proposal
   acc[i] <- 1 }  #keep a count of acceptances
  else
   {x[i]  <- x[i-1]} #reject proposal
 }
hist(x[500:100000],100)
```

Before turning to the results, a few additional comments on the code can be made. To be consistent with the horizontal proposal line in Figure 4.5, we simulate move proposals randomly along a line with a `linelength` of 1. Also, *R* runs slowly inside loops. So the acceptance probabilities, a_i, given by `accprob` are calculated in advance. The proposed moves, `propmove`, are also precalculated. Notice that the uniformly distributed `propmove` has 0.5 subtracted away from it. We need to allow the proposals to be both positive and negative to allow the moves to go to both the left and the right. Since the uniform distribution goes from 0 to the `linelength` of 1, subtracting 0.5 from it makes the proposal distribution symmetric around zero (i.e., symmetric around the current location).

A histogram of the results (Figure 4.6) shows that, after a 500-iteration burn-in, x seems to closely follow a standard normal distribution. Using the `mean` command on `x` gives −0.005, a value close to 0 as would be appropriate. However, `sd(x[500:100000])` gives a value of 0.988, which is slightly on the low side for our large value of `n.iter`. Our implementation of the Metropolis algorithm seems to be correct, but we can do better in terms of accuracy. It turns out that the problem is one of efficiency. We are performing 100,000 simulations but it is as if we only have a small number of simulations when it comes to accuracy.

Zooming in on the trace of the Markov chain by entering the command `plot(ts(x[501:1000]))` shows that the chain's behavior can be described as leisurely or timid (Figure 4.7). The majority of the moves

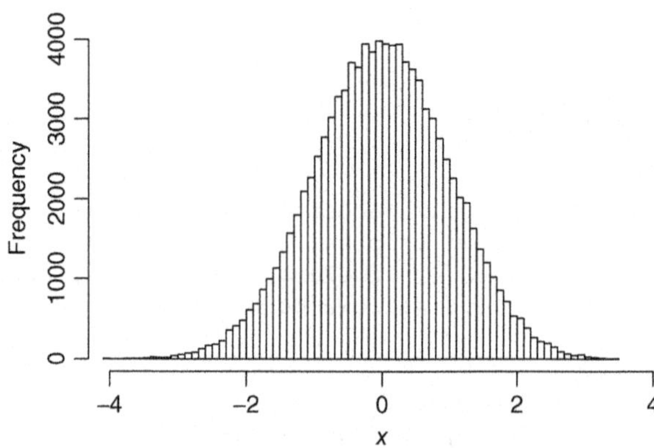

Figure 4.6 Standard Normal Distribution using the Metropolis Algorithm

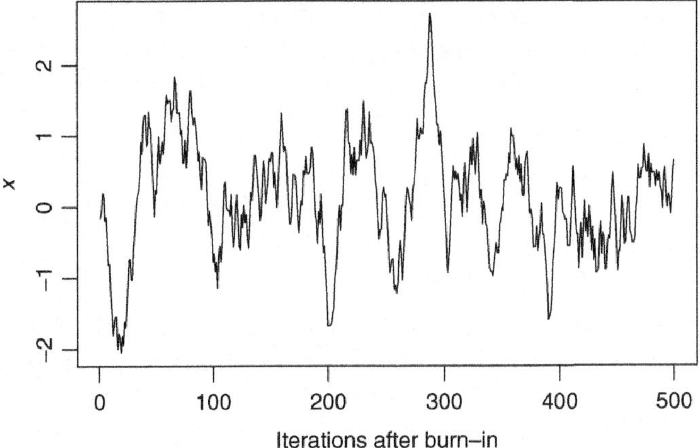

Figure 4.7 Trace Plot of Metropolis Markov Chain

are small, and the chain tends to meander in a local vicinity for numerous iterations before moving on to a substantially different area of the target distribution. Entering `mean(acc)` shows that over 90% of the moves were accepted. The slow movement of the Markov chain is termed *slow mixing* and the consequence of slow mixing is that our estimates are slower to approach their true values. We could always raise the number of iterations but this can become prohibitively slow in larger problems. The Metropolis algorithm guarantees that, with a large enough sample size, we will eventually be able to estimate the parameters as accurately as we require. However, with a poorly tuned algorithm, this may take much longer than we want or, even worse, be impossible with the finite computer memory available on a given system.

It is better to address the slow mixing directly by tuning the algorithm to have better performance. The way to do this is to expand the width of the jumping proposal distribution. This is analogous to expanding the width of the horizontal proposal line in Figure 4.5. Expanding the width will increase the chance of larger jumps being proposed. If the width of the proposal distribution is expanded greatly, then very large jumps to low probability areas will be proposed too often. These proposals will tend to be rejected, and the resulting chain may not move at all for many iterations. A balance is needed to ensure the best mixing. Research by Gelman et al. (1995) shows that Metropolis acceptance rates should be between 23% and 44% for best performance depending on the number of variables. In the case of one variable, a 44% acceptance rate produces the best mixing, while for five or more variables a 23% acceptance rate is recommended. Under these conditions, the algorithm will be making some bold proposals but not so many that the chain fails to move for long periods.

We can make our proposal distribution wider by making edits on two lines of code and then resubmitting the block of code. This gives the following:

```
...
linelength <- 5 #length of proposal line in Fig 4.4
...
propmove <- propmove - 2.5
...
```

where ... denotes the code that has not been changed. The resulting trace plot appears in Figure 4.8. We can see that the chain mixes more quickly after the changes. The Metropolis acceptance rate is now 55.7%, which is closer to the optimal range. The post burn-in standard deviation of x improves to 0.995, while for the mean we also have an improved estimate of −0.0002. At some point, if we were to continue to make the proposal distribution even wider, the trace would begin to appear blocky. The chain would have long runs where no proposals are accepted and the trace would show horizontal lines punctuated by an occasional move up or down. At extremely high widths, the chain would be very unlikely to move and the trace could appear as a long horizontal line. In these situations, we will be getting a very poor return on our computational investment, since a large number of iterations will have done very little to help us characterize the posterior. To get a better return with regard to accuracy, we would want to tune the algorithm to have a more appropriate acceptance rate.

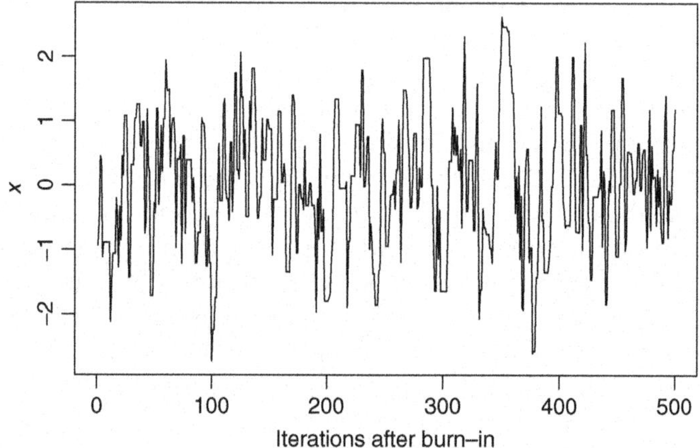

Figure 4.8 Trace Plot of Metropolis Markov Chain with Wider Proposal

4.6.2 In Practice: Regression Analysis Using the Metropolis Algorithm

In Section 4.4, we saw that some work was required in order to find the full conditional distributions needed for the Gibbs sampler in the context of regression analysis. The Metropolis algorithm does not impose this requirement on us. All we need is the product of the prior and the likelihood to proceed. We have already obtained this in (4.4). Since we will need to calculate (4.4) repeatedly, it is worthwhile to write an *R* function that calculates this for us. The product term in most posterior densities involves a product of many small numbers that are close to zero. The resulting product term for the posterior density will therefore be an extremely small number, and for large datasets it can become so small that the computer cannot represent it adequately. A way to avoid this problem is to use logarithms. Then, to find the ratio in (4.13) that the Metropolis algorithm requires, we can take the difference in the logarithms and then exponentiate the result. That is because, by the rules of logarithms, $a/b = \exp\left(\log(a) - \log(b)\right)$. We will also use the associated rule that $a * b = \exp\left(\log(a) + \log(b)\right)$. With these rules in mind, we can create a function for the log posterior density.

```
mylogpost <<- function(b0,b1,tau)
 { logpost1 <- ((n/2)-1)*log(tau)
  resid  <- y - b0 - b1*xdif
  logpost2 <- -(tau/2) * sum(resid^2)
  return(logpost1 + logpost2)
 }
```

A few comments on this function are in order. The function is named `mylogpost` and the use of the command `<<-` allows the function to see the other variables in *R*'s memory. This saves us from having to inform the function about the other variables needed in the calculation, such as `n`, `xdif`, and `y` (good programming practice would be to avoid modifying these "global" variables inside the "local" function, and we follow that practice here). We will be supplying the latest values of `b0`, `b1`, and `tau` to the function. It helps matters to process (4.13) piece by piece. This reduces the chance of errors and makes checking easier. The first term is the one involving τ. By the rules of logarithms, $\log(a^b) = b\log(a)$, which produces the right-hand side of `logpost1`. We need the sum of squared residuals for the second term, so we find `resid` based on the current values of `b0` and `b1`. We then set `logpost2` equal to the second term of (4.13). From the rules of logarithms, the sum of our two `logpost` terms is the desired result so this value is returned to the main program. Note that we do not perform the exp calculation on this sum yet, as it will be done in the main program.

One possible issue you might have considered is whether we can use the Metropolis algorithm for τ even though τ itself does not have a symmetric distribution. The Metropolis algorithm does produce the appropriate (asymmetric) posterior distribution for parameters such as τ because of the generality of the method. Another possible complication for τ is that we need to accommodate the fact that the precision cannot be negative. As τ approaches zero, our symmetric proposal will extend into negative values, which are not meaningful for τ. A solution is to use a *reflected* random walk proposal distribution. In this proposal distribution, negative proposed values for τ are set to their absolute values. This leads to a symmetric proposal distribution because the density from the negative values is folded back in the appropriate manner. For now, we will continue to use the Gibbs sampler for `tau`. This will be revisited when we discuss the Metropolis–Hastings algorithm in Section 4.7.

With the `mylogpost` function now available, we can implement the Metropolis algorithm with the following code:

```
n.iter <- 100000
y <- c(6,7,8,12, 3,6,9,8); x <- rep(c(1,2,3,4),2)
n <- length(y); xdif <- x-mean(x);
tau <- b0 <- b1 <- rep(0,n.iter)

tau[1] <- .5; b0[1] <- 0; b1[1] <- 0 #initial values

b0accprob <- runif(n.iter,0,1) #acceptance probs
b1accprob <- runif(n.iter,0,1)
b0acc <- b1acc <- rep(0,n.iter) #tuning check
b0length <- 5; b1length <- 5;  #change for tuning
b0move <- runif(n.iter, 0, b0length) - b0length/2
b1move <- runif(n.iter, 0, b1length) - b1length/2

for (i in 2:n.iter)
 { if(i%%10000==0) cat(i, " iterations complete","\n")
 #sample tau using Gibbs
  error <- (y - b0[i-1] - b1[i-1]*xdif)
  tau[i] <- rgamma(1, n/2, sum(error^2)/2)
 #simulate b0 posterior
  prop <- b0[i-1] + b0move[i]
  propratio <- exp(mylogpost(prop,  b1[i-1],tau[i-1])
         - mylogpost(b0[i-1],b1[i-1],tau[i-1]))
  if (propratio > b0accprob[i])
   {b0[i]  <- prop   #accept proposal
   b0acc[i] <- 1 }   #count of acceptances
  else
   {b0[i]   <- b0[i-1]} #reject proposal
 #simulate b1 posterior
  prop <- b1[i-1] + b1move[i]
  propratio <- exp(mylogpost(b0[i],prop,  tau[i-1])
```

```
        - mylogpost(b0[i],b1[i-1],tau[i-1]))
  if (propratio > b1accprob[i])
   {b1[i]   <- prop    #accept proposal
   b1acc[i] <- 1 }    #count of acceptances
  else
   {b1[i]    <- b1[i-1]} #reject proposal
}
```

Most of the ideas from the code appearing in Section 4.6.1 reappear in the above code with the main difference being that we need an additional section of code for each additional variable. The program begins with defining the variables and setting the initial values. Then the looping portion is reached. In Section 4.6.1 we only simulated from the standard normal distribution, so the Metropolis ratio had two `dnorm` terms in it. Now we are simulating from the posterior distribution, so we replace `dnorm` with `mylogpost`. The `propratio` calculations now use the properties of logarithms to help ensure numerical accuracy. In each ratio, we supply the proposed value along with the current values to `mylogpost`. Then we supply the current values to `mylogpost`. The difference of these is exponentiated to obtain the Metropolis ratio.

The acceptance probabilities for the β parameters are both near 36%. These are within a reasonable range. Inspection of the trace plots suggests that posterior convergence was reached before iteration 500. The posterior means $(\beta_0 = 7.373, \beta_1 = 1.856, \sigma = 1.743)$ and standard deviations $(\beta_0 = 0.649, \beta_1 = 0.596, \sigma = 0.619)$ are similar to the values obtained by Gibbs sampling in Table 4.1.

4.7 HASTINGS' EXTENSION OF THE METROPOLIS ALGORITHM

The Metropolis algorithm relies on a symmetric jumping proposal distribution. Hastings (1970) showed that the requirement of a symmetric distribution could be relaxed with a modification of (4.13). His modification permits the use of an asymmetric distribution for a jumping proposal distribution, $j(\cdot)$. Hastings' variation of (4.13) is to accept a move when

$$\frac{p(\hat{\theta})j(\theta^{(s-1)})}{p(\theta^{(s-1)})j(\hat{\theta})} > a^{(i)}. \tag{4.14}$$

Using an asymmetric proposal can help enhance MCMC's performance when the target distribution is asymmetric, such as for precision and variance parameters. If we can match the proposal distribution closely to the target distribution, this approach can become very efficient. Because Hastings' extension is so valuable, it is common for researchers to refer to this as the *Metropolis–Hastings algorithm*. In fact, this has become the default name for the algorithm even when Hastings' extension is not actually being used in a given context.

In addition to the random walk proposal distributions we have seen previously, another type of proposal distribution is the *independence* proposal. When used in the context of the Metropolis algorithm, we have the *independence sampler* Metropolis algorithm. In the random walk, the proposed value depends on the current value. Successive values tend to be correlated with each other, and this reduces the overall efficiency of the algorithm as we have seen in Figure 4.7. With the independence sampler, values are proposed independently, again contributing to greater efficiency. A caveat of this approach is that we need to make sure that our proposal distribution has heavier tails than the target distribution (Gilks et al., 1996). If the tails of the proposal distribution are lighter than the target, the algorithm will not venture into the tails as often as required. This will cause parameter estimates to be inaccurate. On the positive side, the independence proposal does not need to be tuned. We happen to know what our posterior distribution for τ looks like from previous analyses. Figure 4.9 shows the posterior along with three possible proposal distributions for an independence sampler. The dotted line displays the density of a Gamma(1, 1) distribution. This would be a suitable proposal distribution as there is plenty of weight on both the left and right sides as compared to the target. The dashed line displays the density of a Gamma(2, 2) distribution. This, by contrast, would not be a desirable proposal distribution since the values close to zero are

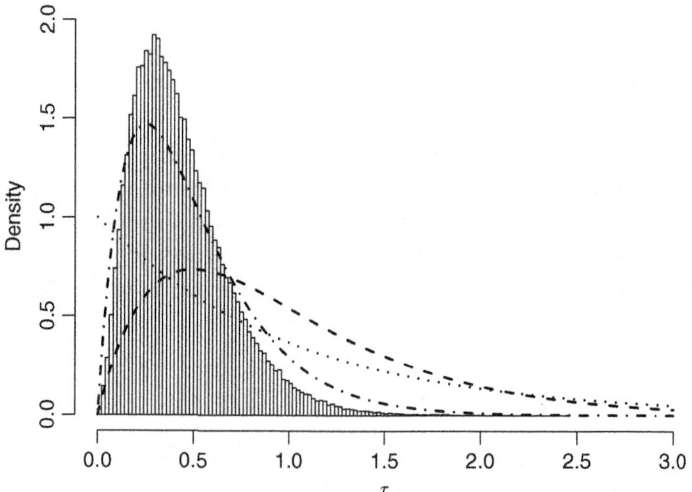

Figure 4.9 Target Posterior Distribution and Three Hastings Proposal Distributions

underrepresented by the Gamma(2, 2) distribution as compared to the target. The dash-dotted line shows the Gamma(2, 4) density. This is also suitable as there is slightly more weight on both the left and right sides as compared to the target. Because the Gamma(2, 4) density is closer to the target posterior than the Gamma(1, 1) density, it will be more efficient in terms of proposing suitable values to the algorithm.

It is convenient here that we already know what the posterior distribution looks like, but in real life, typically, we will not have this information. To obtain this information, we could consider doing a pilot run using an independence sampler with a reasonably flat proposal such as the Gamma(1, 1) here. After the pilot run, we would respecify the proposal distribution based on what we had learned. While this may appear to be more time consuming than the regular Metropolis algorithm, recall that for the random walk Metropolis we will more than likely have to tune the algorithm. In effect, this means that a pilot run (or two or three) will be needed for the random walk approach as well.

4.7.1 In Practice: The Metropolis–Hastings Algorithm

In the code below, the asymmetric Gamma(1, 1) distribution will be used as an independence sampler proposal distribution. We could have also considered the Gamma(2, 4) distribution here, but we will use the Gamma(1, 1) for the time being. By adding a few lines of code to the Metropolis algorithm of Section 4.6.2 and rewriting the tau sampling section (all changes in **boldface**), we implement an independence sampler Metropolis–Hastings algorithm as follows.

```
...
b1accprob <- runif(n.iter,0,1)
tauaccprob <- runif(n.iter,0,1)
b0acc <- b1acc <- tauacc <- rep(0,n.iter)
tauprop <- rgamma(n.iter, 1, 1)
...
 { if(i%%10000==0) cat(i, " iterations complete","\n")
 #sample tau using Metropolis-Hastings
  Hratio  <- dgamma(tau[i-1],1,1)/dgamma(tauprop[i],1,1)
  propratio <- exp(mylogpost(b0[i-1],b1[i-1],tauprop[i])
         - mylogpost(b0[i-1],b1[i-1],tau[i-1]))
  if ((propratio*Hratio) > tauaccprob[i])
   {tau[i]  <- tauprop[i]  #accept proposal
   tauacc[i] <- 1 }  #count of acceptances
  else
   {tau[i]   <- tau[i-1]} #reject proposal
 #simulate b0 posterior
...
```

We have added `tauaccprob` and `tauprop` to the set-up portion of the code. These create the Metropolis–Hastings acceptance probabilities and the Gamma(1, 1) independence samples, respectively. In the main loop we have added `Hratio` as the Hastings term. The numerator of `Hratio` has the Gamma(1, 1) density at the proposed value of `tau`, and the denominator has the Gamma(1, 1) density of `tau`'s current value. The `if` statement then checks the product of `propratio` and `Hratio` following (4.14).

Inspection of the trace plots suggests that posterior convergence was reached before iteration 500. The posterior means ($\beta_0 = 7.373, \beta_1 = 1.849, \sigma = 1.731$) and standard deviations ($\beta_0 = 0.662, \beta_1 = 0.588, \sigma = 0.666$) are again similar to the two previous sets of results. The code for `tauacc` is not essential because we do not need it for tuning, but it may be of interest to see the proposal acceptance rate. Here, 42.1% of the proposals were accepted as found by entering `mean(tauacc[501:100000])`.

4.7.2 The Relationship Between the Gibbs Sampler and the Metropolis–Hastings Algorithm

In the previous section, we saw that our Gamma(1, 1) proposal led to 42.1% of the proposals being accepted. What if we had used one of the workable alternatives such as the Gamma(2, 4)? Taking a look at Figure 4.9 suggests that this would be a better approach. The Gamma(2, 4) tends to overemphasize the tails and underemphasize values in the vicinity of 0.3–0.5, but still appears to be an improvement over the more conservative Gamma(1, 1).

It is easy to modify our Hastings sampler to try this out. The `tauprop` specification of `rgamma` needs to be changed from the Gamma(1, 1) distribution to the Gamma(2, 4) by substituting `2, 4` for `1, 1` in that line. A similar change needs to be made for `Hratio` in the `rgamma` portions. Rerunning the code with these changes gives an acceptance probability for `tau` of 71.0%. This is an improvement, but can we do even better? In fact, we can ensure that the jump is always accepted with 100% probability with the appropriate choice of a proposal distribution. Referring back to the Hastings' acceptance ratio of (4.14), note that, if our proposal distribution is the full conditional for the parameter of interest θ_P, namely $p(\theta_P|y, \theta_1 \ldots \theta_{P-1})$, we have

$$\frac{p(\hat{\theta})\, j(\theta^{(i-1)})}{p(\theta^{(i-1)})\ j(\hat{\theta})} > a^{(i)}$$

$$\frac{p(\hat{\theta})\, p(\theta_P^{(i-1)}|y, \theta_1^{(i)} \ldots \theta_{P-1}^{(i-1)})}{p(\theta^{(i-1)})\, p(\theta_P^{(i)}|y, \theta_1^{(i)} \ldots \theta_{P-1}^{(i)}))} > a^{(i)}$$

$$\frac{p(\theta_P^{(i)}|y, \theta_1^{(i)} \ldots \theta_{P-1}^{(i)})\, p(\theta_P^{(i-1)}|y, \theta_1^{(i)} \ldots \theta_{P-1}^{(i-1)})}{p(\theta_P^{(i-1)}|y, \theta_1^{(i)} \ldots \theta_{P-1}^{(i-1)})\, p(\theta_P^{(i)}|y, \theta_1^{(i)} \ldots \theta_{P-1}^{(i)})} > a^{(i)}$$

$$1 > a^{(i)}. \tag{4.15}$$

On line 2 of (4.15) we set the jumping distribution to be the full conditional. On line 3 we explicitly indicate that the target distribution we wish to simulate from is the full conditional. We then see these terms cancel. Summing up (4.15) in words, this means that with a full conditional proposal our acceptance rate is 100% as desired. This is because 1 in the last line of (4.15) is greater than or equal to any random uniform number, $a^{(i)}$, we simulate. This proves that the Gibbs sampler is a special case of the Metropolis–Hastings algorithm which occurs when the jumping distribution $j(\cdot)$ is the full conditional.

Either the independence proposal or the random walk proposal can be used with the original Metropolis algorithm, and, additionally, either can be used with the Metropolis–Hastings algorithm. Thus we have identified four different variations of this kind of MCMC method here. Thinking more broadly for a moment, we see that all the different MCMC methods described here have individual strengths and weaknesses. There is no silver bullet approach—instead, we will have to consider the trade-offs and use which one is best for a given scenario.

4.8 SUMMARY

This chapter has examined some of the technical details associated with MCMC methods. We have seen the following:

- The posterior is proportional to the prior multiplied by the likelihood. In some simple cases, we can find an exact expression for the posterior by doing this multiplication and working out the details. This gets more difficult as the complexity goes up.
- Fortunately, we can simulate from the posterior using MCMC methods. MCMC methods can be shown to produce simulations from distributions of interest.
- There are two main MCMC methods used in the literature: the Gibbs sampler and the Metropolis–Hastings algorithm. It turns out that the Gibbs sampler is a special case of the Metropolis–Hastings algorithm.
- The advantage of the Gibbs sampler is that it can be somewhat faster to code when we are writing our own MCMC sampler programs. It will usually run faster than Metropolis–Hastings since there are no rejection calculations. The disadvantages is that we must find the full conditionals, which sometimes are available and sometimes are not. Typically, conjugate priors need to be used for Gibbs samplers.
- Another advantage of the Metropolis–Hastings algorithm is that it can be implemented with almost any prior and that we do not need to explicitly find the full conditional distributions. The disadvantage is that it requires tuning (or pilot runs) and additional programming code compared to Gibbs sampling.

4.9 EXERCISES

1. Explain in a sentence or two what a Markov chain is.
2. Explain in several sentences why it is necessary to perform output analysis when using MCMC.
3. Explain the differences between the random walk Metropolis algorithm and the independence sampler Metropolis algorithm. What are the pros and cons of using each?
4. Modify the random walk Metropolis algorithm code at the beginning of Section 4.7.1 so that `linelength` is 20. What acceptance probability do you obtain?
5. As in the above exercise, modify the random walk Metropolis algorithm code on page 81 so that `linelength` is 20. Plot the trace for the chain. Describe the appearance of the trace and contrast this with the appearance of the trace in Figure 4.8.
6. Explain why the Gibbs sampler is actually a special case of the Metropolis algorithm.
7. Compare the conditional distribution for τ in (4.2) with the marginal distribution for τ in (3.14). What term(s) change and by what quantity(ies)?
8. Starting with (4.6) and (4.9), show that the full conditional posterior for β_1 is as stated in (4.12).
9. Consider the Hastings algorithm and modify the car recall regression code of Sections 4.6.2 and 4.7.1. Modify the code to use a Gamma(2, 20) proposal distribution. Comment on the results.

5

ESTIMATING BAYESIAN MODELS WITH *WinBUGS*

We have seen how to perform Bayesian inferential calculations by hand using algebra and some common statistical distributions. Yet it would be attractive to have software that could assist with these tasks so that we could be free to focus on more important modeling issues. The freely available software *WinBUGS* (Lunn et al., 2000) does precisely this for a very large family of models. The release of *WinBUGS* (as well as its DOS-based predecessor *BUGS*) has been an important practical milestone for Bayesian methods, and it is no overstatement to credit *WinBUGS* as a contributor to the current popularity of Bayesian methods. *WinBUGS* can be downloaded from `http://www.mrc-bsu.cam.ac.uk/bugs/winbugs/contents.shtml`. The download URL contains detailed installation instructions but (at the time of writing) to summarize, it is necessary to download three things: the main executable file, a patch to upgrade the executable file to version 1.4.3, and the free registration key.

Various extensions to *WinBUGS* are available. These include the *GeoBUGS* extension, which permits spatial statistics and statistical mapping; the *Jump* interface, which allows for model selection/model comparison using reversible jump Monte Carlo Markov chain (MCMC); and the *PKBUGS* interface for pharmacokinetic models. Additional resources that come with *WinBUGS* are an online manual accessible by pressing `F1` in the software and two sets of examples comprising 34 worked analyses complete with data, code, and associated commentary.

Bayesian Methods for Management and Business: Pragmatic Solutions for Real Problems,
First Edition. Eugene D. Hahn.

Yet another worthwhile resource is the *BUGS* discussion mailing list at `http://www.jiscmail.ac.uk/`. Once you are at this URL, type `BUGS` in the search box to find the list. Subscribing is worthwhile to enter the ebb and flow of *WinBUGS*-related questions and developments, but perhaps even more so is searching through the extensive back archive for information on topics of relevance to your current models of interest.

As described in Section 2.7, in Bayesian inference the model consists of the priors, the likelihood, and the functional form. The advantage of *WinBUGS* is that the user only needs to supply these three things to the program (along with the data and initial values for the simulation). *WinBUGS* then automatically samples from the appropriate posterior distributions. This eliminates the need to manually find the posterior distribution as was done in Chapters 2, 3 and 4. *WinBUGS* accomplishes this with an expert system which inspects the priors and the likelihoods to determine the best way to sample from a parameter's conditional posterior distribution. Two of the sampling methods *WinBUGS* uses are Gibbs sampling and Metropolis–Hastings sampling (Chapter 4), but others are also available depending on the expert system's determination. The user then compiles the model and tells *WinBUGS* to begin MCMC sampling. In the absence of any error messages, the program proceeds and then *WinBUGS*'s output analysis tools can be used to examine and summarize the posterior distributions.

In this chapter, we will gain experience with the core functionality of *WinBUGS*. We will also use *WinBUGS* to estimate the commonly used statistical models including those from the linear regression family.

5.1 AN INTRODUCTION TO *WinBUGS*

A *WinBUGS* session begins by launching the *WinBUGS* executable file with a double-click of its icon. The program loads and displays the license agreement. Figure 5.1 shows the menu bar of *WinBUGS*. At far right is the `Help` menu. Under this menu, select `About` to ensure that the latest version of *WinBUGS* (which is version 1.4.3) is installed. At left, the `File` menu brings up the standard file operations such as `New`, `Open`, `Save`, and `Close`. *WinBUGS* files are saved with the `.odc` extension, and an alternative way to open a *WinBUGS* `.odc` file is to drag it onto the *WinBUGS* icon and drop it there. To the right of `File`, the `Tools` menu is infrequently used except when upgrading *WinBUGS* to the latest version. `Edit` gives standard commands such as `Copy` and `Paste`, while `Attributes` allows for font formatting to be changed.

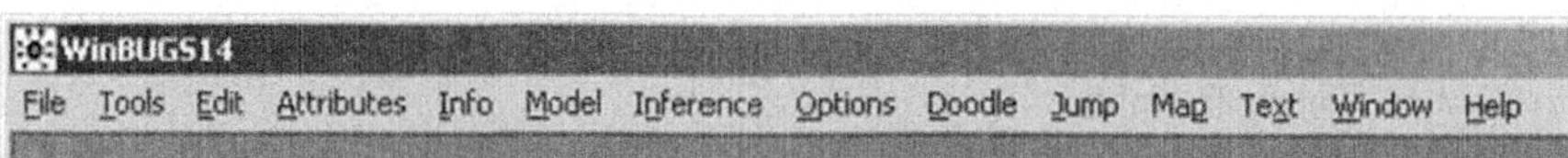

Figure 5.1 *WinBUGS'* Menu Bar

Toward the right of the menu bar is the `Text` menu. This menu provides the `Find` and `Replace` options for searching through text and making changes. `Window` provides window management, while `Help` can be used to call up the user manual and *WinBUGS* program examples. All the menus we have seen so far are standard fare for many programs and will likely seem familiar. These menus will enable us to type in a model for *WinBUGS* to process and allow us to perform necessary edits and file operations.

The remaining menu options are specific to *WinBUGS*. The two most regularly used menus are the `Model` and the `Inference` menus. Items under the `Model` menu allow the completed text specification of the model to be submitted to *WinBUGS* for checking. If *WinBUGS* can interpret the model without any errors, additional items under this menu will allow the MCMC sampling to begin. The `Inference` menu allows the monitoring of parameters and can be used to produce summary statistics, graphical depictions, and diagnostic information. In the event of trouble, the `Info` menu allows logs to be created as well as more detailed information about *WinBUGS'* sampling operations. The `Doodle` menu provides a way to specify a model graphically with little to no code. It invokes *DoodleBUGS* and the *Doodle Editor*, which then allow the user to draw ovals and arrows to indicate parameters and relationships. The full power of *WinBUGS* is not accessible through *DoodleBUGS* however, so we do not use it in this book. The `Map` menu is for the spatial statistics capabilities provided by the *GeoBUGS* extension and is an advanced option.

5.2 IN PRACTICE: A FIRST *WinBUGS* MODEL

Code for *WinBUGS* has many resemblances to code for *R* because both programs have a common ancestor in terms of their programming languages. Specification of the prior, likelihood, and functional form all occurs within the context of a `model{ ... }` section. Separate sections are used for the data and the initial values. We begin with a reanalysis of the Lenz and Engledow (1986) gross sales data in Section 3.6 and use *WinBUGS* to estimate the normal mean and standard deviation. The likelihood function therefore involves the normal distribution. We can omit the functional form in this analysis because we do not create any additional functional relationships among the parameters. Finally, the prior for μ will be a vague normal prior with mean zero and precision 0.0000001, while for the τ we use a Gamma(0.001, 0.001) prior.

To obtain a fresh document for our program, click on `New` in the `File` menu in *WinBUGS*. The code below can then be entered.

```
model
{
#likelihood
```

```
  for (i in 1:n)  {y[i] ~ dnorm(mu, tau)}
#priors
  mu   ~ dnorm(0, 0.0000001)
  tau ~ dgamma(0.001, 0.001)
#calculated quantities
  sigma <- 1/sqrt(tau)
}
```

A *WinBUGS* model code statement begins with `model {` and ends with a closing `}`. Comments can be placed after the # sign. The `for` loop applies our normal likelihood function to the `n` observations in the data `y`. Indexing for the loop has been specified here with the loop counter `i`. This will allow us to reference a particular value of a column of data in `y` by using the index in brackets; that is, `y[i]` picks out the `i` th value of `y`. As we saw in Chapter 2, ~ means "is distributed as," and the tilde sign ~ is used to specify this relationship in *WinBUGS*. Note that the normal density (`dnorm`) is parameterized in terms of the precision (instead of the variance or the standard deviation) as well as the mean. As a result, we will need to place a prior on the unknown mean parameter `mu` and the unknown precision parameter `tau`.

In addition to the normal distribution and the gamma distribution, *WinBUGS* recognizes 21 other types of distributions. These include all the distributions discussed in this and previous chapters: the uniform distribution (`dunif`), the beta distribution (`dbeta`), the binomial distribution (`dbin`), the Poisson distribution (`dpois`), the chi-squared distribution (`dchisqr`), and the t-distribution (`dt`). The full list of distributions natively supported by *WinBUGS* can be accessed by calling up the User Manual under the `Help` menu, and then clicking on the word `Distributions`.

Our code indicates the prior for `mu` is a normal prior with a mean of zero and a small precision of 0.0000001. Recalling the definition of the precision, this prior precision is equivalent to a prior variance of 1,000,000 or a prior standard deviation of 1000. Here our prior specifications use proper priors (i.e., they are valid statistical distributions) whereas previously, in Section 3.5, we used the improper priors of (3.10). This in part reflects the design strategy of *WinBUGS* , where the use of proper priors is encouraged so as to ensure that posterior distributions are proper. The prior for `mu` is very spread out and flat compared to the data and so this prior should lead to results that are similar to the strictly flat prior for μ of (3.10). This kind of prior is sometimes called a *just proper* prior because it so closely resembles an improper prior. Similarly, our code for `tau` is a gamma distribution with small values (0.001) for both of its parameters. Looking back at Figure 3.3, we see that this kind of prior will be very flat across most of its range with a near-vertical spike at zero. This again provides a just proper prior that will satisfy *WinBUGS* with results similar to the flat prior of (3.10) in many cases. The last line of the code before the braces is a calculated quantity we may be interested in, the posterior standard deviation `sigma`. We can have *WinBUGS* calculate this quantity as it runs, and thereby obtain its posterior distribution.

After the end of the `model` statement, we can add the data and initial values as follows:

```
#data
list(y=c(4.5, 2.1, 30.0, 65.8, 7.8, 26.5, 7.5, 5.2, 6.3, 2.3),
n=10)
#inits
list(mu=0, tau=1)
```

The data and initial values are supplied using list statements. Vectors such as `y` appear inside collections `c(...)`, while scalars such as `n` and `mu` can stand alone.

With the data typed in, the Specification tool is accessed from the `Model` menu (Figure 5.2). Selecting the initial word `model` at the top of the code window and clicking on the `check model` button tells *WinBUGS* to review the code. If *WinBUGS* finds an error, it will beep and print a small error message at the very bottom left of the application window. *WinBUGS* will also make an educated guess about the location of the error and will place the cursor there. Some of the more commonly encountered errors include variable or distribution misspellings, inappropriate use of commas or brackets, and inappropriate use of ~ versus `<-`. Starting with small programs and reusing portions of programs that are known to work are two ways to minimize the introduction of errors. If *WinBUGS* is satisfied with the model syntax, it will display the

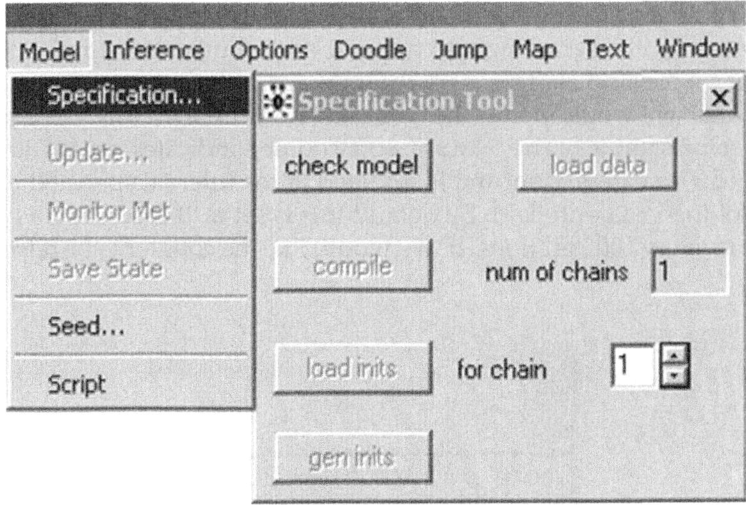

Figure 5.2 *WinBUGS'* Specification Tool

message 'model is syntactically correct' in the status bar at the bottom left portion of the application window (Figure 5.3).

After success with check model, highlight the word list in the data portion of the code window just prior to the actual data and just after the line labeled #data. Clicking load data in the specification tool will instruct *WinBUGS* to read and load the data. *WinBUGS* will display the message "data loaded" in the status bar if the data follows *WinBUGS*'s format requirements. Next the compile button is pressed. The status bar will contain the message "model compiled" if this step is successful. Finally, we need to load the initial values. The word list in the portion of the code window that contains the initial values should be highlighted. Then the load inits button should be pressed. Alternatively, we can ask *WinBUGS* to generate its own initial values for us. *WinBUGS* will do this by generating a random value from each parameter's prior and then using it as the initial value. When a parameter has a vague prior, this randomly generated value can be extremely far from the posterior distribution. If a particular model exhibits slow mixing (Section 4.6.1), then it may take a very long time for the effect of the extreme initial value to diminish. Convergence occurs sooner if sensible initial values are specified, so providing initial values to *WinBUGS* is a recommended practice. In the absence of specific knowledge about the posterior distributions, zero can be tried as an initial value for a pilot run of normal distribution parameters, while a value of 1 can be tried for precision/variance parameters. Once all steps have been completed, *WinBUGS* shows the message "model is initialized" in the status bar.

The Update tool is used to produce MCMC samples based on the provided model, data, and initial values (Figure 5.4). The number of desired samples (or updates) can be entered in the field next to the word updates. Each time the update button is pressed, *WinBUGS* will proceed with the sampling and append that number of samples to the current set of samples. This is useful if we are in the midst of the analysis and decide we would prefer the increased numerical accuracy of a larger number of samples. The iteration counter indicates what iteration *WinBUGS* is currently on, and is updated at the interval specified by the value of the refresh field. By default this is set to 100, so the counter is updated every 100 iterations. If you would like the counter to be refreshed

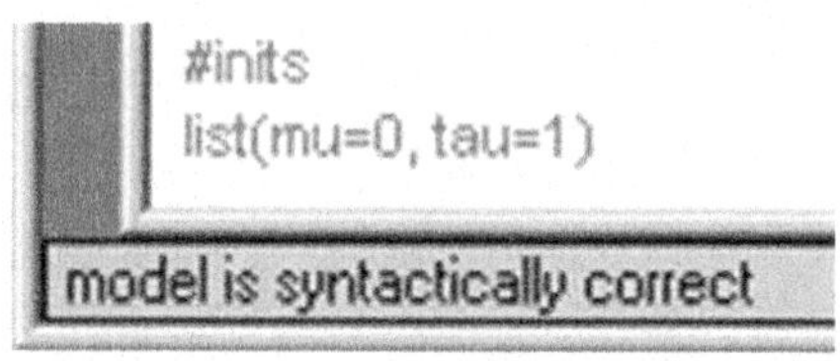

Figure 5.3 *WinBUGS'* Status Bar

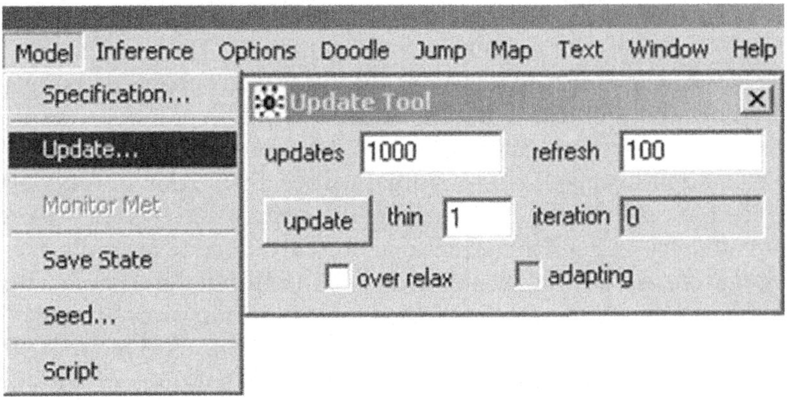

Figure 5.4 *WinBUGS'* Update Tool

more regularly, the value of `refresh` can be reduced. However, this comes at a considerable computational cost (try setting the value to 1, for example, and observe the slower performance compared to 100) and so should be left at 100 or even set higher to speed up processing. Table 5.1 shows the performance impact of different refresh rate values for the current model running on a laptop computer. We see that adding a zero or two to the default 100 refresh rate gives an easily obtained performance improvement. In general, the performance improvement will vary depending on the run lengths and the model specification, and may be less noticeable for short runs of more computationally intensive models.

Other options include the `thin` and `over relax` checkboxes. These can be selected while attempting to reduce the autocorrelation of the chain (Section 6.2) at the expense of longer run times. Finally, the `adapting` checkbox will be checked when *WinBUGS* is performing an adaptive tuning step for Metropolis sampling or another kind of MCMC called *slice sampling*. Any samples produced while this box is checked will be ignored by *WinBUGS*. Very briefly, this behavior of the `adapting` process occurs as a consequence of Markov chain sampling theory and cannot be overridden.

TABLE 5.1 Run Times for 100,000 Iterations of Gross Sales Data Model

Value of `refresh`	*WinBUGS* Run Time (s)
10	624
100	62
1000	6
10,000	<1

In a typical *WinBUGS* session, the Update tool would be used just after *WinBUGS* indicates the model has been initialized. These initial updates would serve as the burn-in iterations for the Markov chain. Although the required number of burn-in iterations depends on the model and the initial values, burn-in periods of 1000–5000 iterations are often sufficient for simpler models and well-chosen initial values. Inspection of the Markov chain trace plots is useful for assessing the appropriateness of a particular burn-in period. For the purposes of our current code example, we use 1000 iterations as burn-in, so the value `1000` (without the comma) should be entered in the Update tool and the `update` button pressed.

Once the burn-in iterations are complete, the *WinBUGS* Sample Monitor tool (Figure 5.5) is used to tell *WinBUGS* which parameters to monitor. To monitor a parameter, type its name into the `node` field in the Sample Monitor tool. Once the parameter's name has been typed, the `set` button becomes clickable. Clicking the `set` button causes *WinBUGS* to monitor the output of the Markov chain for that parameter. Only parameters that have been defined in the model code can be monitored. So if by chance a parameter of interest has not been specified, the model will have to be corrected and resubmitted. For our current example, we monitor `mu`, `tau`, and `sigma`. Once a parameter has been entered, we can click on the drop-down button at the right end of the `node` field to review what parameters are available. If needed, we can click on one of the parameters in the drop-down menu and its name will appear in the `node` field. Parameters that have been `set` can be removed with the `clear` button. In our current example, we are satisfied with the three parameters listed, so `clear` does not need to be used.

When the parameters of interest have been set, the Update tool is used to continue the run of the Markov chain. In our example model, we use 10,000 iterations for analysis. The number `10000` is entered into the `updates` field, and the `update` button is clicked. The status bar displays the message "`model is updating`." Once the updates are

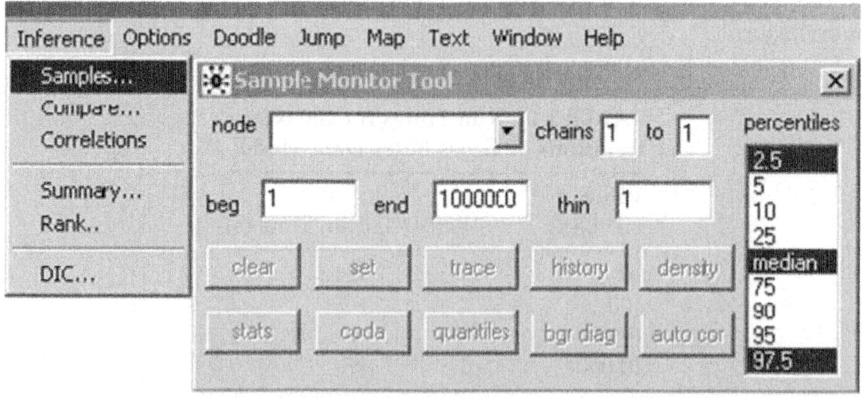

Figure 5.5 *WinBUGS'* Sample Monitor Tool

complete, the Sample Monitor tool is selected. Typing an asterisk in the `node` field will cause subsequent button clicks to provide results for all parameters. Alternatively, if we prefer to look at one parameter at a time, the drop-down menu can be used to select a parameter of interest. We review the behavior of the Markov chain by clicking on the `trace` button on the Sample Monitor tool. This button provides a zoomed-in view of the most recent behavior of the Markov chain (Figure 5.6). At this zoomed-in level, the chain appears to be moving around readily, suggesting that reasonably good mixing is occurring. This output is called the *dynamic trace* in *WinBUGS* because it will be continuously refreshed in real time if the Update button is pressed. The scale of the y-axis of the dynamic traces also contains useful information. We see in Figure 5.6 that the y-axis for `sigma` ranges from 0 to 100 despite the recent values being in the range of 20–40. This suggests that the chain for `sigma` has had some large excursions toward higher values at some point in the past. Also, the trace for `mu` suggests that it may have had some large excursions toward lower values.

The trace tool displays only 200 iterations, and this is not enough information to gauge the overall mixing of the chain from its start to its finish. We would also like to find out more about the large excursions for `sigma`. Clicking on the `history` button on the Sample Monitor tool gives a zoomed-out view of the entire Markov chain. The trace history begins at iteration 1000 when monitoring is initiated. The plots indicate good mixing. The occasional excursions for `sigma` are consistent with the notion that `sigma` has a long upper tail. By contrast, `mu` appears to have one substantial downward excursion near iteration 9000. With a total run of 10,000 iterations, it seems conceivable that such an excursion could have happened by chance. As an aside, it is often more informative to

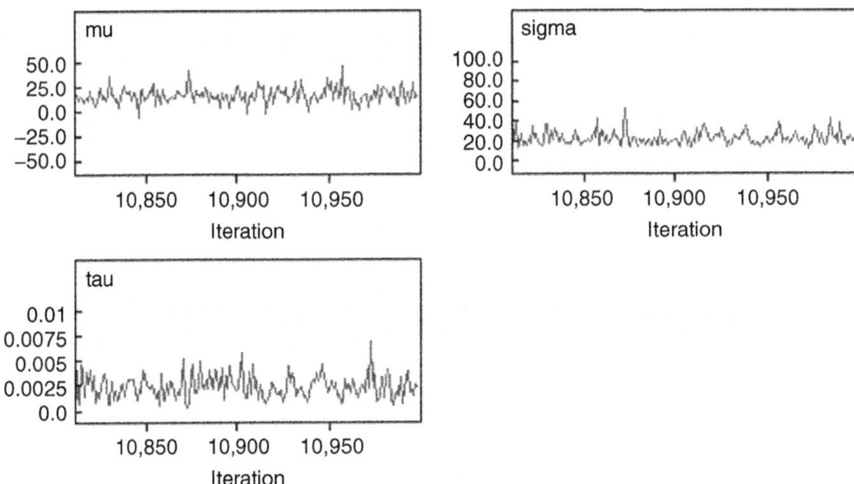

Figure 5.6 Dynamic Trace Output

look at the entire trace by using the trace plot provided by the `history` button, so for the rest of this book we refer to this output as the "trace plot" whereas we refer to the output of the `trace` button as the "dynamic trace" (Figure 5.7).

The evidence reviewed so far suggests that the Markov chains have achieved their stationary distributions for all parameters. We can now use other *WinBUGS* tools to display and summarize the distributions of the parameters. Clicking the `density` button on the Sample Monitor tool after typing an asterisk in the node field produces graphical displays of the parameters' densities (Figure 5.8). The *WinBUGS* default values for the density plots are typically acceptable, but customization of the plots is possible. Right-clicking on a plot brings up a Plot Properties dialog box where graphical options for the plot can be changed. The `Margins` tab allows greater or lesser amounts of whitespace to be placed around each plot. The `Axis Bounds` tab allows the limits of the x- and y-axes to be changed. In our current example, Figure 5.8 shows that the x-axis

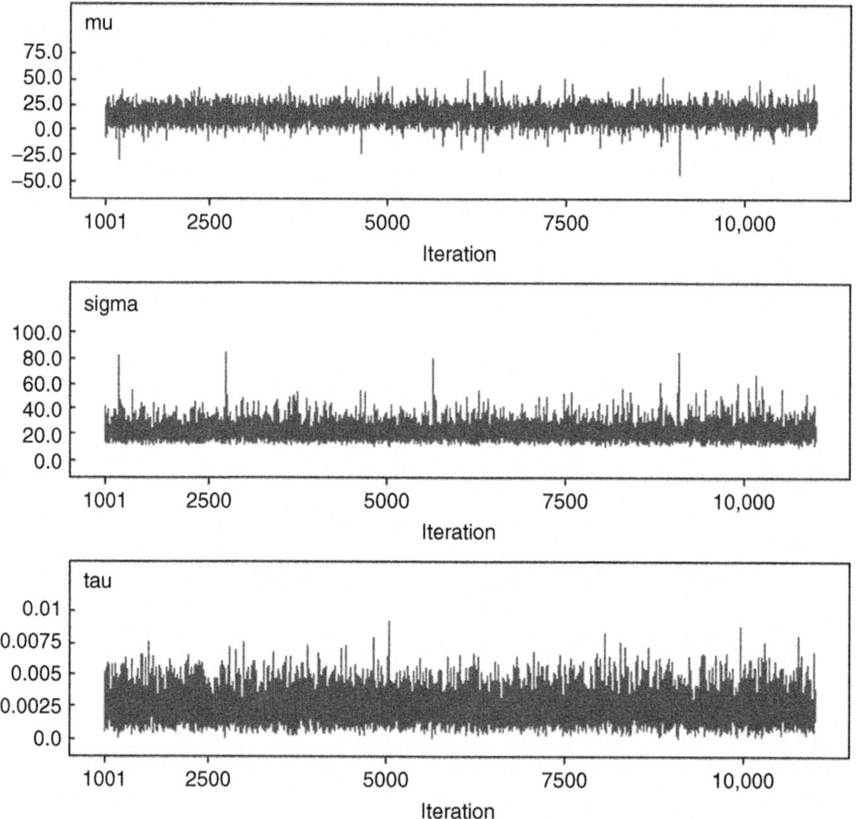

Figure 5.7 *WinBUGS'* Trace History Output

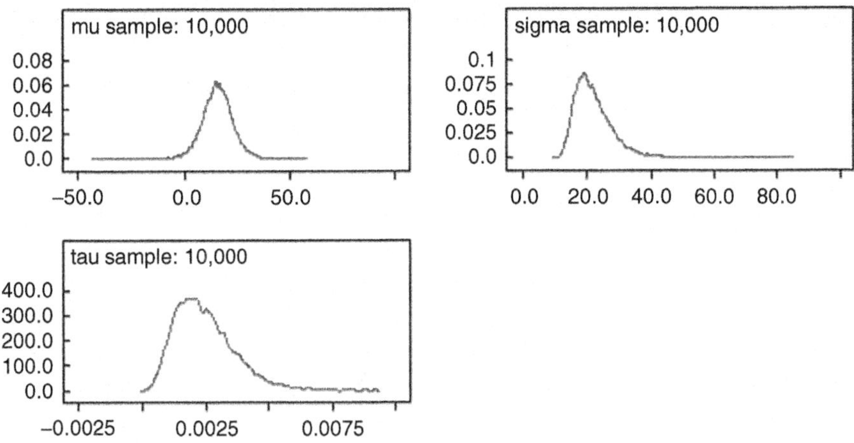

Figure 5.8 *WinBUGS* Density Plots

for `tau` extends to −0.005, an impossible value for precision. We might want to alter the `x-min` value under `Axis Bounds` for `tau` to be zero. The `Titles` tab allows the plot titles such as "mu sample: 10000" to be changed to other text. Alternatively, the title text can be deleted if desired. The `All Plots` tab allows changes to be made to all plots simultaneously (which is useful if a large number of parameters is being monitored), while the `Fonts` tab permits changes in the axis and title fonts. There is also a `Special` button which allows the smoothing parameter for the densities to be changed. Raising the value above the default value of 0.2 makes the density curve appear rougher, while lowering it makes the density appear more perfectly smoothed.

The goal of our analysis is to support inference regarding our unknown parameters μ and σ. We also report on results for τ for illustrative purposes. Clicking on the `stats` button provides summary statistics for the parameter distributions. This output will appear in a new window titled `Node statistics`. Portions of this output for the current example appear in Table 5.2. By default, *WinBUGS* prints output with four decimals of precision, but this can be increased by clicking on the `Options` menu, selecting `output options`, and increasing the precision. The results can be compared with the results of Section 3.6.3, where the same model was estimated using the Jeffreys' prior, which we have approximated here. The posterior mean of μ and its 95% credible interval appearing in Table 5.2 are similar to the values calculated in Section 3.6.3 (note that two different approaches were used to get the marginal distribution for μ in Section 3.6.3—the *WinBUGS* results are consistent with both). *WinBUGS* estimates the posterior mean as 15.73 versus 15.79 for the direct approach based on Equation (3.13) and 15.80 based on integrating out τ. *WinBUGS* gives the 95% credible interval for

μ as (1.289, 30.31), while the two approaches of Section 3.6.3 give (1.320, 30.17) and (1.366, 30.24), respectively.

The `Node statistics` window also lists the posterior medians for the parameters. For skewed distributions, it may be that the median is of interest at times. Here, the distribution of σ is skewed, and we observe that the posterior median is 20.94, which is lower than its posterior mean. As an aside, the posterior median also arises as the natural Bayesian estimator in certain advanced decision-theoretic contexts (Bernardo and Smith, 1994, p. 257). The default credible interval in the `Node statistics` window is the 95% interval. However, at the far right of the Sample Monitor tool, other percentiles can be selected so as to provide credible intervals of different widths such as the 90% credible interval. We can also change the beginning and the ending value of the Markov chain by changing the values of the `beg` and `end` in the Sample Monitor tool. Usually, these can be left alone because we would like to take advantage of all of our simulations, but sometimes it is useful to raise the value of `beg` if we believe our burn-in period was not long enough.

5.3 IN PRACTICE: MODELS FOR THE MEAN IN *WinBUGS*

In Section 5.2, we estimated the normal mean and variance at the same time in a single sample. This model is known as the *single-sample t-test* in classical statistics. By changing our *WinBUGS* code, we can implement other related models. In the following, we examine the Bayesian models corresponding to the classical single-sample z-test and the two-sample t-test.

5.3.1 Examining the Single-Sample Mean

If the variance or standard deviation is known, then in classical statistics the single-sample t-test for the mean reduces to the single-sample z-test. The z-test is well known in classical statistics, but note that in the Bayesian approach we do not need to consider the z-score for making decisions about the mean. Instead, if we would like to assess whether a particular *parameter value* is credible or not, we can find the posterior credible

TABLE 5.2 Results for Gross Sales Data

Parameter	Mean	Std. Dev.	95% Credible Interval
μ	15.73	7.313	(1.289, 30.31)
σ	22.07	6.084	(13.79, 36.64)
τ	0.00246	0.00116	(0.000746, 0.00228)

interval for the parameter and see whether the value is inside the credible interval or not (Lindley, 1965, ch. 5, 6, Zellner, 1971, ch. 2.6).

Suppose as an example that the standard deviation for the data of Section 5.2 is 20. For a Bayesian analysis, the known value for the standard deviation can be supplied to *WinBUGS* by placing it in the data listing, as shown below. Also, suppose we like to calculate the standardized z-score associated with our unknown sample mean, not for inferential purposes but just out of curiosity. The formula for the standardized z-score is $z = (\mu - \mu_r)/(\sigma/\sqrt{n})$, where μ_r is the reference value. In many instances $\mu_r = 0$, and we can begin with that value here. Supplying the code for calculated values or data transformations to *WinBUGS* allows us to sample these quantities as long as they have an unknown or stochastic component. Here, entering the code `z <- (mu - mu.r)/(sigma/sqrt(n))` will allow us to sample from `z` because `mu` is unknown. The following code implements the single sample test for the mean with the option for calculating the standardized z-value for arbitrary μ_r.

```
model
{
#likelihood
  for (i in 1:n) {y[i] ~ dnorm(mu, tau)}
#priors
  mu ~ dnorm(0, 0.0000001)
#calculated quantities
  tau <- 1/(sigma*sigma)
  z <- (mu - mu.r)/(sigma/sqrt(n))
}
#data
list(y=c(4.5, 2.1, 30.0, 65.8, 7.8, 26.5, 7.5, 5.2, 6.3,
   2.3), sigma=20, mu.r=0, n=10)
#inits
list(mu=0)
```

Sampling can be initiated in *WinBUGS* by following the steps described in Section 5.2. Here we ask *WinBUGS* to monitor `mu` and `z` by using the Sample Monitor tool. Selected results appear in Table 5.3. The value of μ is now more precisely estimated when we assume that σ is known and here it corresponds exactly to the empirical mean, 15.8. The estimate of standard deviation of μ (6.319) is now quite close to the standard error value which can be calculated using the standard error formula $\sigma/\sqrt{n}$ for our current values of σ and n.

Although there would be no reason to report it in a Bayesian analysis, *WinBUGS* gives the mean value of z as 2.499, which is almost exactly equal to the classical standardized z-score (2.498) generated by the same dataset. The principal use of this parameter is pedagogical—we see that Bayesian and classical methods can give very similar numerical results in

TABLE 5.3 Single-Sample Mean Test for Gross Sales Data with Standard Deviation Known

Parameter	Mean	Std. Dev.	95% Credible Interval
μ	15.8	6.319	(3.314, 28.27)

certain kinds of models when vague priors are used. Note also that *WinBUGS* will compute a standard deviation and 95% credible interval for z. This illustrates a difference between Bayesian and classical statistics. In particular, we see that in the Bayesian approach our uncertainty about μ carries over to z.

5.3.2 The Two-Sample *t*-Test

The two-sample comparison of means (or *t*-test in classical statistics) extends the single-sample mean analysis of Section 5.2, and the *WinBUGS* code for this model follows this extension. The two-sample *t*-test model described here is also known as the *independent samples t-test*. The *t*-test can be used when the data contains one outcome variable where the mean is of interest. The outcome variable is typically called the *dependent variable* because we suspect the outcome variable depends on or is influenced by other variables. The influencing variable in the *t*-test is membership in one or the other group. Thus we expect group membership to be influential in affecting outcomes, which in turn are summarized by the group outcome means. The influencing variable is typically called the *independent variable* because in this model we assume that group membership affects the mean outcome. The independent variable in the *t*-test is called a *categorical* (or *nominal*) variable. This kind of variable is a grouping variable that carries no other information except as a label to distinguish the groups from one another. The dependent variable in the *t*-test is a *quantitative* or metric variable. Quantitative variables, such as the amount of money a firm has, differ from categorical variables because they convey information about measurable quantities and magnitudes.

The data in Table 5.4 originally appeared in Mascarenhas and Aaker (1989, Table 4), who examined the effect of market barriers to entry on firm performance in the petroleum drilling industry. Firm performance was measured as return on drilling assets (RDA). Group 1 consisted of companies that had a strategic focus on shallow onshore drilling, while Group 2 consisted of companies that focused on deep onshore drilling. Mascarenhas and Aaker (1989) hypothesized that greater barriers to entry for deep drilling should allow Group 2 companies to generate more profits than Group 1. Table 5.4 lists the mean performance of these two groups of companies by year.

TABLE 5.4 Mean Performance (RDA) by Strategic Focus

Year	Group 1	Group 2	Year	Group 1	Group 2
1974	0.242	0.184	1979	0.285	0.187
1975	0.476	0.218	1980	0.260	0.226
1976	0.732	0.221	1981	0.330	0.231
1977	0.496	0.222	1982	0.204	0.140
1978	0.427	0.252			

In developing the model for *WinBUGS*, the standard t-test is based on the assumption that the dependent variable is normally distributed. The likelihood for our model is therefore the normal distribution. In terms of the functional form, each group is hypothesized to have its own mean. If we use the unequal variances t-test, each group is also hypothesized to have its own separate variance. We adopt vague priors for both groups. This leads us to consider the following specification:

$$y_{i,1} \sim \text{Normal}(\mu_1, \tau_1)$$
$$y_{i,2} \sim \text{Normal}(\mu_2, \tau_2)$$
$$\mu_1, \mu_2 \sim \text{Normal}(0, 0.0000001)$$
$$\tau_1, \tau_2 \sim \text{Gamma}(0.001, 0.001).$$

Here, i indexes the 9 years of available data. The focal point of the t-test is the difference in the means and whether this difference is nonzero. We can calculate the difference in *WinBUGS* , and this will gives us the posterior distribution of the difference. We can then examine this distribution and see whether the value of zero seems plausible. The following *WinBUGS* code (see also **WinBUGS Code 5.2 two sample t-test.odc**) implements the unequal variances t-test.

```
model
 {
 #likelihood
  for (i in 1:n)
    { y1[i] ~ dnorm(mu1, tau1)
      y2[i] ~ dnorm(mu2, tau2) }
#priors
  mu1 ~ dnorm(0, 0.000001)
  mu2 ~ dnorm(0, 0.000001)
  tau1 ~ dgamma(0.001, 0.001)
  tau2 ~ dgamma(0.001, 0.001)
#other calculated parameters
  diff <- mu1-mu2
  sigma1 <- 1/sqrt(tau1)
  sigma2 <- 1/sqrt(tau2)
 }
```

```
#data
list(y1 = c(0.242, 0.476, 0.732, 0.496, 0.427, 0.285, 0.26,
    0.33, 0.204), y2=c(0.184, 0.218, 0.221, 0.222, 0.252,
    0.187, 0.226, 0.231, 0.140), n=9)
#inits
list(mu1=0, mu2=0, tau1=1, tau2=1)
```

To estimate this model, we enter this code into a *WinBUGS'* code window and follow the *WinBUGS* processing steps in Section 5.2. A 1000-iteration burn-in period was used, and model inferences were based on a subsequent 10,000 iterations of the Markov chain. Inspection of all trace plots (using the `history` button in the *WinBUGS'* Sample Model tool) suggested that the chains had reached their stationary distributions during the burn-in period. Selected results appear in Table 5.5. We see that, contrary to the study hypothesis, Group 1's mean RDA performance (μ_1) is higher than that of Group 2, despite Group 2 having greater barriers to entry. The 95% credible interval for the posterior distribution of $\mu_1 - \mu_2$ (using `diff` in *WinBUGS*) excludes the value of zero. Thus the value of zero would not be considered credible. We can also calculate the classical *t*-test using *R* and its `t.test` command. We might be concerned about *t*-test assumption violations given the smaller sample sizes and the dependence of the classical *t*-test on the central limit theorem. For example's sake, we proceed with the classical *t*-test. *R* reports the sample means as $\mu_1 = 0.384$ and $\mu_2 = 0.209$, which are very close to the *WinBUGS* results in Table 5.5. *R* gives a *t* statistic of 3.07 with an associated *p*-value of 0.0141, again broadly similar to the findings from *WinBUGS* that contradict the study hypothesis.

5.3.3 An Alternative Parameterization of the Two-Sample *t*-Test

We can also conceptualize the two-sample *t*-test model of Section 5.3.2 in terms of an overall mean and group-specific departures from the mean. This approach emphasizes what is common to both groups through an overall mean parameter μ, and also what is different about the groups through a difference parameter δ. Although the emphasis is different, mathematically the results are equivalent. We consider this model for three reasons here. First, this approach gives us some experience with writing the functional form of the model in an alternative way. As

TABLE 5.5 Results for RDA Performance Data

Parameter	Mean	Std. Dev.	95% Credible Interval
μ_1	0.383	0.0642	(0.254, 0.513)
μ_2	0.209	0.0143	(0.181, 0.238)
$\mu_1 - \mu_2$	0.174	0.0658	(0.0417, 0.307)

a consequence, it illustrates that the t-test is a *linear model* where the predicted value of a data point can be expressed as a sum of weighted terms. This sum of weighted terms is called the *linear predictor*. Second, it gives us the opportunity to look at some of the interpretational aspects of writing the model in an alternative form. Third, we will see in future chapters that consideration of alternative ways of specifying the functional form can lead to better MCMC performance.

The specification that we now consider is as follows:

$$\begin{aligned}
y_{i,1} &\sim \text{Normal}(\mu_1, \tau_1) \\
y_{i,2} &\sim \text{Normal}(\mu_2, \tau_2) \\
\mu_1 &= \mu + \delta \\
\mu_2 &= \mu - \delta \\
\mu, \delta &\sim \text{Normal}(0, 0.0000001) \\
\tau_1, \tau_2 &\sim \text{Gamma}(0.001, 0.001).
\end{aligned}$$

We see that the likelihood for the data is specified in the same way as in the previous model. However, in this model the predicted value for Group 1, μ_1, is influenced by two parameters. The first is an overall mean term μ. This term is shared for both groups and hence the predicted value of Group 1 will be influenced partly by Group 2, and vice versa. Therefore, μ can be interpreted as the average performance of both groups in the industry sample. The second term in the predicted value for Group 1 is the group-specific offset δ. Increasing the values of δ positively influences the predicted value of Group 1. Thus if the overwhelming bulk of the posterior distribution for δ was found to be above zero, it would indicate that Group 1 outperformed the industry average, as indicated by the industry sample.

The predicted value for Group 2, μ_2, is also influenced both by the overall mean μ and by the group-specific offset δ. However, increasing the values of δ negatively influences the predicted value of Group 2. For both groups, we see that the linear predictors μ_1 and μ_2 take the form of a weighted sum. The weight for δ is positive for Group 1 and negative for Group 2, while the weight for μ is positive for both groups.

The following *WinBUGS* code (see also **WinBUGS Code 5.2 two sample t-test.odc**) implements the model just discussed. We again use the RDA data of Mascarenhas and Aaker (1989) with this model for comparative purposes. In doing so, we have omitted the data statement below because it is the same as that of the previous two-sample t-test model.

```
model
 {
 #likelihood
```

```
 for (i in 1:n)
   { y1[i] ~ dnorm(mu1[i], tau1)
     y2[i] ~ dnorm(mu2[i], tau2)
 #functional form
     mu1[i] <- mu + delta
     mu2[i] <- mu - delta
   }
#priors
  mu ~ dnorm(0, 0.000001); delta ~ dnorm(0, 0.000001)
  tau1 ~ dgamma(0.001, 0.001); tau2 ~ dgamma(0.001, 0.001)
#other calculated parameters
  sigma1 <- 1/sqrt(tau1); sigma2 <- 1/sqrt(tau2);
 }
#data: same as previous
#inits
list(mu=0, delta=0, tau1=1, tau2=1)
```

The results in Table 5.6 were obtained from a run of 10,000 MCMC iterations after a 1000-iteration burn-in had occurred. The overall RDA for all firms considered was estimated as 29.6%. The overall difference in firm performance resulting from different strategic orientation was estimated by δ as 8.72%. This parameter had a 95% credible interval which excludes the value of zero; hence we again conclude that different strategic orientations lead to different levels of performance. Group 1 firms performed significantly above the overall mean by 8.72%, while Group 2 firms performed significantly below the overall mean by 8.72%. Comparing these findings to those of Table 5.5, we see that doubling the value for δ, 8.72%, gives 17.4%, which equals the estimated value of $\mu_1 - \mu_2$ in Table 5.5.

Comparing the parameters' standard deviations in the two sets of results also reveals some minor differences. In Table 5.6, the standard deviations for the parameters μ and δ are for all practical purposes identical at an approximate value of 0.033. This is because the entire dataset is being used to estimate both μ and δ in the model corresponding to Table 5.6. In Table 5.5, the standard deviation for Group 1's μ_1 (0.064) was considerably larger than that of Group 2's μ_2 (0.014). In the model corresponding to Table 5.5, each group's mean parameters are being estimated separately. To summarize, our two analyses of the RDA data provide identical overall conclusions but have slightly different nuances in terms of interpretation. In our first analysis, the model specification

TABLE 5.6 Results for RDA Performance Data: Alternative Parameterization

Parameter	Mean	Std. Dev.	95% Credible Interval
μ	0.296	0.0335	(0.229, 0.362)
δ	0.0872	0.0334	(0.0196, 0.153)

treated both groups as being entirely independent of each other, while in our second analysis the model emphasized that both groups shared a common overall component but also had some individualized characteristics. The choice of which model to use therefore comes down to which of the two interpretations is more relevant for a particular situation and research question.

5.4 EXAMINING THE PRIOR'S INFLUENCE WITH SENSITIVITY ANALYSIS

Bayes' theorem shows how the prior and the likelihood come together to form the posterior distribution. Given the potential for the prior to substantially influence the posterior, it is worthwhile to try multiple priors and compare the different sets of results obtained with an eye toward any changes or differences. This practice is known as *sensitivity analysis*. In sensitivity analysis we hope to uncover whether the results are sensitive to the influence of the prior. Alternatively, we may find that our results are robust to a certain degree of alterations in the prior if we find the results change only trivially under these differing priors. We can distinguish between two types of situations where prior sensitivity analysis would be undertaken: when informative priors are used, and when vague or noninformative priors are used.

5.4.1 Sensitivity Analysis with Informative Priors

In most real-world situations, we have at least some prior information about the business situation at hand. We may have past sales histories for similar products, awareness of competitors' likely strategies, or information from early adopters, beta testers, and/or pilot runs. If data from the past is available, it can be used to create an *informative* prior. For example, we could calculate the mean and the variance (or precision) of last year's data and use these values to form a prior distribution for next year's data. The data distribution may happen to resemble a common distribution such as the normal distribution and, if so, we may be satisfied with a prior that uses that common distribution where the mean and the variance have been matched. This technique is known as *moment matching*.

If the distribution of the data does not resemble a common distribution, we may instead use the histogram prior. Entering our data into *R* and using *R*'s `hist` command might produce a result such as that of Figure 5.9. To find the probability of each bin, we take its observed frequency and divide this by the total number of observations, which is 20 here. We can then use these bins and probabilities as a prior for our model. If we have a discrete prior such as when the data are only integers, a discrete histogram prior can be implemented in *WinBUGS* (see the instructions entitled "Specifying a discrete prior on a set of values" under the Advanced Use of the BUGS Language Section of the manual). It is important to note, however, that events that have a zero prior probability will also have a zero posterior probability by Bayes' theorem. Hence, with reference to Figure 5.9, we should add some bins and use expert judgment if we believe that the posterior distribution might be outside the range 8–12.5. A data-based informative prior is often termed an *objective* informative prior as long as we acknowledge that the choice of which data to use as relevant for a given situation is itself a subjective decision.

If we do not have hard data with which to form an informative prior, we may also form an informative prior using expert judgments. Such a prior is called a *subjective informative prior*. This process typically requires some time investment because converting subjective beliefs to probability distributions is not a familiar task for most experts. Moreover, people may make mistakes and may be inconsistent in their judgments (Chaloner, 1996).

A thorough process for eliciting expert judgments involves at least four stages (Garthwaite et al., 2005). In an initial stage, the experts are chosen and then given training about representing their judgments probabilistically. Next, the second stage involves the experts rendering their judgments. A distribution is then fitted to these judgments in the third stage. In the fourth stage, the outputs of the third stage are reviewed with a focus on adequacy and fidelity of the representation provided by the fitted distribution. If the output is found wanting, the second (or potentially even the first) stage is restarted and refinements are conducted.

In the third stage, we may want to use a common distribution such as the normal distribution and try to fit this distribution to the judgments. Or, we may find it easier to let the experts provide a histogram of judgments, and then smooth the histogram to obtain the fitted distribution

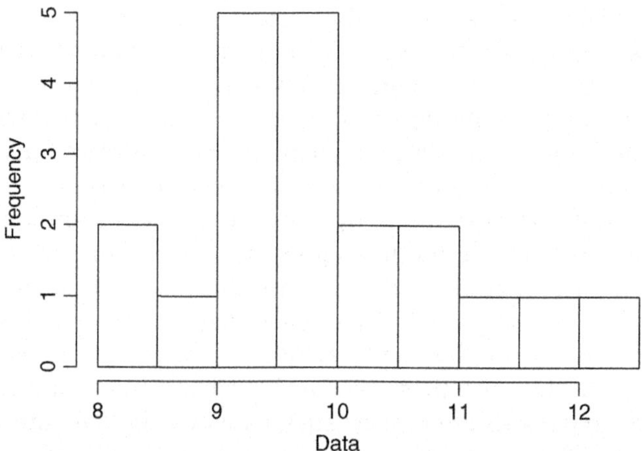

Figure 5.9 Histogram Distribution

(Press, 2003, ch. 2). An expert who does not have hard data may also create a histogram prior and express his or her beliefs using it directly. A probability concept familiar to most experts is the odds ratio as used in betting (e.g., there might be a 2 : 1 odds for a certain team being the winner in a sports event). A histogram prior can be formed on the basis of a series of odds ratio judgments (Hahn, 2006), and then the prior can used for inference with MCMC methods.

Once we believe we have an appropriate informative prior (whether using hard data or expert judgments), sensitivity analysis with informative priors often involves comparing the results under the informative prior with the results under a noninformative prior. Consumers of the research may not always hold the same prior beliefs as the expert, or may simply wish to see what the results would be like without the influence of the prior so as to better gauge the informative prior's effect.

In business scenarios, there may even be situations where we would simultaneously use multiple priors for the same data and same model. As an example, we might be interested in knowing whether a new experimental product will exhibit some outcome or characteristic (such as eventually breaking even, as measured by projected sales volumes). We may start with a "data-derived" prior that captures the notion that the product will break even based on data from other projects. We might also produce an "enthusiastic expert" prior as well as a "skeptical expert" prior for product parameters (Spiegelhalter et al., 1994). Finally, we could also use a noninformative prior. We collect data as the product is tested and goes to market. If at some point the posterior probability of success (such as breaking even) goes below a predefined point, changes are made such as managerial reorientation or product termination. With Bayesian inference and multiple priors, we can have multiple real-time perspectives on the new product that run in parallel from parallel analyses. If the product begins to appear promising even under the skeptical prior, this would provide additional evidence about for the viability of the product and any beta-testing or premarket testing can be concluded early so that full product launch can begin. Conversely, in business there is often the problem of knowing when to abandon an idea that is not delivering on its promise. But if it seems unlikely that desired outcomes will be obtained even with the enthusiastic prior, this would be compelling evidence to discontinue the product early without going into a full product rollout. Similar kinds of analyses occur routinely in the medical clinical trials literature, where an expensive trial may be able to be concluded early based on especially favorable (or unfavorable) outcomes.

5.4.2 Sensitivity Analysis with Noninformative Priors

The use of a carefully constructed informative prior can have important benefits such as reducing time and cost in beta-testing, but the use of a poorly constructed informative prior will undermine the believability and relevance of the results. Given the time investment required for a carefully constructed informative prior, it is much more commonplace to see noninformative priors used in practice (as we have used in this book). There are a number of ways to produce noninformative priors besides the Jeffreys' prior. These include reference priors (Bernardo, 1979), unit-information priors (Kass and Wasserman, 1995), and *g*-priors (Zellner, 1986). However, all the methodological variety in effect illustrates that there is no single universally accepted way to produce a noninformative prior for all possible circumstances and all possible models. Because there is no automatic solution for choosing a noninformative prior, we must choose a prior on a case-by-case basis and assess its impact using sensitivity analysis.

5.4.3 In Practice: Pre-sensitivity Analysis: Graphically Examining a Mean Parameter's Prior and Posterior Distribution

As we have seen in Section 2.4, a prior that is a constant value (or nearly so) across a broad range that includes the posterior distribution will tend to have little effect on the posterior distribution. Thus we will often choose distributions that have this property. As an initial check (which can be called *pre-sensitivity analysis*), a graphical comparison of the prior and posterior distribution can be performed. A graphical comparison usually can be performed both by the person conducting the research analysis and by the consumers of the research. Hence it is worthwhile to invest the time on this check. In Figure 5.10, we compare the posterior distribution for μ_1 as analyzed in Section 5.3.2 against its prior using the following *R* code:

```
mu1prior <- rnorm(10000, 0,      1/sqrt(0.000001))
mu1post  <- rnorm(10000, 0.383, 0.0658)
```

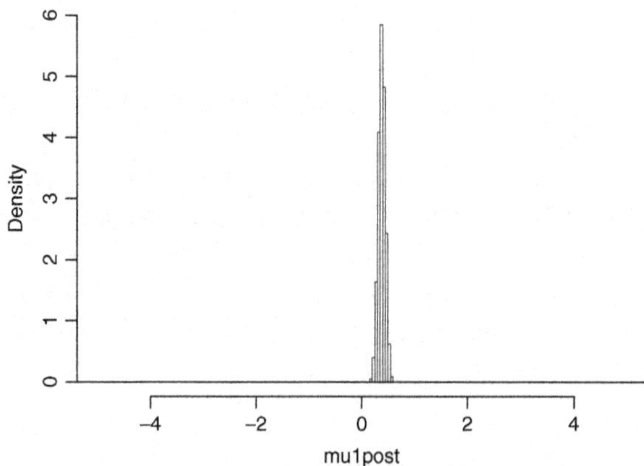

Histogram: posterior distribution; solid line:prior distribution (indistinguishable from x-axis)

Figure 5.10 Prior and Posterior Distributions for μ_1 of Section 5.3

```
hist(mu1post,prob=TRUE, xlim=c(-5,5))
lines(density(mu1prior))
```

The prior distribution, mu1prior, is simulated according to the prior mean and prior precision given in the *WinBUGS* code from Section 5.3. Similarly, the posterior distribution, mu1post, is simulated using moment matching with the posterior mean and standard deviation for μ_1 as listed in Table 5.5. We ask *R* to plot a histogram of mu1post in such a way that some room to the left and the right of the bulk of mu1post is retained. This will allow us to view any curvature or nonlinearity in mu1prior. Figure 5.10 shows that the posterior for μ_1 is highly concentrated around its posterior mean of 0.383. By contrast, the prior distribution cannot be distinguished from the horizontal x-axis. We conclude that this prior distribution probably will not affect the posterior distribution very much when we perform the sensitivity analysis in the next step. Note that, technically speaking, the distribution of mu1post is a t-distribution, while in the *R* code above we have approximated this by using the normal distribution. As we are doing an informal graphical check of the prior, this approximation is acceptable here. If we were to use a t-distribution for the posterior of μ_1, the histogram in Figure 5.10 would be very similar and our overall conclusions would be the same.

Next is an example where a graphical review of the prior and the posterior distributions could reveal an issue about the informativeness of a prior. Suppose we came across an analysis where a researcher used a prior with a mean of zero and a precision of 0.001. The researcher might have believed that his prior was vague because the precision might appear to be small. While this value of the precision may be small for some problems, it might not be for others, because the prior's standard deviation $\sigma = 1/\sqrt{\tau}$ is itself a rather narrow 31.6. As a result, most of the prior's weight will be within the range −100 to 100. The following point is therefore emphasized: *a vague prior for one model and dataset is not necessarily a vague prior for another model and/or dataset*. This is true in real life as well. Having a vague idea about the amount of money in your wallet will not necessarily be the same as having a vague idea about the size of the national deficit or having a vague idea about the number of years the universe has existed. Phrased differently, you might be pleasantly surprised to find an extra 1000 dollars (or euros or pounds) in your wallet, whereas the discovery that the national deficit was a 1000 dollars more than anticipated is not likely to surprise anyone given the magnitude of the spending involved. This indicates that a 1000-dollar difference may be informative in one context but not in another.

Returning to the researcher, suppose he indicated his posterior distribution was approximately normal with posterior mean −100.5 and posterior standard deviation 32. Figure 5.11 displays a plot of these two distributions, which we can create by simulation in *R*. We see that the prior

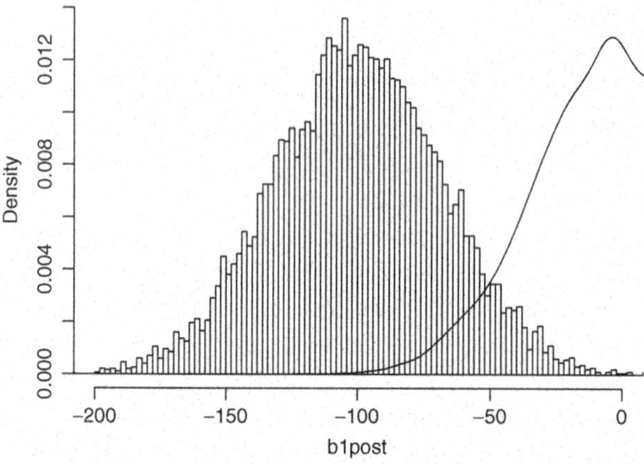

Figure 5.11 Prior and Posterior Distributions Where the Prior is Likely Informative

distribution is not approximately flat across the range of the posterior, but rather is concentrated to the right of it. This would have the effect of exerting a pull on the posterior so that it becomes closer to zero. If the researcher's sample size were small, this pull would be strong and would have an appreciable impact on the results compared to a classical analysis. In the event that the researcher's sample size was very large, the effect of this pull would be small, although it might be influential enough to make research consumers wonder about its effect. To summarize, we would conclude that, if a noninformative prior was desired for this analysis, there are alternatives that might fit the bill more closely. We also point out that, since it is possible for us to do an after-the-fact "audit" of the researcher's choice of prior given the posterior distribution he reported, we should exercise some thought as to what our "auditors" might think of our choice of priors.

It is natural to want to focus on the posterior distribution when undertaking a graphical examination of the prior, because we want to see whether the posterior distribution has been influenced by the prior. A limitation of this focus in the above graphical techniques is that some experimentation with the scale on the posterior x-axis is required in order to ensure that the plot does not omit important information. For example, the flatness of the prior is more evident in Figure 5.10 when the x-axis is expanded well beyond the bulk of the posterior. Similarly, the discrepancy between the posterior and the prior in Figure 5.11 is clearer when the x-axis of Figure 5.11 is changed to $(-300, 100)$. Hence we propose a variation of the above graphical approach, which is to plot the difference of the prior and the posterior vis-á-vis the prior. We can do this in *R* with a code such as the following:

```
hist(mu1prior-mu1post,prob=TRUE)
lines(density(mu1prior))
```

Running this code for the two examples discussed in this section produces the plots in Figure 5.12. On the left of the figure, the histogram of the difference between the prior and the posterior is very similar to that of the prior density. On the right of the figure, we see the histogram of the difference between the prior and the posterior does not correspond well with the prior. This gives an indication that this prior may not be the best candidate if a noninformative prior is desired. The general principle behind the prior difference plot is that, if the range of the posterior is small compared to the range of the prior, subtracting the posterior away from the prior will not have a substantial impact on the prior. The approach is most effective with symmetric unbounded posteriors and priors such as seen here. Note that, in extreme cases of prior/posterior discrepancy or for unusually shaped posterior distributions, it may be necessary to adjust the x-axis for the difference plot as well. Also, the effectiveness of any particular way of visualizing information depends on the situation and the viewer, so use the graphical method and settings that work best for the intended audience.

5.4.4 In Practice: Pre-sensitivity Analysis—Graphically Examining a Precision Parameter

We can also examine precision parameters graphically in *WinBUGS*, although the options are more limited than in *R*. After looking at the top row of Figure 3.3, we can infer that a Gamma(0.001, 0.001) distribution will be highly concentrated at zero and will be flat elsewhere. We do not have the ability to overlay two density plots in *WinBUGS* as we do in

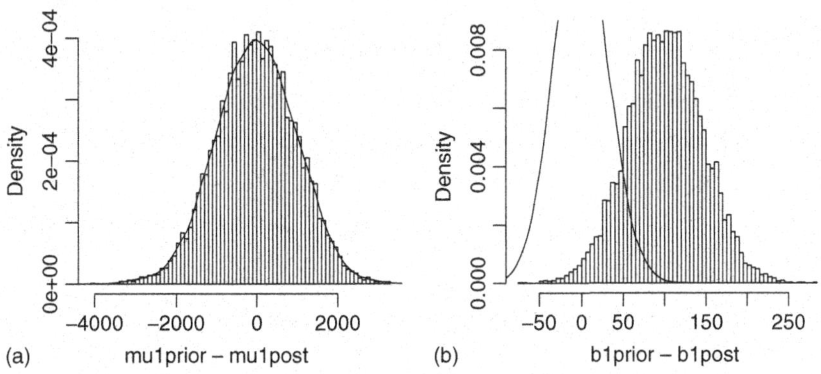

(a): plot where prior is noninformative; (b): plot where prior is likely informative.
Histogram: prior difference distribution; solid line: prior distribution.

Figure 5.12 (a,b) Prior Difference Plots

R. However, we can inspect a posterior distribution plot and compare it with the known properties of the gamma distribution. For example, Figure 5.13 displays the posterior distribution for τ_1, and we can compare this with our knowledge of the Gamma(0.001, 0.001) distribution as suggested by Figure 3.3. The posterior distribution in the figure is no7t concentrated in the vicinity of the prior's zero emphasis, but instead is far from zero. This might give some preliminary evidence that the prior was not very influential. As always, any one diagnostic provides only one piece of information, so a thorough sensitivity analysis would involve multiple priors and multiple examinations.

It is also possible to export our MCMC simulations from *WinBUGS* and read them into *R*. Once the simulations from the posterior distribution are in *R*, we can generate an overlay plot of the prior and posterior. A workflow for exporting MCMC simulations from *WinBUGS* to *R* appears in Section 5.10.

5.4.5 In Practice: Sensitivity Analysis for a Mean Parameter

To conduct sensitivity analysis for our prior on μ_1, we try two alternative priors. We increase and decrease the precision parameter by a multiple of 100. Note that these are only examples of possible changes to the prior's specification. We could also explore other multiples as well as changes in the mean. As before, since a vague prior for one model and dataset is not necessarily a vague prior for another model and/or dataset, so too will a sensitivity analysis need to be customized to the model and data at hand.

Table 5.7 shows the posterior distribution of μ_1 after two additional *WinBUGS* runs in which the prior has been changed. It also includes the original results for reference purposes. We see that posterior inference for μ_1 does not change over the range of priors used. This provides some evidence that, over a range of priors, the results are not affected (at least not to four significant digits).

5.4.6 In Practice: Sensitivity Analysis for a Precision Parameter

There are a number of possible alternative priors we could try for our precision parameter. We could change the values of the prior to obtain the

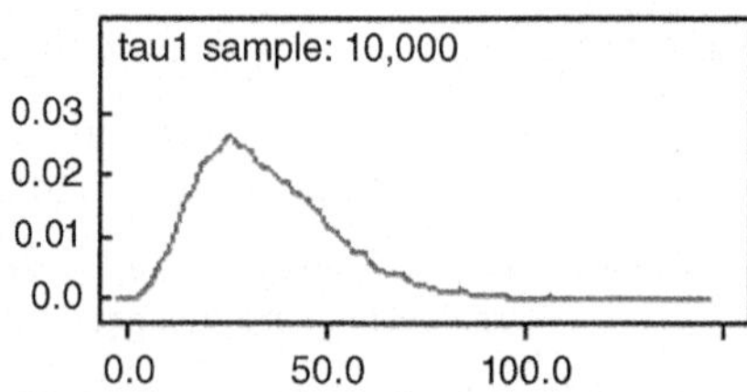

Figure 5.13 Posterior Distribution of τ_1

TABLE 5.7 Sensitivity Analysis Results for μ_1 Using Different Priors

Prior for μ_1	Mean	Std. Dev.	95% Credible Interval
`dnorm(0, 0.000001)` (original)	0.383	0.0642	(0.254, 0.513)
`dnorm(0, 0.00000001)`	0.383	0.0642	(0.254, 0.513)
`dnorm(0, 0.0001)`	0.383	0.0642	(0.254, 0.513)

somewhat less flat Gamma(0.01, 0.01). A different type of gamma prior is the Gamma(1, 0.001) prior. This prior differs from the previous gamma priors in that it does not have a large spike at zero. Rather, it is approximately flat in any given small range but gently slopes downward as it increases. This prior can be attractive if we *a priori* believe that τ might plausibly be small such as in random-effects models. If the data indicate that τ is small, the Gamma(0.001, 0.001) prior might become informative as τ approaches the spike. The Gamma(1, 0.001) prior, however, would not have a large local change in its informativeness in such a situation.

Instead of placing a prior on the precision, we could also place a prior on the standard deviation, variance, or some other related quantity of interest. For example, in Section 3.5.3 we discussed that the Jeffreys' prior for the precision corresponded to a uniform prior on $\log \tau$. Our gamma priors described earlier in the book were designed as an approximation to this, but we can enter this prior directly into *WinBUGS*. To do so, we create a new parameter and then place the desired prior on it. Then we backtransform the new parameter to obtain τ_1. For example, in *WinBUGS* code this could be implemented as follows:

```
logtau1 ~ dunif(-10, 10)
tau1 <- exp(logtau1)
```

Here, the `logtau1` prior is restricted to lie within the range −10 to 10 on the log scale. The backtransformed prior for `tau1` will be restricted to the range e^{-10} to e^{10} (i.e., 0.000045–22026). If we felt a broader range was necessary, we could alter these values.

Another possible flat prior is a uniform prior on the standard deviation. Here we examine the prior and backtransformation:

```
sigma1 ~ dunif(0,1000)
tau1 <- 1/(sigma1*sigma1)
```

The range for `sigma1` can again be changed as needed. Note that we will also have to change the *WinBUGS* initial values to correspond to the `logtau1` and `sigma1` priors as appropriate, so that *WinBUGS* will have a starting value for these chains. We use the value of 1 for both these initial values. Before leaving this section, it is worth pointing out that some so-called noninformative priors can actually have an undesirably substantial effect on inference. For example, results from Gelman (2006) indicate that putting certain (inverse-)gamma priors on certain standard deviations and precisions in hierarchical models can negatively affect inference. Performing sensitivity analysis for precision parameters is therefore especially recommended.

Table 5.8 shows the posterior distributions of τ_1 and σ_1 after additional *WinBUGS* runs in which the relevant prior and associated transformations have been changed. We find that the `dgamma(0.001, 0.001)` prior, the `dgamma(0.01, 0.01)` prior, and the `logtau1` prior all give similar results as expected. The `dgamma(1, 0.001)` prior produces a slightly larger value for τ_1 (and smaller mean value for σ_1). This makes sense because the mean of this prior (mean = 1000) is much higher than the mean of the previous two priors (which is equal to 1). By contrast, the `dunif(0, 1000)` prior on σ_1 produces a slightly larger mean value for σ_1, again displaying the impact of a large prior mean. This slightly larger mean is consistent with results in Gelman (2006), where it was noted that this prior may tend to cause posterior estimates to be slightly larger than other priors. Despite these differences, the 95% credible intervals for σ_1 are all quite similar across the various priors. As a result, our σ_1 inferences are not especially sensitive to changes in the prior for the priors we have considered here.

5.5 IN PRACTICE: EXAMINING PROPORTIONS IN *WinBUGS*

In Section 2.2, we first discussed proportion data. The proportion is of interest when we have a certain number of events that can result in yes/no or success/failure outcomes. The number of events (or "trials") is denoted

TABLE 5.8 Sensitivity Analysis Results for τ_1 and σ_1 using Different Priors

Parameter	Prior	Mean	Std. Dev.	95% Credible Interval
τ_1	`dgamma(0.001, 0.001)` (original)	35.28	17.76	(9.725, 77.16)
σ_1		0.187	0.0570	(0.114, 0.321)
τ_1	`dgamma(0.01, 0.01)`	32.76	16.48	(9.034, 71.57)
σ_1		0.194	0.0591	(0.118, 0.333)
τ_1	`dgamma(1, 0.001)`	44.43	20.05	(14.58, 92.6)
σ_1		0.163	0.0414	(0.104, 0.262)
τ_1	`logtau1~dunif(-10, 10)`	35.57	17.95	(9.66, 78.9)
σ_1		0.186	0.0550	(0.113, 0.323)
τ_1		31.52	16.49	(7.83, 70.9)
σ_1	`dunif(0, 1000)`	0.200	0.0626	(0.119, 0.358)

by n, and the number of successes is denoted by y. Then the proportion π of successes is equal to y/n. The binomial distribution (2.1) is the likelihood function that is typically used for π in these situations. When $n = 1$, the binomial distribution is typically called the *Bernoulli distribution* after Jacob Bernoulli who made important early discoveries in probability theory. Both the binomial distribution and the Bernoulli distribution are available in *WinBUGS*. For the binomial distribution, *WinBUGS* expects the successes y to be nonnegative integers less than or equal to n. *WinBUGS* also expects n to be an integer. The binomial distribution can then be specified with `dbin(pi,n)`. For the Bernoulli distribution, *WinBUGS* expects the data to take the form of a binary (0/1) variable where successes are indicated with 1's. The Bernoulli distribution can then be specified in *WinBUGS* with `dbern(pi)`.

5.5.1 Analyzing Differences in Proportions

We use data originally reported by Loeb (1971, Table 1) for our *WinBUGS* analysis of proportions. Loeb (1971) investigated the ethical behavior of accountants using a methodology originally designed to investigate the ethical behavior of lawyers. A screening sample of practicing accountants in Wisconsin was asked to review a list of other practicing accountants in the local area and identify which accountants on the list were highly ethical. This information was used to create a sample of accountants who were considered ethical by their peers (Group 1).

In addition, the screening sample was asked to identify which accountants on the list were unethical. This information, combined with reputation information and information about which accountants had been sanctioned by ethics boards as a result of complaints, was used to create a sample of unethical accountants (Group 2). Both Group 1 and Group 2 were provided with a questionnaire which asked about ethical conflict situations. The questions were designed so as to refer to borderline or gray area issues. The responses to this questionnaire formed the pretest data appearing in Table 1 of Loeb (1971). Loeb reports the sample size of unethical ($n = 19$) and ethical ($n = 22$) accountants and provides percentages of unethical responses by group. Based on this information, we may calculate the needed values of y for *WinBUGS* that appear below.

A natural question of interest for Loeb was to determine whether the pretest questions did indeed discriminate between the two groups of accountants. If no differences could be found, then it would imply that either the pretest questions were not sufficiently discriminating or else that there were no differences between the two groups in terms of their ethical responses. Hence, we reexamine the data to find whether there was a difference between the two groups of accountants in terms of providing ethical responses.

For the *WinBUGS* model, we use the binomial likelihood for each group's responses. In this model, we begin with a functional form which indicates that each group has a separate overall propensity to respond unethically, i.e., we have separate values of π for each group. A uniform prior on both π parameters is used to represent vague prior information. If we let i index the 13 pretest questions, we arrive at the following specification:

$$\begin{aligned} y_{i,1} &\sim \text{Binomial}(\pi_1, n_1) \\ y_{i,2} &\sim \text{Binomial}(\pi_2, n_2) \\ \pi_1, \pi_2 &\sim \text{Uniform}(0, 1). \end{aligned}$$

The *WinBUGS* code appears below (and also in **WinBUGS Code 5.5 Proportions 1.odc**). On Question 6 for Group 1, the percentage is given by Loeb as 19%. Since 19% times the sample size of 22 is slightly discrepant from an integer value, we assume that there may have been a typographic error. Accordingly, we assume this was 18%, which would indicate a corresponding y value of 4.

```
model
{
#likelihood
for (i in 1:n)
   {y1[i] ~ dbin(pi1, n1)
    y2[i] ~ dbin(pi2, n2) }
#priors
 pi1 ~ dunif(0, 1); pi2 ~ dunif(0, 1)
#other calculated parameters
 diff <- pi2-pi1
}
#data
list(y1=c(13,8,8,4,3,4,7,2,5,1,1,7,11), y2=c(17,10,10,9,
    12,8,9,5,10,6,8,14,14), n=13, n1=22, n2=19)
#inits
list(pi1=0.5, pi2=0.5)
```

After a 1000-iteration burn-in of the Markov chain, results were obtained by sampling the parameters for an additional 10,000 iterations. Inspection of the trace plots suggests that convergence to the posterior distribution occurred during the burn-in period. The results appear in Table 5.9. We see that Group 1 accountants did have a lower proportion of unethical answers ($\pi_1 = 0.260$ versus $\pi_2 = 0.534$) on average across the entire survey. More importantly, the 95% credible interval for the difference in proportions ($\pi_2 - \pi_1$) excludes the value of zero by a wide margin. Thus we conclude that the Group 2 accountants were more likely to give unethical answers to the survey than the Group 1 accountants.

We can also examine the item-by-item discriminability of the individual questions with minor changes to the code. For questionnaire item i, we now assume that responses will depend on the unknown value of $\pi_{1,i}$ for

TABLE 5.9 Results for Accounting Behavior Survey: Model 1

Parameter	Mean	Std. Dev.	95% Credible Interval
π_1	0.260	0.0259	(0.210, 0.312)
π_2	0.534	0.0316	(0.472, 0.596)
$\pi_2 - \pi_1$	0.275	0.0408	(0.193, 0.354)

Group 1 accountants, whereas it will depend on $\pi_{2,i}$ for Group 2 accountants. We can also compute the difference in proportions for each item and use this to assess the items' discriminating power. The *WinBUGS* code will need to be modified so that the likelihood functions involve $\pi_{1,i}$ and $\pi_{2,i}$ (see **WinBUGS Code 5.5 Proportions 2.odc**).

It will also be convenient to place the priors in a loop. In addition, we may also calculate the item differences in a loop. Since all of these loops will loop over the 13 item questions, we may place all of these statements inside the same loop to economize on the code. Making these changes produces our Model 2 code below. The `data` code does not need to be changed, and so it has been omitted. However, the `inits` code must be updated because the initial values are no longer scalars. Instead, the initial values need to be supplied in vector format since we are using a loop for the priors. We can do this using the collection statement `c(...)` in the same way we have been doing this for the data.

```
model
{
#likelihood
for (i in 1:n)
  {y1[i] ~ dbin(pi1[i], n1)
   y2[i] ~ dbin(pi2[i], n2)
#priors
   pi1[i] ~ dunif(0, 1); pi2[i] ~ dunif(0, 1)
#other calculated parameters
   itemdiff[i] <- pi2[i] - pi1[i]
  }
}
#inits
list(pi1=c(0.5,0.5,0.5,0.5,0.5,0.5,0.5,0.5,0.5,0.5,
   0.5,0.5,0.5), pi2=c(0.5,0.5,0.5,0.5,0.5,0.5,0.5,
   0.5,0.5,0.5,0.5,0.5,0.5))
```

Results from this model appear in Table 5.10. We see that the items vary in terms of their ability to discriminate between the groups. Item 5 has the highest ability to discriminate between the groups. Conversely, several of the item difference scores have 95% credible intervals that do not exclude the value of zero, casting some doubt on their discriminability in this small sample.

TABLE 5.10 Results for Accounting Behavior Survey: Model 2

Parameter	Mean	Std. Dev.	95% Credible Interval
$\pi_{2,1} - \pi_{1,1}$	0.273	0.123	(0.0266, 0.510)
$\pi_{2,2} - \pi_{1,2}$	0.147	0.144	(−0.141, 0.425)
$\pi_{2,3} - \pi_{1,3}$	0.149	0.146	(−0.142, 0.427)
$\pi_{2,4} - \pi_{1,4}$	0.268	0.132	(0.005, 0.522)
$\pi_{2,5} - \pi_{1,5}$	0.453	0.127	(0.190, 0.687)
$\pi_{2,6} - \pi_{1,6}$	0.220	0.135	(−0.049, 0.475)
$\pi_{2,7} - \pi_{1,7}$	0.143	0.142	(−0.140, 0.417)
$\pi_{2,8} - \pi_{1,8}$	0.160	0.117	(−0.066, 0.391)
$\pi_{2,9} - \pi_{1,9}$	0.273	0.137	(0.002, 0.538)
$\pi_{2,10} - \pi_{1,10}$	0.251	0.116	(0.029, 0.490)
$\pi_{2,11} - \pi_{1,11}$	0.347	0.118	(0.117, 0.577)
$\pi_{2,12} - \pi_{1,12}$	0.380	0.134	(0.103, 0.623)
$\pi_{2,13} - \pi_{1,13}$	0.216	0.139	(−0.060, 0.484)

5.5.2 Predicting Customer Behavior: Part 2 Revisited

We return to our problems of Sections 3.4.1 and 3.4.2. There, we were interested in examining whether women or men were more likely to purchase an item. Fourteen out of 37 women purchased an item, whereas 3 out of 22 men purchased the item. We also wanted to forecast how many sales would be made to women assuming that there were 500 women shoppers in a given month. We previously used *R* to examine the customer purchase data. Now we examine the data using *WinBUGS*. A few modifications of our *WinBUGS* code above produce the following model (see also **WinBUGS Code 5.5 Customer Behavior.odc**):

```
model
{
#likelihood
  y.w ~ dbin(pi.w, n.w)
  y.m ~ dbin(pi.m, n.m)
#priors
  pi.w ~ dunif(0, 1); pi.m ~ dunif(0, 1)
 #other calculated parameters
  diff <- pi.w - pi.m
  sales.w ~ dbin(pi.w, 500)
}
#data
list(y.w = 14, y.m=3, n.w=37, n.m=22)
#inits
list(pi.w=0.5, pi.m=0.5, sales.w=250)
```

The results from *WinBUGS* are essentially the same as those we found using *R* in Sections 3.4.1 and 3.4.2. The posterior mean of `sales.w` is almost the same at 192.1. The posterior standard deviation of `sales.w` is 39.81, while the 95% credible interval is (116, 272).

In Section 3.4.3, we wanted to extend the results of Section 3.4.1 to a situation where the number of women was unknown but could be forecasted. We can use *WinBUGS* to handle this situation. Suppose we forecast that instead of an exact 500 female shoppers, the number of female shoppers has a Poisson distribution with a mean of 500. Let us call this quantity `cust.w`.

We will add a line of code to our *WinBUGS* program that simulates from `cust.w`. We will also change `sales.w` to depend on `cust.w`. The following two lines of code make the needed changes (see also **WinBUGS Code 5.5 Proportions Customer Behavior.odc** for a full listing):

```
...
 sales.w ~ dbin(pi.w, cust.w)
 cust.w  ~ dpois(500)
...
```

We see that `sales.w` now depends on `cust.w`, as desired. In turn `cust.w` has the appropriate Poisson distribution. We use a 1000-iteration burn-in and 10,000 MCMC iterations for estimation. The posterior mean of `sales.w` is virtually unchanged at 191.9. However, the standard deviation is somewhat wider, at 40.96. The 95% credible interval (115, 275) is also somewhat wider.

We can give other distributions than the Poisson to `cust.w`. However, we need to make sure `cust.w` takes on only integer values. Suppose we thought that the distribution of women shoppers was approximately uniform from 400 to 600. The standard *WinBUGS* uniform distribution is continuous. We can, instead, use the *WinBUGS* `round` function to give only integer values. For example, we could make the following changes:

```
...
 cust.unif  ~ dunif(400,600)
 cust.w     ~ round(cust.unif)
...
```

Now `cust.w` will take on only integer values. *WinBUGS* will then sample from `sales.w` appropriately.

5.6 ANALYSIS OF VARIANCE MODELS

The t-test model of Section 5.3 is designed for the scenario where there are only two groups with means to compare. If there are more than two groups and we only have qualitative independent variables, then the model is called *analysis of variance* (*ANOVA*). Extending the *t*-test to an arbitrary (and perhaps large) number of groups leads us to introduce additional capabilities of *WinBUGS* such as new ways for reading data, new *WinBUGS* graphical options, and new ways to program parameter comparisons in *WinBUGS*.

5.6.1 In Practice: One-Way ANOVA

To motivate the use of ANOVA in practice, we examine flows of consumer credit in the United States during the period 2007–2011 as tabulated by the Federal Reserve (Board of Governors of the Federal Reserve System, 2012). The flow of consumer credit is the year-on-year change in credit extended to individuals such that positive values indicate the existence of more credit extended in a given year as compared to the previous year. We examine flows among major nongovernmental holders of consumer credit. Values are in billions of U.S. dollars. The time period covers the period of the 2007–2008 financial crisis and its aftermath. There were five categories of major nongovernmental holders of consumer credit. Annual data was examined and the total sample size was 25. An ANOVA will allow us to examine whether the average credit flow differed from year to year during this period.

Table 5.11 shows that we have data over time for the different types of credit holder. We would expect that credit flows exhibit some correlation over time. However, ANOVA does not model this correlation. Therefore, any results we obtain should be regarded as preliminary. We will proceed with our preliminary analysis and then revisit this example in Section 8.1.2 when we have a better way for accounting for this correlation.

Descriptive Preanalysis In both Bayesian methods and classical methods, it is almost always worthwhile to check over the data using

TABLE 5.11 2007–2011 U.S. Consumer Credit Flows by Major Holder (in Billions of $)

	Year				
Type	2007	2008	2009	2010	2011
Depository institutions	58.1	70.1	−64.5	−70.3	7.1
Finance companies	40.2	−17.9	−82.4	−27.0	−17.8
Credit unions	0.5	−0.4	0.9	−10.7	−3.4
Nonfinancial business	2.2	−3.0	−4.6	−2.3	0.0
Pools of securitized assets	34.1	−40.1	−39.7	−50.3	−8.3

exploratory data analysis before going onward to perform inference. The goal is to make sure the data is accurate and the existence of any problems is revealed before proceeding further. Excellent discussions of exploratory data analysis can be found in Tukey (1977), and a number of the *R* procedures that can be used for these tasks have been described by Venables and Ripley (2002). A detailed treatment of this area would move us away from our main topic of Bayesian inference; however, as a brief example we take a descriptive look at the current data so as to make sure it is a good candidate for ANOVA. Figure 5.14 contains descriptive box-and-whisker plots (Tukey, 1977) from *R*, which visually provide an overview of the data (the *R* code appears in **WinBUGS Code 5.6 one-way ANOVA.odc**). The group medians appear as heavy solid lines within the boxes, while the boxes and whiskers show the dispersion. Extreme values (outliers) are displayed as circles.

We see there is a likely outlier in the 2008 data because depository institutions expanded their credit flows to consumers by about $70 billion in that year despite contractions among other types of creditors. When we encounter outliers, some extra thought is required. Outliers may arise from typographic data errors, respondents' failure to follow instructions, or other data quality reasons. Presumably, the Federal Reserve reviews the data for these kinds of issues so the data quality can be considered high. Another issue is that we may have an apples-and-oranges situation, with depository institutions being the orange in 2008 amongst four other apples. Since the mean can be strongly influenced by outliers, one approach would be to remove the outlying orange from the data. A second approach that we will explore later in the book is to use models that reduce the impact of the outlier. We will investigate these models

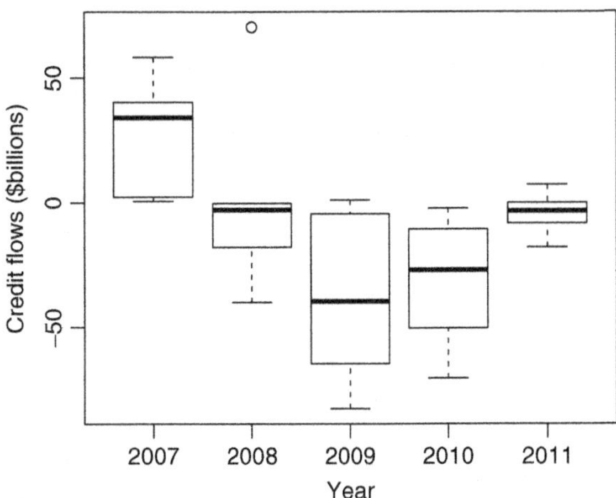

Figure 5.14 Descriptive Analysis (Data Box-and-Whisker Plots) of U.S. Consumer Credit Flows: 2007–2011

in Chapter 10. A third approach (which we will adopt here) would be to acknowledge the effect of the outlier on the mean, retain the outlier, and proceed with the originally planned analysis. We take the third approach here for reasons to be described in Section 5.6.3.

Inferential Analysis The one-way ANOVA extends the t-test, and so the model specification follows this extension. We first consider what we will need for the likelihood function. The t-test uses data from two groups, $y_{i,1}$ and $y_{i,2}$. The ANOVA uses data from J groups. We let j index the J groups so that j takes on values from 1 to J. Each group will have its own mean, μ_j, so we will estimate J means in total. The dependent variable for each group is assumed to be normally distributed. The conventional way of performing ANOVA is to assume that all the groups have the same variance. This is called a *homogeneous variance* assumption and most classical textbooks and software packages use this assumption. For priors, we again use vague normal priors for the means and a vague gamma prior for the precision. The functional form does not need to be explicitly written out since the functional form is that each group has its own mean. Bearing all this in mind, a model specification for one-way ANOVA is as follows:

$$
\begin{aligned}
y_{i,j} &\sim \text{Normal}(\mu_j, \tau) \\
\mu_j &\sim \text{Normal}(0, 0.0000001) \qquad (5.1) \\
\tau &\sim \text{Gamma}(0.001, 0.001).
\end{aligned}
$$

When translating the model specification into *WinBUGS*, we will need an efficient way to deal with data that is indexed by both i and j at the same time. So far, we have only seen collections or columns of data (i.e., vectors) such as `y = c(1, 2, 3)`. *WinBUGS* also allows us to supply data in matrix form using the `structure` option. Putting together the model specification with the data gives the program code below (see also **WinBUGS Code 5.6 one-way ANOVA.odc**):

```
model
{
#likelihood
for (i in 1:holders) {
     for (j in 1:years) {
       y[i,j] ~ dnorm(mu[j], tau)    }
    }
#priors for mu's and tau
 for (j in 1:years) { mu[j] ~ dnorm(0, 0.000001) }
 tau ~ dgamma(0.001, 0.001)
#other calculated parameters
 sigma <- 1/sqrt(tau)
 for (i in 1:holders) {
```

```
    for (j in i:years) {
      mudiff[i,j] <- mu[i]-mu[j] }
   }
}
#data
list(years = 5, holders = 5,
  y = structure(.Data =
   c(58.1, 70.1, -64.5, -70.3, 7.1,
     40.2, -17.9, -82.4, -27, -17.8,
      0.5, -0.4, 0.9, -10.7, -3.4,
      2.2, -3, -4.6, -2.3, 0,
     34.1, -40.1, -39.7, -50.3, -8.3),
  .Dim = c(5, 5))
   )   #(rows, columns)
#inits
list(mu=c(0,0,0,0,0), tau=1)
```

This program code has several new aspects worth mentioning. First, we have defined the likelihood in terms of two loops: one for the creditor index *i*, and one for the year index *j*. When using multiple nested loops, a common source of errors involves the incorrect placement of the brackets so it is worth reexamining the code to see exactly where each loop begins and ends. Second, we have specified the priors for the μ parameters in a loop to save time and economize on code. Consequently, the initial values for these priors need to be given in vector format. We were also able to further economize on code by specifying the priors within a preexisting loop. Third, we would like to be able to compare which μ parameters are different from which. Therefore, another nested loop is used to calculate a series of `mudiff` parameters. Observe that the outer loop variable, `i`, goes from 1 to `years`. However, the inner loop variable, `j`, begins with whatever the current value of `i` is and then goes to `years`. This approach allows us to calculate, for example, $\mu_1 - \mu_2$ without wasting computer (and human) time on a redundant analysis of $\mu_2 - \mu_1$. Finally, we see the `structure` usage. The dimensions of the matrix need to be supplied in the `.Dim` portion. Rows are supplied first and then columns, as is indicated by the comment after the data statement.

After a 1000-iteration burn-in period, parameters were estimated based on 10,000 iterations of the Markov chain. *WinBUGS* can create a posterior box plot which is a useful way of visualizing the posterior distributions of the `mu` vector. To create the box plot, click on the `Inference` menu option in *WinBUGS*, and then select `Compare`. In the Comparison tool (Figure 5.15), enter the posterior vector variable in the box labeled `node`. Then click `box plot`. A similar plot having a different orientation can be created by clicking `caterpillar plot`). It is possible to have the box plot based on a subset of the Markov chain if desired. This is done by entering the beginning iteration of the subset in the `beg` box and the ending iteration in the `end` box of the tool.

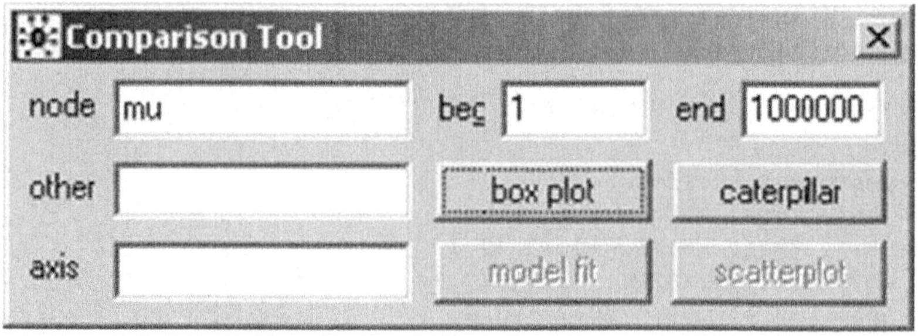

Figure 5.15 *WinBUGS'* Comparison Tool

The box plot by the *WinBUGS'* Comparison tool appears in Figure 5.16. The posterior mean of a parameter is indicated by a solid line inside of a box. The box itself indicates the posterior interquartile range (the 25th percentile to the 75th percentile of the distribution). The whiskers indicate the 95% central credible interval, with the lower whisker at the 2.5th percentile and the upper whisker at the 97.5th percentile of the distribution. Comparing the credible interval with any given value on the y-axis allows us to determine whether this value is in the credible interval or not. For example, we can see that 0 is outside the 95% credible interval for μ_3, so zero would not be a credible value for μ_3. Comparing any

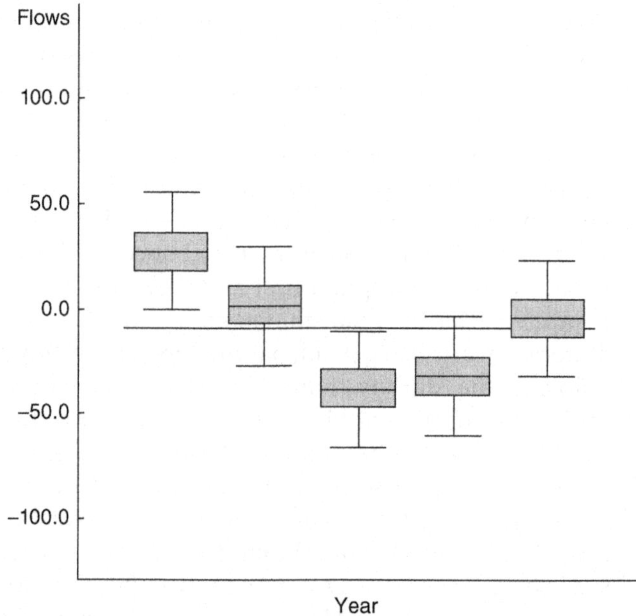

Figure 5.16 Posterior Distribution Box Plots for U.S. Consumer Credit Flow Means: 2007–2011

two parameters to one another is best accomplished with the `mudiff` parameters estimated separately. The reason why this is true can be illustrated with a short example. Suppose two parameters were estimated, and it happened that the upper whisker for one parameter was exactly parallel to the other parameter. Suppose also that the posterior distributions of the two parameters are uncorrelated with one another. Then the joint probability of the two distributions overlapping is a much lower $2.5\% \times 2.5\% = 0.0625\%$. Further complicating matters is the fact that distributions may be correlated, so it is best to look at the posterior distribution of the differences to make statements about parameter differences.

WinBUGS provides some basic options for customizing the box plots in addition to those discussed in Section 5.2. For the box plot, clicking the `Special` button after right-clicking on the plot and selecting `Properties` allows the display of a baseline horizontally across the plot. Here, we see that *WinBUGS* has drawn this line at about −9 on the y-axis, which approximately equals the overall average of the posterior distributions displayed in the plot. Orientation and box color are among the other options available by clicking the `Special` button.

Results for all estimated parameters appear in Table 5.12. A main question of interest was whether or not average consumer credit flows remained the same over the time period. Results show that the credit flows significantly declined from 2007 to 2009, as indicated by the $\mu_1 - \mu_3$ parameter. Similar declines occurred between 2008 and 2009 as well as between 2007 and 2010. As a side note, we see in Table 5.12 that all of

TABLE 5.12 One-Way ANOVA Results for Consumer Credit Flows

Parameter	Mean	Std. Dev.	95% Credible Interval
μ_1	26.91	14.14	(−0.42, 55.75)
μ_2	1.71	14.21	(−26.65, 29.66)
μ_3	−38.12	14.10	(−66.14, −10.43)
μ_4	−32.19	14.15	(−60.19, −3.79)
μ_5	−4.56	13.88	(−32.08, 22.90)
$\mu_1 - \mu_2$	25.20	19.97	(−14.23, 65.96)
$\mu_1 - \mu_3$	65.03	20.02	(25.41, 104.7)
$\mu_1 - \mu_4$	59.10	20.11	(18.99, 99.22)
$\mu_1 - \mu_5$	31.47	19.87	(−8.41, 70.83)
$\mu_2 - \mu_3$	39.82	20.07	(0.32, 80.17)
$\mu_2 - \mu_4$	33.90	20.13	(−6.56, 73.89)
$\mu_2 - \mu_5$	6.27	19.96	(−34.18, 45.38)
$\mu_3 - \mu_4$	−5.93	19.82	(−45.85, 32.88)
$\mu_3 - \mu_5$	−33.56	19.83	(−72.17, 5.818)
$\mu_4 - \mu_5$	−27.63	19.87	(−67.14, 11.36)
σ	31.28	5.23	(22.86, 43.45)

the posterior standard deviations for μ_1 to μ_5 are about the same because of the homogeneous variance assumption. The homogeneous variance assumption can also be seen in the width of the box plots in Figure 5.16.

5.6.2 In Practice: One-Way ANOVA with Effects Coding

In Section 5.3.3, we considered an alternative parameterization of the *t*-test. This alternative parameterization can be extended to ANOVA as well. To do this, we retain parts of our current model specification and add a functional form to it. The μ_j parameters are redefined so that $\mu_j = \overline{\mu} + \delta_j$, where $\overline{\mu}$ is the overall mean and δ_j is the year-specific deviation from the overall mean. We need to make sure that all of the δ_j parameters sum up to zero. This is often called *effects coding* and is done by setting one of the δ parameters to minus the sum of the others. Here we set δ_1 to the necessary sum. Since δ_1 is now a deterministic sum of other parameters, it does not receive a prior distribution. The final model specification is then as follows:

$$
\begin{aligned}
y_{i,j} &\sim \text{Normal}(\mu_j, \tau) \\
\mu_j &= \overline{\mu} + \delta_j \\
\delta_1 &= -\sum_{j=2}^{5} \delta_j \\
\overline{\mu} &\sim \text{Normal}(0, 0.0000001) \\
\delta_2, \ldots, \delta_5 &\sim \text{Normal}(0, 0.0000001) \\
\tau &\sim \text{Gamma}(0.001, 0.001).
\end{aligned}
$$

The program code for this model appears below (see also **WinBUGS Code 5.6 one-way ANOVA.odc**). The primary differences are the name changes (from `mu` to `mu.overall` and `delta`) and the code for calculating `delta[1]`. Note that the initial value for `delta[1]` has been set to `NA` (not applicable). This is because `delta[1]` is no longer a random parameter. *WinBUGS* will generate an error message if `NA` is not used for this initial value. In our effects-coded ANOVA we can still compare each year with another just as is done in Table 5.12. The differences in δ parameters (`deltadiff`) will be equivalent to the differences in the previous μ parameters, excluding small variations due to Monte Carlo error. In addition, the δ parameters give us some additional insight by simultaneously letting us see whether any particular year was above or below the overall average $\overline{\mu}$.

```
model
{
#likelihood
```

```
for (j in 1:years) {
  for (i in 1:holders) {
      y[i,j] ~ dnorm(mu[j], tau) }
 #linear predictor
  mu[j] <- mu.overall + delta[j]
 }
#priors for mu.overall, delta and tau
 mu.overall ~ dnorm(0, 0.000001)
 for (j in 2:years) {
    delta[j] ~ dnorm(0, 0.000001)  }
 tau ~ dgamma(0.001, 0.001)
#other calculated parameters
 delta[1] <- -sum(delta[2:years])
 sigma <- 1/sqrt(tau)
 for (i in 1:holders) {
    for (j in i:years) {
      deltadiff[i,j] <- delta[i]-delta[j]  }
   }
}
#data - same as previous
#inits
list(mu.overall=0, delta=c(NA,0,0,0,0), tau=1))
```

Table 5.13 presents selected results from the MCMC run for our ANOVA with effects coding. The selected results show us the additional information from effects coding that was not apparent in the previous analysis. The 95% credible interval for δ_1 indicates that the consumer credit flows in 2007 were significantly different from the average flows during the time period. We also see that the 95% credible interval for $\overline{\mu}$ includes the value of zero. Hence we cannot reject the notion that overall flows were zero during the time period.

5.6.3 In Practice: One-Way ANOVA with Unequal Variances

The t-test discussed in Section 5.3.2 allowed unequal variances whereas the usual ANOVA assumption is to have equal variances for all groups.

TABLE 5.13 Additional Results from One-Way Effects-Coded ANOVA for Consumer Credit Flows

Parameter	Mean	Std. Dev.	95% Credible Interval
δ_1	36.34	12.57	(11.01, 60.95)
δ_2	10.91	12.72	(−14.42, 36.15)
δ_3	−28.93	12.66	(−53.67, −3.32)
δ_4	−23.01	12.42	(−47.62, 1.55)
δ_5	4.685	12.55	(−20.72, 29.12)
$\overline{\mu}$	−9.195	6.36	(−21.88, 3.31)

Unequal-variance t-tests are recommended as being more conservative than equal-variance t-tests when variances are dissimilar (Ott, 1993; Hogg and Tanks, 1997), so it would be interesting to use an unequal variance ANOVA. Fortunately, the flexibility of *WinBUGS* is such that we can easily do this.

We begin with model specification. We extend our previous ANOVA model specification from Section 5.6.1 so that each group has its own variance. The precision parameters are given a subscripted j to indicate that they are now estimated separately on a group-by-group basis. After making the change, we have the following model:

$$y_{i,j} \sim \text{Normal}(\mu_j, \tau_j)$$
$$\mu_j \sim \text{Normal}(0, 0.0000001)$$
$$\tau_j \sim \text{Gamma}(0.001, 0.001).$$

Now that the variances are allowed to be unequal, the variances are said to be heterogeneous. We have J precision parameters to estimate for our group-specific variance terms. The *WinBUGS* program code can be updated so that the priors for the `tau` parameters are defined in a loop. Also, the initial values for `tau` are changed so as to be a vector of five initial values. A full listing appears in **WinBUGS Code 5.6 one-way ANOVA.odc** and the program code appears below.

```
model
{
#likelihood
for (j in 1:years) {
  for (i in 1:holders) {
      y[i,j] ~ dnorm(mu[j], tau[j])
    }
 #priors for mus and taus
  mu[j]   ~ dnorm(0, 0.000001)
  tau[j] ~ dgamma(0.001, 0.001)
 #other calculated parameters
 sigma[i] <- 1/sqrt(tau[i])
 }
for (i in 1:holders) {
    for (j in i:years) {
      mudiff[i,j] <- mu[i]-mu[j]   }
 }
}
```

Posterior box plots of the means as well as the standard deviations for this model appear in Figure 5.17. In part (a) of the figure, we see that the height of the box plots for the means now varies noticeably between years. The 2008 data with its outlier has a large variance, as does the 2009 data. By 2011, credit flows seem to have stabilized in the sense that all major holders were issuing approximately the same amount of credit as they had the year before. This has led to a much smaller standard deviation during this period. In Figure 5.17(b), the 95% credible intervals indicate that the standard deviation for 2011 was considerably smaller than that in other years. If desired, a formal test of size differences among the standard deviations would involve calculating a `sigmadiff` variable in the same way that `mudiff` was calculated previously (Section 5.11). Since the standard deviation parameters are highly skewed as usual, comparing standard deviations using a normal-based test would not be attractive. By contrast, Monte Carlo simulation as in *WinBUGS* is an effective way of examining the differences of skewed variables.

Table 5.14 displays selected results from the unequal-variance ANOVA. The unequal-variance ANOVA leads to one different conclusion from the equal-variance ANOVA found in Table 5.12. The values in the "mean" column are about the same in the two tables. However, the parameter standard deviations are considerably different. The standard deviations are larger, particularly if the comparison involves Years 2 and 3. In Table 5.12, the conclusion from $\mu_2 - \mu_3$ was that Year 2 was significantly different from Year 3. However, in Table 5.14 the 95% credible interval for $\delta_2 - \delta_3$ is dramatically wider. Consequently, there is no evidence of a significant difference between these two years.

The reasons for this can be found in the lower part of Table 5.14. With only a sample size of 5 per year, the estimates for the σ parameters are subject to a large degree of error. This large degree of error carries over into the δ estimates and then to the differences between them. In the equal-variance ANOVA, all 25 data points were contributing to the

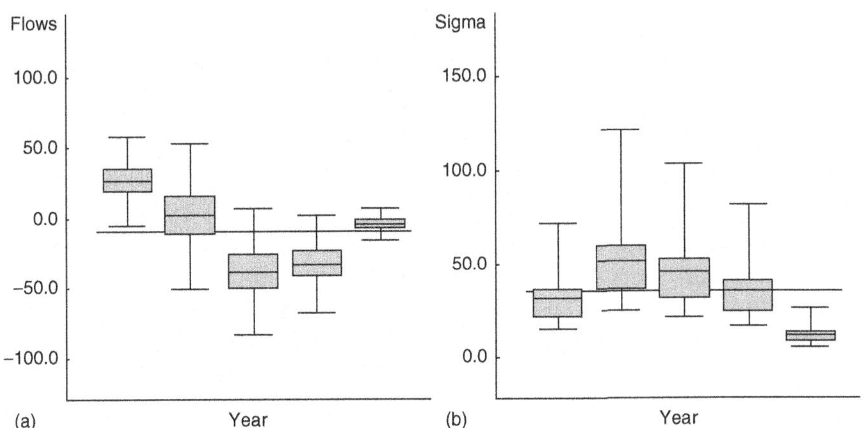

Figure 5.17 Posterior Box Plots for Credit Flow Means and Standard Deviations: 2007–2011, Heterogeneous Variances ANOVA. (a) Means. (b) Standard Deviations

TABLE 5.14 One-Way Unequal-Variance ANOVA Results for Consumer Credit Flows

Parameter	Mean	Std. Dev.	95% Credible Interval
$\delta_1 - \delta_2$	25.13	30.41	(−35.74, 84.79)
$\delta_1 - \delta_3$	65.22	28.02	(10.04, 121.5)
$\delta_1 - \delta_4$	59.46	23.41	(14.07, 105.5)
$\delta_1 - \delta_5$	31.56	16.81	(−1.919, 65.03)
$\delta_2 - \delta_3$	40.09	34.68	(−29.85, 108.8)
$\delta_2 - \delta_4$	34.33	31.15	(−25.39, 95.61)
$\delta_2 - \delta_5$	6.44	26.64	(−45.56, 58.40)
$\delta_3 - \delta_4$	−5.76	29.51	(−65.3, 51.60)
$\delta_3 - \delta_5$	−33.66	24.09	(−82.34, 13.47)
$\delta_4 - \delta_5$	−27.90	18.59	(−64.42, 8.30)
σ_1	31.4	16.24	(15.15, 72.07)
σ_2	51.71	26.86	(25.16, 120.6)
σ_3	45.62	24.02	(21.45, 104.9)
σ_4	35.09	18.06	(16.94, 78.95)
σ_5	11.73	6.002	(5.58, 26.90)

estimation of σ. The posterior standard deviation of σ in that model is therefore much smaller. This is the trade-off of using a more comprehensive (and more parameterized) model—we will need more data to better estimate the additional parameters. Overall, the unequal-variance ANOVA is probably more realistic for decision making in situations where groups have substantially different variances.

5.6.4 Indexing Parameters by Group Membership Variables

The credit flows data was already organized in a tabular format so it was easy to use that particular structure in *WinBUGS*. However, we may find that the data we wish to analyze is not organized in that manner. In this situation, it can be useful to inform *WinBUGS* on the fly about which group an observation belongs to. This can be accomplished by adding a group indicator variable to the *WinBUGS* data and then referencing it in the code. Suppose we call this variable `group`. Then for observation *i*, the value of `group` is `group[i]`. As an example, we might use the following code segment for our heterogeneous variance ANOVA:

```
#likelihood
for (i in 1:n) {
      y[i] ~ dnorm(mu[group[i]], tau[group[i]])
   }
```

This code ensures that `y[i]` contributes to the estimation of `mu` and `tau` for the appropriate group. Another situation where the above code is useful is when the sample size for each group is different. This is known as *unbalanced* data. Unbalanced data can be handled with nested loops with some organization and forethought but the above code makes it easy to handle unbalanced groups without the need for reorganizing the data.

A somewhat related scenario is where we would like to weight data. For example, we may have a marketing study where 50% of our sample's respondents indicated they were in Group X but we know that in the population 40% of our customers are in Group X. In normal distribution models such as ANOVA, the precision τ characterizes information about how precise our estimates are. Multiplying τ by a nonnegative weight changes the amount of "evidence" that a particular piece of data conveys. We can use this fact to apply population weights (or other kinds of weights) to our data. For our marketing study, we see that 40%/50% = 0.8 and so a first pass at *WinBUGS* code might be `weighted.tau[i] <- 0.8 * tau`. However, this will apply the downweighting to all groups equally. A more general approach is

```
weighted.tau[i] <- popWT[group[i]]/sampWT[group[i]] * tau
```

where the *WinBUGS* dataset would need to contain the appropriate data for variables `popWT` and `sampWT` (e.g., 40% would appear among the values in `popWT` while 50% would appear among the values in `sampWT`). Having calculated our weighted τ, we would use it in the likelihood, i.e.,

```
y[i] ~ dnorm(mu, weighted.tau[i])
```

This will cause `mu` to be estimated with the data weights applied, which should cause `mu` to more closely resemble its population value. A brief illustration of sample weighting can be found in **WinBUGS Code 5.6.4 sample weighting.odc**.

5.7 HIGHER ORDER ANOVA MODELS

We often have more than one grouping variable that is relevant for understanding an outcome. For example, we may be able to group companies by size into large versus small, and also to simultaneously group them by ownership type into publicly traded versus privately held. A two-way ANOVA is the name of the model when there are two grouping variables used in an ANOVA. Generally speaking, higher order ANOVA models have multiple grouping variables as compared to the one-way ANOVA with its single grouping variable.

In order to provide the right functional form for higher order ANOVA, we need to review how many parameters we can estimate per grouping variable in ANOVA. Imagine we have grouping Variable 1 and grouping Variable 2 for our two-way ANOVA. Suppose there are J groups for Variable 1 and K groups for Variable 2. We will be able to estimate $J - 1$ parameters for Variable 1 and $K - 1$ parameters for Variable 2. We can also estimate one additional overall parameter as a reference parameter. To do so, one approach is to designate one subgroup as the *reference category*. The reference category could be, for example, data that belongs to both the first group in Variable 1 as well as the first group in Variable 2. We are free to choose whichever group we want among the groups in Variables 1 and 2 so as to select the subgroup we want as the reference category. The reference parameter will then pertain to the reference category and the remaining parameters are interpreted as differences (or offsets) from the reference category. Another approach would be to estimate the overall mean as the reference parameter. We have seen this approach in Section 5.6.2, where the reference parameter was $\overline{\mu}$. If we use this approach, we can still estimate only $J - 1$ parameters, with the final group's parameter being calculated as in Section 5.6.2.

Implementing higher order ANOVA is now a matter of modifying the functional form of the model so that the linear predictor contains all the parameters of interest. The linear predictor μ gives us the predicted value of a given observation. The reference parameter can be written as α_0. The remaining variables are added to the linear predictor to give the functional form. Writing out the functional form, we have $\mu_{i,j,k} = \alpha_0 + \alpha_j + \beta_k$, where $\alpha_1 = \beta_1 = 0$. Here we see that the predicted value for the ith observation in the jth group of Variable 1 as well as the kth group of Variable 2 is equal to the sum of α_0 and the pertinent offsets. Completing our normal model with a precision term τ and priors for all parameters leads to the following model specification:

$$
\begin{aligned}
y_{i,j,k} &\sim \text{Normal}(\mu_{i,j,k}, \tau) \\
\mu_{i,j,k} &= \alpha_0 + \alpha_j + \beta_k \\
\alpha_1 = \beta_1 &= 0 \\
\alpha_0, \alpha_2, \ldots, \alpha_J &\sim \text{Normal}(0, 0.0000001) \\
\beta_2 \ldots, \beta_K &\sim \text{Normal}(0, 0.0000001) \\
\tau &\sim \text{Gamma}(0.001, 0.001).
\end{aligned}
$$

The data can be supplied to *WinBUGS* using `structure` as in Section 5.6.1. The dimension of `structure` is increased from 2 to 3 to accommodate the new variable. In order to use `structure` effectively in higher dimensions, we need to understand how to assign the dimensions properly through the use of *WinBUGS'* `dim` specification. The first dimension in the `dim` specification is the one that changes the slowest as we proceed through the data. Conversely, the last dimension is the one that changes the fastest as we proceed through the data. Bear this in mind when supplying data to *WinBUGS* because preexisting data may be in a dimensional format that needs to be changed to make it more convenient or logical for use in *WinBUGS*.

Fortunately, *WinBUGS* allows you to inspect its internal representation of the data. It is highly recommended for data quality reasons to review this information when using many dimensions in a `structure`. To do so, load the data into *WinBUGS* as usual (compiling and further steps are not needed). Next, click on the `Info` menu and select `Node info`. Type the name of the variable into the dialog box (e.g., `y` in our dataset) and then click `values`. *WinBUGS* will list out the entire data for the variable in a new window called `log`. The `log` output will show which dimensions and values are changing faster or slower according to *WinBUGS'* reading of the data `structure`. Check the listing to ensure that *WinBUGS* has loaded the data the way you wanted. Then, if necessary, respecify the data's dimensionality as needed. The dimensional format of the data can be changed using other software such as Excel. Alternatively, it is often easy enough to change the dimensional format in *WinBUGS* . To do this, we write a short piece of code to read in a variable (e.g., `y`) in its preexisting format and then create a new variable (e.g., `y.new`) where the dimensions have been remapped (see **WinBUGS Code 5.7 two-way ANOVA.odc** for an example). After inspecting the new variable with `Node info`, we can copy the output for use in the dataset of a second piece of *WinBUGS* code or make further changes if needed.

5.7.1 In Practice: Two-Way ANOVA with `structure` Data

We reexamine the Federal Reserve data on credit flows using an additional variable reported in the dataset. Annual credit flow data was also available subdivided into two types of credit: revolving credit (e.g., credit cards), and nonrevolving credit. Nonrevolving credit as defined by the Federal Reserve includes loans for automobiles, boats, mobile homes, education, or vacations (but does not include loans secured by real estate which are treated separately). Since nonrevolving credit typically is for longer term loans involving larger amounts of money, we may expect that these credit flows might differ from revolving credit flows. A two-way ANOVA will allow us to see if there were differences by credit type as well as by year. The *WinBUGS* program code for the two-way ANOVA appears below (see also **WinBUGS Code 5.7 two-way ANOVA.odc**). Only a portion of the data appears for brevity.

```
model
{
 for (i in 1:years) {
   for (j in 1:holders) {
     for (k in 1:types) {
       y[i,j,k] ~ dnorm(mu[i,j,k], tau)
       mu[i,j,k] <- alpha0 + alpha[j] + beta[k]}
     }
   }
 # Reference group is Year 2007 and revolving credit type
  alpha[1] <- beta[1] <- 0
 #priors and calculated parameters
  alpha0 ~ dnorm(0, 0.000001)
  for (j in 2:years) { alpha[j] ~ dnorm(0, 0.000001)}
  for (k in 2:types) { beta[k]  ~ dnorm(0, 0.000001)}
  tau ~ dgamma(0.001, 0.001)
  sigma <- 1/sqrt(tau)
}
#data
list(years=5, holders = 5, types=2, y = structure(.Data =
  c(28.4,29.7,31.9,38.2,-56.2,-8.3,-34.4,-36,-0.8,7.8,
    ...
    40.2,-6.1,-15.5,-24.6,-8,-31.6,-31.9,-18.4,-1.6,-6.7
                  ), .Dim = c(5, 5, 2))
 ) #(Dim: slowest=holders, then years, fastest=credit type)
#inits
list(alpha0=0, alpha=c(NA,0,0,0,0), beta=c(NA,0), tau=1)
```

The results for credit type do not show significant differences across type because the 95% credible interval for `beta[2]` includes the value of zero (Table 5.15). The increased sample size leads to sharper estimates for the yearly α parameters. This can be seen by comparing the standard deviations of the α parameters with the standard deviations of the $\mu_1 - \mu_j$ parameters in Table 5.12. As a consequence, credit flows in Year 5 are now found to be significantly different from credit flows in Year 1 because the 95% credible interval for α_5 is (−31.68, −0.15). This can be compared with the nonsignificant credible interval for $\mu_1 - \mu_5$ in Table 5.12.

5.7.2 Two-Way ANOVA with Group Indicator Variables

The functional form of two-way ANOVA model specification of Section 5.7 is tidy and compact. We can clearly see that our expected value is equal to the sum of our linear predictor, $\alpha_0 + \alpha_j + \beta_k$. This may be the best form for communicating the model to others, but the use of the `structure` option can become cumbersome as the model complexity grows. Also, the notation will get more involved as we move beyond the two-way ANOVA data ($y_{i,j,k}$) into higher order models with additional subscripts. It can be more useful for us in *WinBUGS* to have the model specified in terms of simpler individual observations y_i and the corresponding individual predicted values μ_i.

TABLE 5.15 Two-Way ANOVA Results for Consumer Credit Flows

Parameter	Mean	Std. Dev.	95% Credible Interval
α_0	15.18	6.23	(2.87, 27.44)
α_2	−12.64	7.99	(−28.41, 3.32)
α_3	−32.58	8.05	(−48.30, −16.52)
α_4	−29.70	8.03	(−45.42, −13.65)
α_5	−15.87	7.99	(−31.68, −0.15)
β_2	−3.19	5.11	(−13.40, 6.94)
σ	17.94	5.23	(22.86, 43.45)

For convenience's sake, we continue to suppose that our reference group is both the first group in Variable 1 as well as the first group in Variable 2. Next we need to let the functional form convey the following basic information: "data belonging in group j are assigned to the α_j parameter for estimation." Variable 1 contains group membership data, and suppose we write $x_{1,i}$ as the value of Variable 1 for the ith data point. So, instead of writing α_j we can write $\alpha_{x_{1,i}}$. Similarly, we can write $x_{2,i}$ for Variable 2 and write $\beta_{x_{2,i}}$ instead of β_k. To conclude, for *WinBUGS* we have the functional form of the two-way ANOVA written as

$$
\begin{aligned}
y_i &\sim \text{Normal}(\mu_i, \tau) \\
\mu_i &= \alpha_0 + \alpha_{x_{1,i}} + \beta_{x_{2,i}}, \\
\alpha_1 &= \beta_1 = 0 \\
\alpha_0, \alpha_2, \ldots, \alpha_J &\sim \text{Normal}(0, 0.0000001) \\
\beta_2, \ldots, \beta_K &\sim \text{Normal}(0, 0.0000001) \\
\tau &\sim \text{Gamma}(0.001, 0.001).
\end{aligned}
$$

This notation can be extended to higher order ANOVA by adding additional grouping variables and their parameters to the linear predictor.

In Practice: Two-Way ANOVA with Group Indicator Variables For this model, we have added new variables `year` and `creditType` to the dataset. The `year` variable takes on the values 1 through 5, with 1 indicating 2007 and 5 indicating 2011. For `creditType`, the value 1 indicates revolving credit and 2 indicates nonrevolving credit. We set the reference category as revolving flows from the year 2007. The functional form then has α_0, four α parameters for the nonreference years, and one β parameter for the nonreference credit type. The program code below implements the model for this dataset (see also **WinBUGS Code 5.7 two-way ANOVA.odc**). We have included a calculated parameter called `revol`. This will give us the average revolving credit flows by the year. A similar calculated parameter called `nrevol` gives the average nonrevolving credit flows by the year. We also included a calculated variable `adiff` that gives us the added opportunity to examine possible differences between any two years.

```
model
{
 for (i in 1:n) {
   y[i] ~ dnorm(mu[i], tau) #likelihood
   mu[i] <- alpha0 + alpha[year[i]] + beta[creditType[i]]
   }
  alpha[1] <- beta[1] <- 0 #Reference cat.: revolving 2007
#priors and other calculated parameters
  alpha0 ~ dnorm(0, 0.000001)
  for (j in 2:years) { alpha[j] ~ dnorm(0, 0.000001)}
  for (k in 2:types) { beta[k]  ~ dnorm(0, 0.000001)}
  tau ~ dgamma(0.001, 0.001)
  sigma <- 1/sqrt(tau)
  for (i in 1:years) {
   revol[i]  <- alpha0 + alpha[i]
   nrevol[i] <- alpha0 + alpha[i] + beta[2]
   for (j in i:years) {
    adiff[i,j] <- alpha[i]-alpha[j]  }
  }
}
#data
list(years=5, types=2, n=50,  year=c(1,1,1,...,5,5,5),
creditType = c(1,1,1,...,2,2,2),
y = c(28.4,3.9,3.7,...,-19.9,-5,0,-6.7) )
#inits
list(alpha0=0, alpha=c(NA,0,0,0,0), beta=c(NA,0), tau=1)
```

We see in the functional form that `alpha[year[i]]` allows *WinBUGS* to assign the data to the estimation of the correct value of `alpha` based on the `year` of the data. A similar code exists for `beta`. The increased sample size again gives increased statistical power to detect differences. A review of the results for `adiff` shows the differences in year after controlling for credit type. They indicate that flows significantly fell from 2007 to 2009 and from 2007 to 2010. Flows also significantly fell from 2008 to 2009 and from 2008 to 2010. Finally, flows rose from 2009 to 2011 (see **WinBUGS Code 5.7 two-way ANOVA.odc** for detailed results). The evidence from `adiff` suggests that there were three different periods of credit flow during the time studied: a falling period during the first two years, a bottom period in the second two years, and a recovery period in the final year. Since the current model is equivalent to the model of Section 5.7, again no significant difference was found for credit type after the differences in year were controlled for. Here, β_2 had a 95% credible interval of (−13.32, 7.02) and a slightly negative posterior mean (−3.16) (compare with Table 5.15 where similar results were found). Posterior box plots for the yearly revolving credit flows and yearly nonrevolving credit flows can be created using the `revol` and `nrevol` variables in *WinBUGS* (Figure 5.18).

5.7.3 Using Columnar Data in *WinBUGS*

Many times, data would have already been arranged into columns, such as data found in a spreadsheet or on a Web site. This type of data can be used directly by *WinBUGS*, saving the user from having to reformat the data. *WinBUGS* will expect this type of data to look like the following:

```
year[] creditType[] y[]
  1         1        28.4
  1         1         3.9
  ⋮         ⋮          ⋮
  5         2        −6.7
END
```

The data columns begin with the variable names such as `year`. All variable names must be followed by square brackets []. If the column is part of an array, the indexing subscript is placed inside the brackets. So, for example, the variable names for a three-column array might look like this:

```
y[,1] y[,2] y[,3]
```

Higher dimensional matrices can be supplied by extending the subscript notation such as `y[,1,1]`. All columnar data must have the

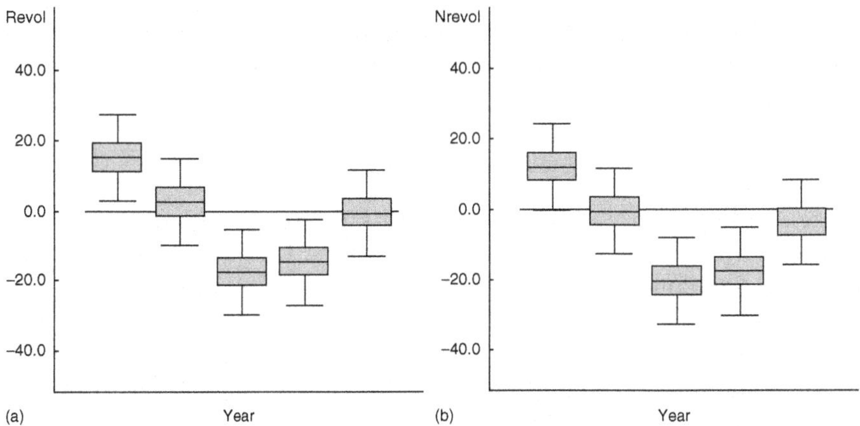

Figure 5.18 Posterior Box Plots for Annual Revolving (a) and Nonrevolving (b) Credit Flows: 2007–2011

statement `END` placed after the data. It is possible to have more than one columnar data listing (as well as more than one `list`-style data listing). It is also possible to use a mixture of both columnar data and `list`-style data. Highlight the first variable name or the word `list` at beginning of each of the data listings and use *WinBUGS'* `load data` button. *WinBUGS* will confirm with the message `data loaded` if it is successful, or display an error message if not. Once a listing has been successfully loaded, highlight the first variable name of the next data listing or the word `list` in the next data listing, and click `load data` again. Continue until all needed data listings have been loaded. It is common to need a short `list`-style data listing along with a columnar one so that *WinBUGS* knows about useful constants such as `n` (see **WinBUGS Code 5.7 two-way ANOVA.odc**). Columnar data can often be pasted directly into *WinBUGS* from spreadsheets or other software as long as you use `Edit > Paste > Paste Special > Plain Text` to ensure that only text is pasted.

5.8 REGRESSION AND ANCOVA MODELS IN *WinBUGS*

Regression models are a fundamental tool for providing insight into business data. In this model we seek to understand how a continuous outcome variable y is associated with one or more continuous predictor variables x. If we find that y is associated with x, regression analysis provides us with the ability to reduce our uncertainty about y by taking advantage of our knowledge of x. For example, we can make sharper predictions about y, conditional on a given value of x than would be possible without x.

As mentioned in Section 4.4, classical regression analysis uses a linear predictor formed from a sum of x variables that have been multiplied by unknown parameters β. Our notation for the linear predictor μ emphasizes that the linear predictor is the expected value of y conditional on x. This parallels the usage of μ in our notation of Section 5.3.3, where μ was the overall expected value for the entire dataset. The linear predictor for regression analysis is

$$\mu_i = \beta_0 + \beta_1 x_{1,i} + \cdots + \beta_k x_{k,i}, \tag{5.2}$$

where i indexes the n observations, and k is the number of predictors. We see that the expected value of observation i is both a function of the unknown β parameters and the particular values of a given x variable for that observation (i.e., $x_{1,i}$). If the value of an x variable (such as price of an item) were to change, then our expected response μ_i (such as demand for the item at price x_i) would also be expected to change. The outcome variable can now be rewritten as

$$y_i = \mu_i + \epsilon_i, \tag{5.3}$$

where ϵ_i is the error term for observation i. The error term is the departure from the expected value that occurs from a particular observation. For example, we may have expected the demand for a product to be 10,000 units at a particular price, but for some reason the observed demand y_i was 9200 units. In this situation, the error term ϵ_i would be −800 given our observed demand and expected value for demand.

In the classical regression model, the error terms are assumed to have a normal distribution, thus giving the likelihood function for the model. The error terms are also assumed to have a constant variance. This is known as the *homoscedasticity* assumption. It turns out that the regression model as described in Equations (5.2) and (5.3) is virtually identical to the ANOVA model as discussed in Section 5.7.2 except for one difference: the regression predictor variables are continuous, while the ANOVA predictor variables are grouping variables. These two models are special cases of what is called the *general linear model* (or sometimes called the *normal linear model*). In the general linear model, either continuous or grouping variables can be used to predict a normally distributed outcome variable using a linear predictor of the form in (5.2).

Our development in Section 4.4 limited our Bayesian treatment of the regression model to the Jeffreys' prior and one x variable. Fortunately, *WinBUGS* allows for much more flexibility with regard to prior selection and the number of variables, as is shown next.

5.8.1 In Practice: Simple Linear Regression Using *WinBUGS*

Being innovative is both a challenge and a necessity for many businesses. Innovation also helps drive the global economy, and as a result businesspeople and political leaders often have an interest in understanding what fosters innovation. Human capital is clearly relevant for innovation, and we might expect there to be a relationship between the number of people innovating and the total number of innovations. For our regression analysis, we examine the relationship between the number of people working in research and development per million residents (x_1) and the total number of patent applications (y) using yearly data over the time period 1999–2008. The data was obtained for two countries, the United Kingdom and France, from World Development Indicators (World Bank, 2012). As a brief aside, the current x_1 variable is a rate of the number of people working in research and development per million residents, but a slight complicating factor is that the population of France (60–64 million) was fractionally larger than that of the United Kingdom (58–61 million) during the time period selected. To make reproducing the data easier, just the original rate variable from World Development Indicators was used for this example. Both the *R* code and the *WinBUGS* code appears in the **WinBUGS Code 5.8 Regression and ANCOVA.odc** file accompanying this book. A plot of the data using *R* appears in Figure 5.19.

A simple linear regression will be the starting point of the analysis. Usually a count data model (see Section 9.3) would be applied to this data but with the large sample sizes the results should be reasonably similar (see Chapter 9 exercises). Note first that the y-axis ranges from 14,000 to 22,000 in Figure 5.19. This tells us that the y intercept, β_0, may be similarly rather large. We will therefore need a widely diffused prior for β_0 if we seek a noninformative prior. We also need a single parameter to implement the homoscedasticity assumption. We therefore estimate the parameter τ as the precision. Since τ does not depend on other parameters, it will be constant throughout the dataset. Bearing in mind the considerations for the prior and the regression assumptions, our specification is as follows:

$$\begin{aligned} y_i &\sim N(\mu_i, \tau) \\ \mu_i &= \beta_0 + \beta_1 x_{i,1} \\ \beta_0 &\sim N(0, 0.0000000001) \\ \beta_1 &\sim N(0, 0.0000000001) \\ \tau &\sim \text{Gamma}(0.00001, 0.00001). \end{aligned} \tag{5.4}$$

We begin with a simple linear regression model while recognizing that a single line will not provide the best fit to the data of Figure 5.19. Running the code in the first Section of **WinBUGS Code 5.8 Regression and ANCOVA.odc** with a 1000-iteration burn-in and 10,000 iterations for estimation produces the first set of results (Model 1) in Table 5.16.

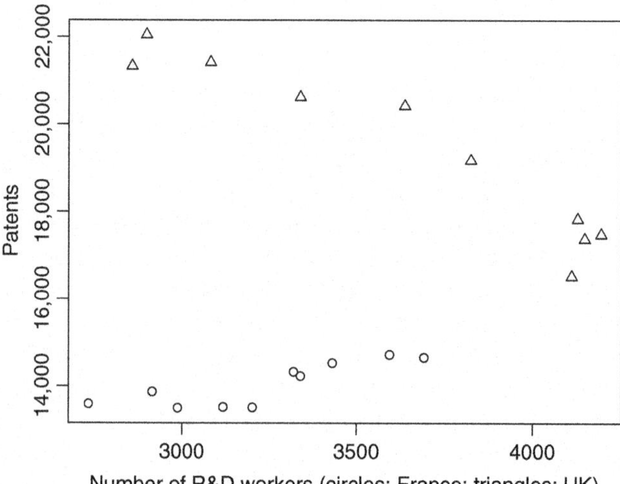

Figure 5.19 Number of R&D Workers per Million and Number of Patents: France and United Kingdom 1999–2008

TABLE 5.16 Regression and ANCOVA Results: Patent Dataset

Model	Parameter	Mean	Std. Dev.	95% Credible Interval
Model 1	β_0	14818	5646	(3579, 26081)
	β_1	0.557	1.634	(−2.701, 3.764)
Model 2	β_0	21613	5646	(17658, 21619)
	β_1	−2.341	0.616	(−3.574, −1.137)
	β_2	6288.5	558.1	(5206.9, 7404.8)
Model 3	β_0	9522.7	1931.1	(5675.8, 13402)
	β_1	1.398	0.595	(0.201, 2.574)
	β_2	22631	2257.3	(18107, 27184)
	β_3	−4.914	0.674	(−6.269, −3.558)

The *WinBUGS* code for our Model 1 is reproduced below, and can be cross-checked against the specification above.

```
model
 { #simple linear regression
  #likelihood + functional form
  for (i in 1:n)
    {patents[i] ~ dnorm(mu[i], tau)
     mu[i] <- beta0 + beta1*rsrch[i]}
  #priors
    beta0 ~ dnorm(0, 0.0000000001)
    beta1 ~ dnorm(0, 0.0000000001)
    tau ~ dgamma(0.00001, 0.00001)
    sigma <- 1/sqrt(tau)
}
```

The results of the Model 1 estimation indicate that the 95% credible interval for β_1 includes the value of zero. Hence under Model 1 we do not have convincing evidence suggesting that there is a positive (or a negative) relationship between the two variables. From Figure 5.19, we see that an explanation for the flat slope of β_1 is that the data for France and the United Kingdom have contrasting patterns. Since β_1 captures the aggregated trend across the two countries in Model 1, a slope of zero in aggregate remains plausible.

5.8.2 In Practice: ANCOVA Models Using *WinBUGS*

Historically, regression models were models in which all variables (predictors and outcome) were continuous. Models that have at least one continuous predictor variable and at least one grouping predictor variable, along with a continuous outcome variable, are known as *analysis of covariance* (*ANCOVA*) models. ANCOVA models are members of the general linear model family along with regression and ANOVA models. Since we have data on two different countries, we might expect that there are country-specific differences with regard to the "patent productivity" of R&D researchers even before plotting the data. Additionally, the data plot itself gives a clear indication that we have two contrasting trends in patent productivity.

We can convert each grouping variable to a continuous variable using something called *dummy coding*. If we have g groups, we create $g - 1$ variables that take on the values 0 and 1. Dummy variable 1 is set to the value 1 when the observations for Group 2 appear, and is set to 0 otherwise. Dummy variable 2 is set to the value 1 when the observations for Group 3 appear, and is set to 0 otherwise. This continues for all remaining groups. The first group will have values of zero for all dummy variables and is typically referred to as the *reference group*. The g parameters estimated for the remaining groups can then be interpreted in terms of a difference between that group and the reference group. The dummy variables thus emphasize the contrasts between the reference group and the other groups.

The second model in **WinBUGS Code 5.8 Regression and ANCOVA .odc** contains an additional variable called `country`. France is the reference group in this dummy variable. In Model 2, we add the dummy variable as x_2 so that our linear predictor is now $\mu_i = \beta_0 + \beta_1 x_{1,i} + \beta_2 x_{2,i}$. Adding this dummy variable to the model will give France and the United Kingdom different intercepts (β_0 for the former and $\beta_0 + \beta_2$ for the latter). However, both countries will share the same slope, β_1. As a result, this kind of ANCOVA model is called the *parallel slopes* ANCOVA model. Having added β_2, we must also specify a prior for it. For this analysis, we give β_2 a prior that is identical to the other β priors. We then run the model with 1000 iterations of burn-in and 10,000 iterations of estimation to produce our results for Model 2 in Table 5.16. For a full code listing, see **WinBUGS Code 5.8 Regression and ANCOVA.odc**.

The posterior mean of β_1 is −2.341 and the 95% credible interval excludes the value zero. The evidence therefore indicates that the more R&D workers a country has, the less the number of patents produced. We also see that β_2 has a posterior mean of 6288.5 and a 95% credible interval that excludes the value zero. This indicates that the patent productivity intercept of the United Kingdom is greater than that of France. We can augment our plot of the data using the *R* code appearing in **WinBUGS Code 5.8 Regression and ANCOVA.odc**. Figure 5.20 shows the data with the Bayesian regression lines formed from the posterior means of β_0, β_1, and β_2 superimposed on the graph. The y-intercept for the regression line of France corresponds to β_0. The y-intercept for the regression line of the United Kingdom corresponds to $\beta_0 + \beta_2$. The slope for both countries' lines corresponds to β_1. Since the slope for both countries is β_1, this

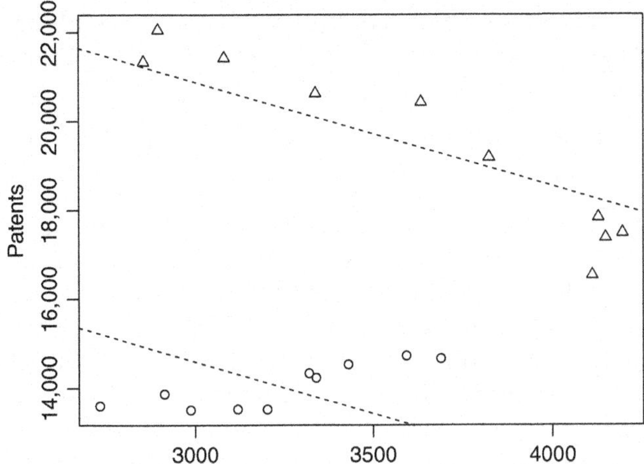

Figure 5.20 Parallel Slopes ANCOVA. Number of R&D Workers and Number of Patents: France and United Kingdom 1999–2008

particular form of the ANCOVA model, where the dummy variable is added to the linear predictor, is sometimes called the *parallel slopes model.*

In Model 1, the 95% credible interval included the value of zero whereas in Model 2 it did not. Just as in classical regression, we observe that omitting relevant variables from the model specification can have an impact on our conclusions about a variable's relationship with the dependent variable. Here we see graphically that, if we allow the two countries to have individual intercepts, the overall trend as measured by β_1 is negative. Yet, Model 2 has the limitation of imposing the same slope on both countries, while the data suggests France's patent productivity may be increasing with greater values of x. Model 3 revises Model 2 to allow each country to have an individual slope and an individual intercept. This can be accomplished by adding an interaction term to the model specification. An interaction term can be created by entering a product of the relevant x variables in the specification such as follows: $\mu_i = \beta_0 + \beta_1 x_{1,i} + \beta_2 x_{2,i} + \beta_3 x_{1,i} x_{2,i}$. Here, the $x_{1,i} x_{2,i}$ product will cause β_3 to be estimated as the difference of the United Kingdom's and France's slopes. It will also cause β_2 to be estimated as the difference of the United Kingdom's and France's intercepts. This ANCOVA formulation allows for unequal slopes and unequal intercepts across the two groups.

Estimates from Model 3 appear in Table 5.16. The results for β_1 indicate that France's patent productivity has been positive such that when France has had more R&D workers, it has produced more patents ($\beta_1 = 1.398$). The 95% credible interval for β_1 excludes the value of zero. Conversely, as the United Kingdom's number of R&D workers increased, the number of patents produced *compared to France* declined ($\beta_3 = -4.914$). Since the 95%

credible interval for β_3 excludes the value of zero, the evidence indicates that the United Kingdom's slope is different from that of France.

5.8.3 In Practice: "Undifferenced" ANCOVA Models Using *WinBUGS*

In the previous section, we estimated standard ANCOVA models using *WinBUGS*. We found that the slope of France's patent productivity was positive and different from zero. We also found that the slope of the United Kingdom's patent productivity was different from that of France. Here, our results permit only a direct comparison of the United Kingdom to France, i.e., the United Kingdom can only be considered in a *relative* or comparative fashion. While in many research contexts this might correspond to a research question of interest, in others they might not. Suppose the research question for the United Kingdom was whether the incremental benefit of more R&D workers on the United Kingdom's patent output was different from zero in an *absolute* sense. Here our research question would lead us to consider a Bayesian ANCOVA model where we estimate the "undifferenced" β_{UK} as opposed to β_3. Note that if you happened to have a research question about β_{UK} and were using classical statistics, it would take some extra work to estimate it in the same model run with France's parameter β_1. Here we will see that estimating both β_{UK} and France's β_1 in the same *WinBUGS* run is straightforward.

One-way to estimate β_{UK} is by variable composition. Since β_3 is the difference between the slopes of the United Kingdom and France, and since β_1 is the slope of France, we see that $\beta_{UK} = \beta_1 + \beta_3$. We can easily add a line of code to our interaction model that calculates this variable. The line of code to add is

```
betaUK <- beta1 + beta3
```

and it can be added just before the code for the priors in the interaction model. Sampling this variable via MCMC allows us to find the posterior of β_{UK}. A complete code listing appears in **WinBUGS Code 5.8 Undifferenced ANCOVA.odc**. We also note in passing that we could also find the intercept for the United Kingdom's regression line by adding `intUK <- beta0 + beta2`. Since a research question for the intercept is not as commonly encountered, this parameter has been omitted.

Performing the MCMC estimation based on a 1000-iteration burn-in and a 10,000 iteration MCMC run gives the results in Table 5.17. There would be little reason to include both the result for β_3 and β_{UK} in the same table because likely only one would involve a research question of interest and because β_3 is partially redundant with the result for β_{UK}. However, both are included here for the purposes of comparison. We see the results for Model 3: Composition in Table 5.17 are identical to those of Model 3 in

TABLE 5.17 ANCOVA Results: Patent Dataset

Model	Parameter	Mean	Std. Dev.	95% Credible Interval
Model 3 (Composition)	β_0	9522.7	1931.1	(5675.8, 13402)
	β_1	1.398	0.595	(0.201, 2.574)
	β_2	22631	2257.3	(18107, 27184)
	β_3	−4.914	0.674	(−6.269, −3.558)
	β_{UK}	−3.515	0.333	(−4.184, −2.866)

Table 5.16 with the exception of the new results for β_{UK}. The 95% credible interval for β_{UK} indicates that, during the sample period, increases in the number of R&D workers were on average associated with declines in the number of patents in the United Kingdom. Note that this conclusion does not hinge on a relative-to-France comparison. Rather, it is a simpler absolute (i.e., non-relative-to-France) conclusion that we may be interested in for our research question.

Given the findings of Table 5.17, we can provide a graphic such as in Figure 5.21. It would also be possible to provide a similar graphic based on the results of Table 5.16 by adding up the slope terms. However, testing classical hypotheses about β_{UK} from the information in Table 5.16 would be less straightforward without some extra work. We could also consider other parameter transformations depending on the data and the research

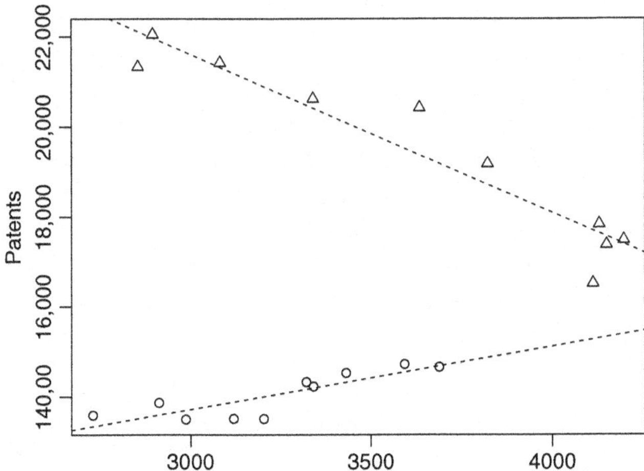

Figure 5.21 ANCOVA with Unequal Slopes and Intercepts. Number of R&D Workers and Number of Patents: France and United Kingdom 1999–2008

questions we had. In a different dataset, we might want to consider the percentage difference between two (positive) coefficients. We could add the following line of code to our *WinBUGS* model:

```
betaPctDiff <- (beta1/(beta1 + beta3) ) - 1
```

This would calculate the ratio of the two group's coefficients and subtract 1 from the result. *WinBUGS* would then obtain the distribution of this quantity, and we could perform inference on the results. This section has demonstrated the flexibility of MCMC estimation. In Bayesian ANCOVA, this flexibility allows us to easily examine research questions that would otherwise be more difficult to examine.

5.9 SUMMARY

This Chapter has introduced the use of *WinBUGS*, its syntax, and a number of its most commonly used dialog boxes and options. We have discussed the idea of prior sensitivity and pointed out how poorly chosen priors can negatively affect our inferences. Finally, we have examined different special cases of the general linear model. Special cases of the linear model include *t*-tests, ANOVA, regression, and ANCOVA models.

5.10 CHAPTER APPENDIX: EXPORTING *WinBUGS* MCMC OUTPUT TO *R*

It is possible to export *WinBUGS'* MCMC output to *R*. In *WinBUGS*, run the simulation as usual and monitor relevant parameters using the Sample Monitor tool. Then, in the Sample Monitor tool select the variable that you wish to export in the node field (or type * in the node field to select all variables). Next click the `coda` button, which is the second button from the left on the bottom row of the Sample Monitor tool (Figure 5.5). *WinBUGS* will produce a file called *CODA index* and a file called *CODA for chain 1* (additional CODA files for chains 2 and higher will be produced if multiple chains are being used). Perform a "Save as" on the CODA index file and select file type Plain text (*.txt) in the dialog box. Then enter a file name with a .ind extension such as

```
"my-run1.ind"
```

making sure to use double quotes as shown. If the file name is not enclosed in double quotes, then a .txt extension will be appended to the file name which will need to be manually removed. Next perform a "Save as" on the CODA for chain 1 file and select file type Plain text (*.txt) in the dialog box. Then enter a file name with a .out extension such as

```
"my-run1.out"
```

making sure to use double quotes as shown. If the file name is not enclosed in double quotes, then a .txt extension will be appended and again will need to be removed.

Once these files have been exported, the information can be read into *R*. There are a variety of ways of doing this. If you are interested in *R* programming, *R* has very powerful tools for reading data. For those who are less interested in *R* programming, the freely available software *BUS*, available at `http://faculty.salisbury.edu/~ edhahn/bus.htm`, can be used to assist. After downloading and running *BUS*, click its Browse button to browse to the location that contains both the .ind and .out files. Click "Read" to have *BUS* read these files and click "Write" to have it write out a file called `bugsout.txt`. *BUS* will write the file `bugsout.txt` to the same directory where the input files were located. Click "Exit" to close the program. Next, in *R* enter a command such as the following:

```
mydat <- read.delim(file="C://My Folder//bugsout.txt", header=TRUE,
   sep="\t")
```

You will need to change the path from `C://My Folder//` to the path where your files are located. Note that *R* requires a double forward slash (//) wherever the usual backslash (\) in the file path occurs. To see that the variables that are now available, enter the following:

```
objects(mydat)
```

The variables you have exported should now be listed. To analyze these variables, type the following:

```
attach(mydat)
```

You can now analyze these variables in *R*. For example, if one of your parameter variables was `beta1`, typing `mean(beta1)` would instruct *R* to calculate its mean.

5.11 EXERCISES

1. The annual unemployment rates (in percent) for men in Arizona during the years 1988–1998 were 6.7, 5.3, 5.9, 6.3, 8.5, 6.4, 5.7, 4.7, 5.3, 4.3, and 3.8 (Bureau of Labor Statistics, 2013). The unemployment rates for women in Arizona over the same time period were 5.7, 5.1, 4.9, 5.1, 6.6, 6.2, 7.1, 5.6, 5.8, 5.1, and 4.5. Use a two-sample *t*-test to analyze this data. Is there evidence of a difference in unemployment rates over this 11-year period?

2. Reexamine the data in **WinBUGS Code 5.2 two sample t-test.odc**. Test whether σ_1 is different from σ_2. Create a new variable `sdiff`, which is the difference between these two parameters. Can we conclude that one standard deviation is greater than the other using a 95% credible interval?

3. In the pharmaceutical industry, 112 out of 201 CEOs said that the past quarter was very challenging. By contrast, 78 of 261 CEOs in the energy industry said that the past quarter was very challenging. Use *WinBUGS* to perform a test of the difference in proportions. Can we conclude that there is a difference using a 95% credible interval?

4. Is the standard deviation of the 2011 credit flows data in Section 5.6.3 significantly different from other years? Modify the code in **WinBUGS Code 5.6 one-way ANOVA.odc** so as to test this possibility.

5. Using **WinBUGS Code 5.6 one-way ANOVA.odc**, reanalyze the first ANOVA model and place a uniform prior from 0 to 100 on `sigma`. Add the backtransformation to `tau` as described in Section 5.4. How do the results change?

6. Singapore's total exports of services from 2002 to 2011 were as follows (in billions of Singapore dollars): 49, 54, 68, 76, 93, 111, 126, 118, 137, and 146 (Department of Statistics Singapore, 2013). Use a simple linear regression to fit this data in *WinBUGS*. Generate a predictor variable `year` that ranges from 1 to 10. Estimate the value of the slope. Does the 95% credible interval for the slope exclude the value of zero?

7. Using your model from the previous exercise, forecast services exports for the year 2012 (`year` = 11). Do this by creating a new variable `forecast` in your model. Define `forecast` to be equal to the slope times 11 plus the intercept. What is the expected value and 95% credible interval for your forecast?

8. Reanalyze the patent data of **WinBUGS Code 5.8 undifferenced ANCOVA.odc** with a normal(0, 0.000001) prior on all predictors. What, if anything, changes in the results? What is your interpretation?

6

ASSESSING MCMC PERFORMANCE IN *WinBUGS*

6.1 CONVERGENCE ISSUES IN MCMC MODELING

We seek to avoid wrong conclusions when we send our Markov chains out to map the posterior distributions of our model's parameters. However, our challenge is that we usually will not know what results to expect in advance. Therefore, we will typically not have a known "gold standard" distribution to compare with the distribution arising from our MCMC run. This motivates the use of diagnostic analyses to assess the performance of the Markov chain. Our primary tool so far has been the trace plot, and so far the Markov chains seem to be performing reasonably well in producing samples from the posterior distributions of our parameters after a brief burn-in period. Of course, this does not always occur. For a variety of reasons, we may find that the output of the Markov chain gives us reason for concern. Consider the following *WinBUGS* code (see also the file **WinBUGS Code 6.1 poor convergence.odc**):

```
model
{
#likelihood + functional form
 for (i in 1:n)
    {y[i] ~ dnorm(mu, 1)}
     mu <- bad1 + bad2
```

Bayesian Methods for Management and Business: Pragmatic Solutions for Real Problems, First Edition. Eugene D. Hahn.

```
#priors and calculated parameters
  bad1  ~ dnorm(0, 0.0000001)
  bad2  ~ dnorm(0, 0.0000001)
}
#data
list(y=c(100, 102), n=2)
#inits
list(bad1=1,bad2=1)
```

After allowing a burn-in of 1000 iterations, we sample from the posteriors for 10,000 iterations. A plot of the traces of `bad1` and `bad2` appears in Figure 6.1.

These trace plots are quite different from the ones we have encountered previously. The Markov chain wanders around in a serpentine fashion and does not appear to resemble random draws from a given distribution. We may wonder whether the burn-in period has been sufficient. If 10,000 iterations might not be enough, perhaps the chain will reach its final destination if it is allowed to run further. Suppose we give the chain plenty of time to burn in by running it an additional 1,000,000 iterations. Plots of the resulting traces appear in Figure 6.2. We see that the additional iterations do nothing to address the problem—if anything, the chains have now wandered even farther away from their initial starting points and look even less like a sample from a posterior distribution.

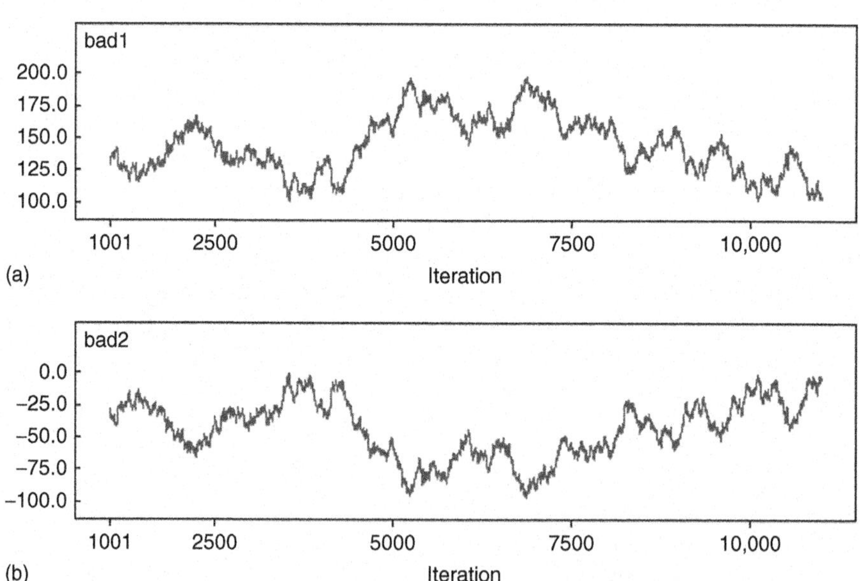

Figure 6.1 Movement of Markov Chain for (a) `bad1` and (b) `bad2` for Iterations 1001–10,000

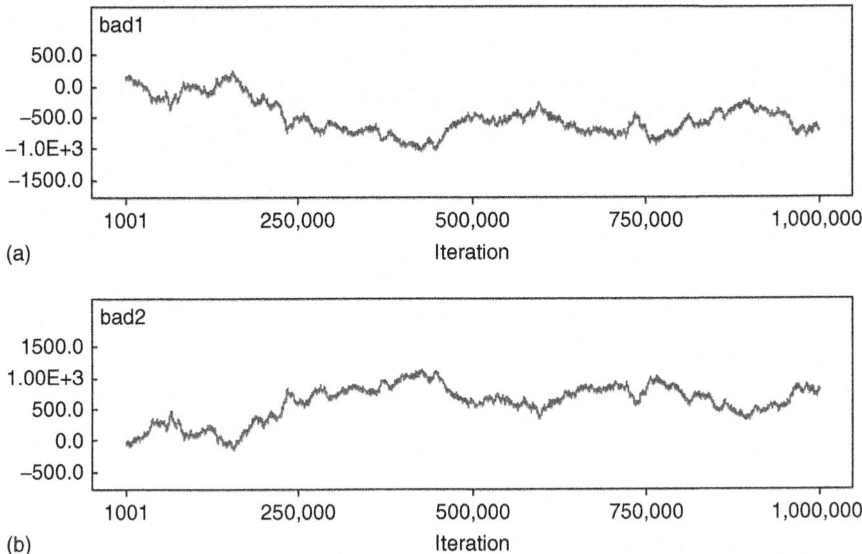

Figure 6.2 Movement of Markov Chain for (a) `bad1` and (b) `bad2` for Iterations 1001–1,100,000

Upon taking another look at Figures 6.1 and 6.2, we see that the trace for `bad2` looks like a mirror image of `bad1`. The *WinBUGS* code uses two parameters to model the expected value of the data `y`. It can be seen that the mean of the data in `y` is 101. Consider the first few iterations of the Markov chain on the left-hand side of the traces in Figure 6.1. The chain estimates `bad1` as being around 140 and `bad2` as being around −40. The average of these two numbers approximates 101, so the Markov chain clearly attempts to model the mean. However, `bad1` and `bad2` are unidentifiable here. Whatever value the Markov chain visits for `bad1`, it will tend toward 101 minus that value in simulating from `bad2`. The converse also occurs. The model is *overparameterized* and the output from the Markov chain for these parameters is not useful for decision making. Since `bad1` and `bad2` appear to be able to wander quite broadly (with no end in sight), we have not been able to learn anything substantial about these parameters or reduce our uncertainty about them. The meandering traces in Figures 6.1 and 6.2 also resemble examples of random walks. This also is a sign that the Markov chain has not reached its stationary distribution. As a result, we will not be able to effectively learn about these parameters with the model and data at hand.

Despite the disappointing results for these parameters, `mu` itself is identifiable and the trace plot for it looks much more satisfactory (Figure 6.3). We have been able to learn about `mu` and thereby reduce our uncertainty

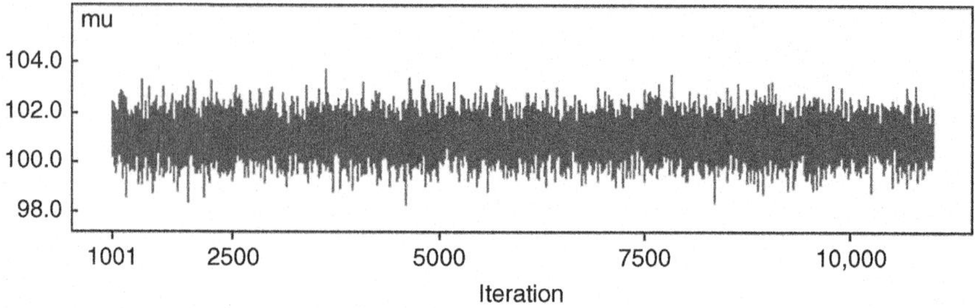

Figure 6.3 Movement of Markov Chain for mu: Iterations 1001–10,000

about it. The results from *WinBUGS* give a 95% credible interval for mu as (99.63, 102.4), so although we may not know mu exactly, we infer that it is likely within this relatively small range.

6.2 OUTPUT DIAGNOSTICS IN *WinBUGS*

Our previous example shows that *WinBUGS* will try its best to do what you tell it to do—whether or not it is sensible to do so! It is therefore critical that we review the output of a MCMC run to see if any problems can be found. This motivates the use of MCMC output diagnostics that have been designed to reveal possible issues. *WinBUGS* includes several diagnostics that help identify some of the more common issues. In this section, we will cover the `quantiles`, `auto cor`, and `bgr diag` tools.

6.2.1 The Quantiles Tool

From Section 4.2, we know that a key consideration in applied Bayesian statistics is determining whether the Markov chain has reached (or converged to) its stationary distribution. If the Markov chain is still progressing toward its stationary distribution at a given point in the run without having reached it, we must omit the current and all previous iterations from our computations as burn-in iterations. Thus one diagnostic to consider is a plot of a moving average of the quantiles and mean of the distribution. The idea behind such a plot is as follows: If we have reached the posterior distribution, our estimate of the mean should be relatively stable over various subsets of the Markov chain. Similarly, our estimates of the quantiles (such as the 95% credible interval) should be stable over various subsets of the Markov chain. Hence, plotting a moving average of the mean and quantiles of the distribution should show relatively little movement over the course of the run if the chain has converged. If the chain has not converged, a moving average of the mean and quantiles should be drifting toward the stationary distribution over the course of the run.

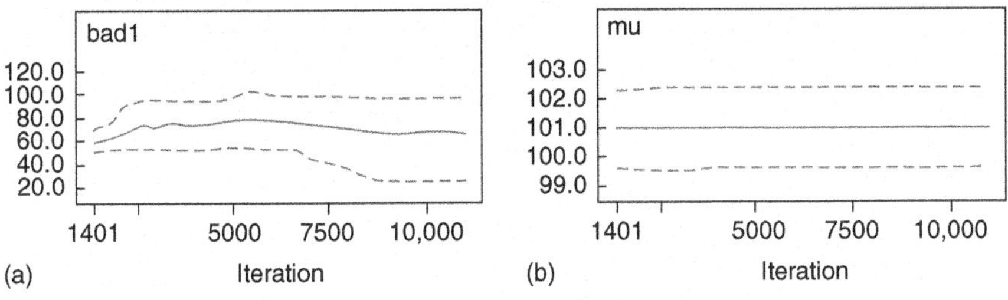

Figure 6.4 Running Quantiles for (a) `bad1` and (b) `mu`

WinBUGS provides diagnostic plots of the running mean and running 95% credible interval. These can be accessed by pressing the `quantile` button on the Sample Monitor tool. Figure 6.4 displays *WinBUGS'* running quantile plots for `bad1` and `mu`. The running mean and quantiles are plotted to correspond with the iteration number on the x-axis. The *WinBUGS* manual is short on details, but from the x-axis it appears that *WinBUGS* is taking a 400-iteration moving average of the quantities plotted. It does not appear to be possible to change *WinBUGS'* moving average to a value other than 400 iterations. However, this value will often produce a graph that is a reasonable starting point.

The plot for `bad1` at left in Figure 6.4 shows drifting in the running mean and quantiles, which would indicate that the chain for this parameter had yet to converge. Conversely, at right we see that the moving averages of the mean and 95% credible interval for `mu` appear stable over the course of the run. Like other convergence diagnostics, the running quantile plots only point to evidence of nonconvergence—they do not confirm that convergence has occurred. Based on the evidence, we would again conclude that any results we might have for `bad1` should be discarded. For `mu`, the evidence does not guarantee convergence but nonetheless seems promising, and we would want to examine other diagnostics for `mu` to make sure it does not fail some other test.

6.2.2 The Autocorrelation Function Tool

Another important output diagnostic is a plot of the autocorrelation function. The autocorrelation function is an idea common from time-series analysis and it is used to display the pattern of correlations over time in a dataset. In applying this idea to MCMC output, we seek to understand how much correlation there is from one iteration to the next over the course of the run. The autocorrelation function graphically plots the correlation between a given iteration and the lagged value of a given iteration over the course of the run. For example, the lag 1 autocorrelation is the correlation between the value of a given iteration and the value of that iteration's most recent predecessor throughout the data, i.e., the correlation between iterations i and iterations $i - 1$ throughout the Markov chain. The lag 2 autocorrelation is the correlation between iterations i and iterations $i - 2$ throughout the Markov chain. Higher lags and their autocorrelations can also be calculated. *WinBUGS* calculates autocorrelations for the first 49 lags. The lag 0 autocorrelation is also calculated by *WinBUGS*. This is always equal to 1 because it is the correlation of a given iteration with itself (i.e., without any lags).

The autocorrelation function gives an idea of the efficiency of the exploration of the Markov chain. In the ideal case, all autocorrelations (except lag 0) are close to zero. In this case, the iterations of the Markov chain are approximately independent over time in the run. Here, each iteration of the chain gives the maximum possible amount of new information about the posterior distributions so as to sharpen our estimates as much as possible. The Monte Carlo error, which quantifies the potential variability in our estimates due to the approximate nature of the Monte Carlo method as well as due to the finite number of samples we have taken, will be as low as possible. As the autocorrelations increase, however, the iterations of the Markov chain become increasingly dependent over time. This means that each iteration of the Markov chain is not contributing as much information toward the sharpening of our estimates. Because of this less than optimally efficient performance, we will need to run the Markov chain for greater lengths in order to reduce the Monte Carlo error to a particular level. It is also possible that the autocorrelations are negative. In this case, the successive iterations are more self-avoiding than would be expected if the iterations were independent of each other. However, large negative autocorrelations are not commonly encountered in practice.

Figure 6.5 shows a plot of the autocorrelation functions for `bad1` and `mu`. The autocorrelation function for `bad1` at left confirms that the Markov chain for this parameter is severely autocorrelated. Double-clicking on the plot in *WinBUGS* and then holding the *Control* key and left-clicking again will instruct *WinBUGS* to list out the actual values of the autocorrelations to a new window. Here, the lag 1 through lag 3 autocorrelations are 0.9988, 0.9976, and 0.9963, revealing that the current location of the

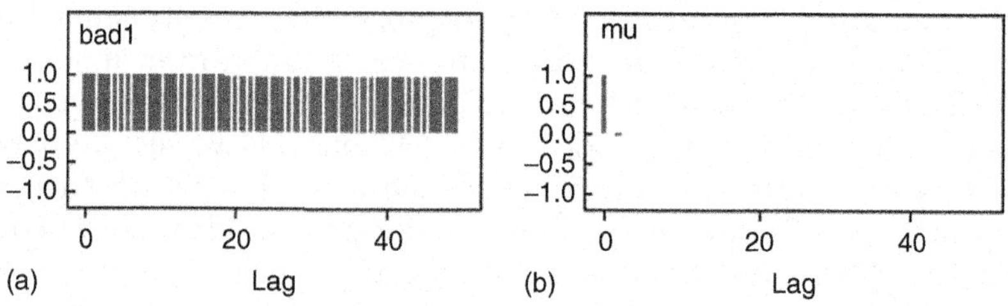

Figure 6.5 Autocorrelation Function Plots for (a) `bad1` and (b) `mu`

Markov chain is almost entirely determined by its previous locations. Even at lag 49, the autocorrelation is over 0.94. By contrast, the autocorrelation plot for `mu1` looks almost picture-perfect. The lag 1 autocorrelation is only 0.0173, and values for successive lags are primarily in the range of −0.015 to 0.015. On clicking on `node statistics` in the Sample Monitor tool, we find that the Monte Carlo error for `mu1` is 0.0067 while for `bad1` it is approximately 300 times greater, at 2.012. When we see autocorrelation functions like the one for `bad1`, it indicates that the Markov chain is having substantial difficulty learning about that parameter with the data at hand. At a minimum, it indicates that we will need to run the chain for a very long time to obtain reasonably precise estimates. It can also indicate (as is the case here) that there is a flaw in the model.

6.3 REPARAMETERIZING TO IMPROVE CONVERGENCE

In some instances, we may be able to rewrite our model so that the Markov chain has an easier time exploring the posterior distributions of the parameters. For example, centering variables around the mean will, for some models, allow the Markov chain to move more freely. The following code (which also appears in **WinBUGS Code 6.3 re-parameterizing.odc**) shows the impact that mean-centering can have on convergence (the data has been omitted for brevity). We first consider the situation where mean-centering has not been used as shown in the following code:

```
model
{
 for (i in 1:n){
   y[i] ~ dnorm(mu[i],tau)
   log(mu[i]) <- beta[1]+beta[2]*x[i]
   }
#priors and data transformations
 for (k in 1:2){  beta[k] ~dnorm(0, .0000001) }
 tau ~ dgamma(0.001, 0.001)
 sigma <- 1/sqrt(tau)
}
#inits
list(beta=c(0,0), tau=1)
```

The model in the above listing is known as a *log-linear model* because the logarithm of the expected value `mu` is assumed to be a linear function of the explanatory variable `x`. In this (nonlinear) model, `beta[2]` tells us how much of a percent change we would expect in y for a one-unit change in x (this is known as *semi-elasticity* in the economics literature). The data consists of 100 simulated values.

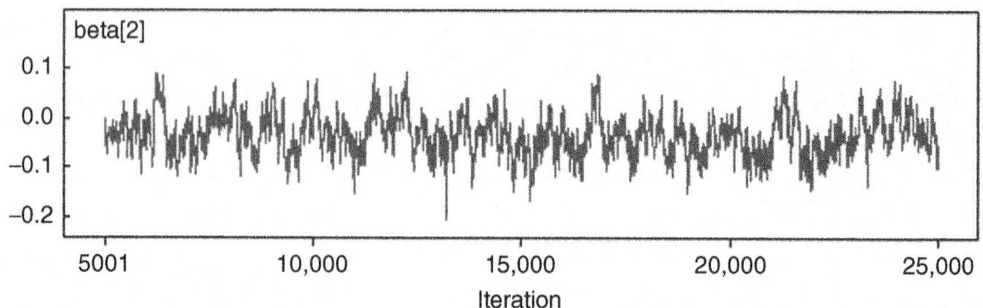

Figure 6.6 Trace Plot for Uncentered `beta[2]` in Log-Linear Model

A 5000-iteration burn-in was used, and the Markov chain was then allowed to run for an additional 20,000 iterations for estimation. The trace plot for `beta[2]` appears in Figure 6.6 (the plot for `beta[1]` looks similar). The trace plot shows evidence of somewhat slow mixing although convergence appears to have occurred.

In general, slow MCMC mixing is more likely when parameters are highly correlated. An extreme version of this was seen in Section 6.1, where the parameters `bad1` and `bad2` are so correlated that they contributed no unique information individually. As a diagnostic, we can request *WinBUGS* to provide a pairwise scatter plot of the values of the Markov chain for two variables. To request pairwise scatter plots, click on the `Inference` menu and then select the `Correlations` option. In the Correlation tool which appears, enter the parameter of interest in the `nodes` text area. One possibility is that you can enter one parameter's name in the first text area and a second parameter's name in the second. Clicking the `scatter` button produces the scatter plot. Alternatively, if the variables are declared as an array (e.g., `beta[1]`, `beta[2]` as we have here), you can enter the array name (`beta` without the indexing in brackets) in the first text area. *WinBUGS* will produce the scatter plots for all members of the array upon clicking the `scatter` button.

Figure 6.7 shows the autocorrelation function for `beta[2]` on the left and the bivariate scatter plot of `beta[1]` and `beta[2]` on the right. The autocorrelation is substantial for `beta[2]`, which means we will need a longer run to obtain a given degree of accuracy in our estimates. The Monte Carlo standard error for `beta[2]` is 0.0026 here. The bivariate scatter plot shows the source of the problem. The two parameters have a strong negative correlation with each other.

For a more detailed look at the reasons behind poor mixing, click on the `Info` menu and then the `Node info...` item to make the `Node tool` appear. Type the word `beta` into the text area and click the `methods` button. *WinBUGS* creates an entry in the `Log` file that the MCMC method being used for `beta[1]` and `beta[2]` is the `UpdaterMetnormal` method. This is a Metropolis–Hastings sampler. Because this sampler updates one parameter at a time based on the parameter's marginal

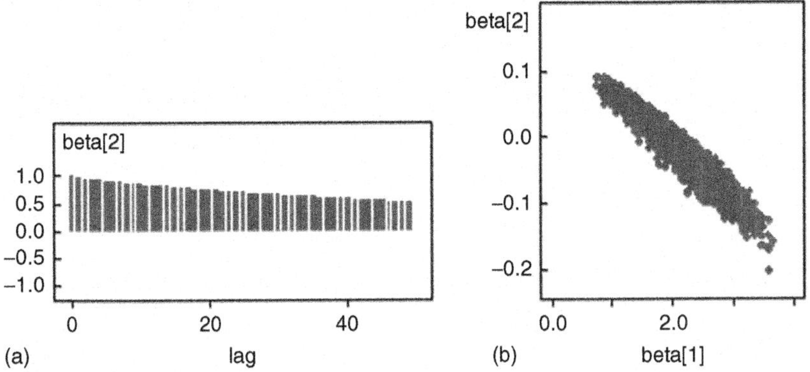

Figure 6.7 Autocorrelation Function and Bivariate Scatter Plot for Uncentered `beta[2]`. (a) Autocorrelation Function. (b) Scatter Plot

distribution, it will attempt to move along either the x-axis or the y-axis of Figure 6.7(b), depending on which parameter is being updated. We see that only very small moves can be made in these perpendicular directions if the sampler wants to stay inside the main density of the posterior distribution. Larger moves will almost always be rejected by the Metropolis algorithm, and in these cases the chain will not move at all. The combination of small moves and rejections (no movement) causes a slowdown in the mixing of the Markov chain and contributes to the high individual-level autocorrelations.

A simple method for addressing this problem is to center the predictor variable around its mean. This change to the functional form makes it much easier for the Markov chain to travel around the joint posterior distribution. We change the log-linear predictor to read

```
log(mu[i]) <- beta[1]+beta[2]*(x[i]-mean.x)
```

We also add a line of code at the end of the listing which reads

```
mean.x <- mean(x[])
```

This instructs *WinBUGS* to calculate the mean of x once and store it in a variable called `mean.x`. Storing the mean in a variable such as `mean.x` is preferable to calculating it in the linear predictor because in the latter case *WinBUGS* will have to recalculate the mean at every iteration, thus slowing down the overall process. Running the Markov chain again with a 5000-iteration burn-in and 20,000 iterations for estimation yields a trace plot as in Figure 6.8. Comparing with Figure 6.6, we see that mixing is better in this run than it was previously.

We may also look at the autocorrelation function for the centered functional form, as well as the scatter plot (Figure 6.9). The autocorrelation

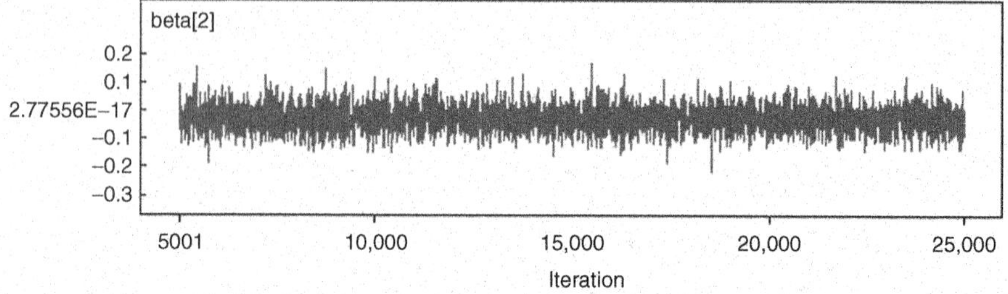

Figure 6.8 Trace Plot for Centered `beta[2]` in Log Linear Model

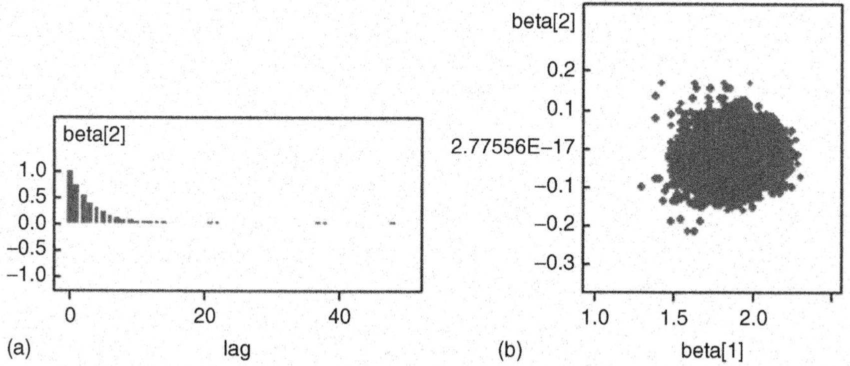

Figure 6.9 Autocorrelation Function and Bivariate Scatter Plot for Centered `beta[2]`. (a) Autocorrelation Function. (b) Scatter Plot

function shows considerable improvement, and the scatter plot indicates that the sampler is much freer to explore `beta[1]` and `beta[2]` in the centered version. The Monte Carlo error for `beta[2]` in the centered version is also over three times smaller (MC error = 0.000763).

When working with a mean-centered functional form, the posterior distribution for the intercept parameter may be considerably different from the posterior distribution for the intercept in the uncentered functional form. Since the intercept parameter does not usually have an important interpretation, the change may not be important; however, in some cases we may want to have the distribution of the original intercept in the uncentered version. We can recover this distribution by examining the two functional forms. Suppose we write the uncentered form as

$$\mu_i = \alpha + \beta_2 x_i \tag{6.1}$$

while in the centered form we have

$$\mu_i = \beta_1 + \beta_2(x_i - \overline{x}). \tag{6.2}$$

Expanding terms and rearranging the centered form (6.2) gives

$$\begin{aligned} \mu_i &= \beta_1 + \beta_2 x_i - \beta_2 \overline{x} \\ &= (\beta_1 - \beta_2 \overline{x}) + \beta_2 x_i, \end{aligned} \tag{6.3}$$

which reveals that the uncentered intercept α is equal to the term in parentheses in (6.3). So, to estimate the distribution of the uncentered intercept, we can create a new parameter in *WinBUGS* which is equal to $\beta_1 - \beta_2 \overline{x}$ and monitor this parameter. It also shows that the intercept and the slope are not independent of each other unless the mean of x is zero. We now have a further understanding of why the two parameters were highly correlated in Figure 6.7(b). The concept behind (6.3) can also be applied for situations with more than one predictor variable. Expanding terms and rearranging will again produce the uncentered intercept for functional forms with more than one predictor variable.

Our discussion so far has presented reparameterization as a change to the functional form of the model. However, this is not the only way to think about it. Gilks and Roberts (1996, p. 92) discuss this as a data transformation from x to $x' = x - \overline{x}$. It is also possible to think about this as a transformation of the priors. In this interpretation, we are transforming the priors from being an independent pair (`beta[1]` and `beta[2]`) to being a dependent pair (`alpha` and `beta[2]`) through the transformation `alpha <- beta[1] - beta[2]*mean.x`. It makes no difference to *WinBUGS* whether the functional form is that of (6.1) or of (6.2), as long as we include code to appropriately calculate α elsewhere in the program listing (see **WinBUGS Code 6.3 re-parameterizing.odc** for an example). If there are many parameters that need to be centered, it can be more convenient to write out a prior transformation on a single line (or data transformations on many lines) rather than flooding the functional form with centering code. Another data transformation that can be useful is to standardize the predictor variables when they have greatly different magnitudes. If the predictor variables are all highly correlated, this collinearity may negatively impact MCMC estimation just as it negatively impacts a frequentist estimation. In this instance, removing certain variables from the model may improve estimation. *R* can be used to calculate the variance inflation factors that quantify the negative effect of variable collinearity so as to help guide the process of variable removal.

6.4 NUMBER AND LENGTH OF CHAINS

6.4.1 Number of Chains

Given that we must use individual judgment to decide whether the Markov chain has converged, one useful way to obtain more information is to run multiple chains independently for the same model and dataset. It is best when we choose starting points for each chain that are very unlike each other. If we begin from very different starting points but end up with similar conclusions, this reinforces the notion that the Markov chain has converged to the posterior despite our efforts to place obstacles in its way. The starting points should additionally be distant from what we believe to be the posterior distribution and should ideally bracket the posterior with at least one starting point being considerably smaller and one considerably larger than the main bulk of the posterior distribution. Such starting points are referred to as being *overdispersed* from the posterior distribution.

One reason for overdispersing the chain's starting points is that some posterior distributions may be multimodal. If this is the case, Markov chains started in different locations may move toward different modes and possibly become trapped near the different modes. The different Markov chains would then be producing different results, which would be clear evidence that the results from any one of them cannot be relied upon exclusively. A second reason is the possibility of slow convergence. As we have seen in Figure 6.7, Markov chains may have very high autocorrelations and therefore only move gradually from iteration to iteration. In taking a look at a small initial portion of the chain, we may not be visually able to detect that the chain is not actually stationary but instead has a slow-moving downward or upward trend. By starting multiple chains in different locations, we will be able to see the relative movement of different chains vis-à-vis each other and it should be easier to spot slow convergence.

WinBUGS permits the user to run multiple chains simultaneously. This is accomplished by using the Specification tool (Figure 6.10). We first begin by accessing the tool and clicking `check model`. Next we place the cursor in the `num of chains` dialog box (see the middle right of Figure 6.10(a)) and enter the desired number of chains. Loading the data and compiling occur as usual. When it comes to loading initial values, *WinBUGS* will display the chain ID number in the `for chain` field at the lower right (Figure 6.10(b)). If you wish to override the ordering and ID supplied by *WinBUGS*, you can use the up–down buttons provided next to the field but this is rarely necessary. In the *WinBUGS* code, select the word `list` that is at the beginning of the first set of initial values, and then click the `load inits` button on the tool. The chain ID number will then be updated, and the process is repeated for subsequent chains. *WinBUGS* will also state at the bottom left of the program screen `chain initialized but other chain(s) contain uninitialized values`. We will need to continue to load sets of initial values while this message is displayed. When the initial values for all chains are entered, *WinBUGS* will be ready to run as usual. *WinBUGS* will confirm that it is

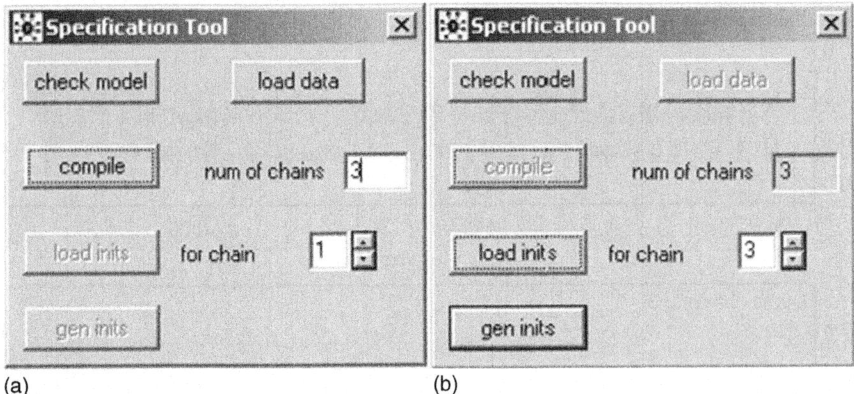

Figure 6.10 Specification Tool for Multiple Chains. (a) Entering Multiple Chains. (b) Loading Multiple Inits

ready by giving the message `model is initialized` at the bottom left of the program screen.

In terms of examples, we first reexamine a variation on the model from the previous section. We examine a regression model instead of a log-linear model by changing the model as follows:

```
mu[i] <- beta[1]+beta[2]*x[i]
```

Three chains are used for this example. Two additional sets of initial values are added to the initial values portion of the code as below:

```
#inits
list(beta=c(0,0), tau=1)
list(beta=c(-200,-100), tau=.1)
list(beta=c(200,100), tau=10)
```

Usually we do not monitor the burn-in period because we wish to omit the burn-in iterations from our estimates. However, in convergence checking we usually do want to monitor all iterations in order to see how the chains move from their starting values. Thus we enter all variables in the `Sample Monitor Tool` before starting the simulation. Multiple-chain trace plots for the `sigma` parameter appear in Figure 6.11. At the top of this Figure we see the first 5000 iterations of all three chains plotted simultaneously. All three chains immediately travel to exactly the same location despite the overdispersed starting values, giving strong evidence that the Markov chain has converged to the true posterior distribution. In Figure 6.11(b), we zoom in for a closer look at the first 100 iterations of the chain (this is accomplished by clicking on the `Sample Monitor Tool` and entering 100 in the `end` dialog box

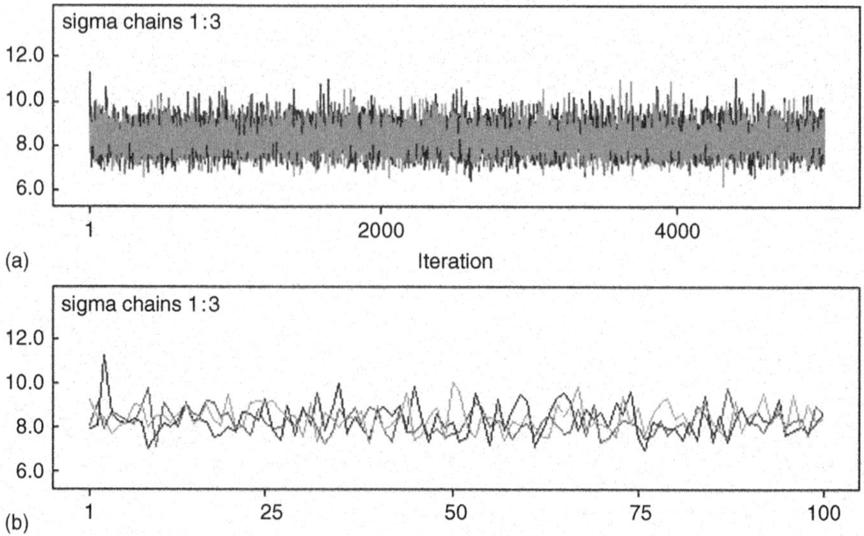

Figure 6.11 (a,b) Multiple-Chain Trace Plot for `sigma` in Simple Regression Model

before clicking `history`). Again, all chains appear to be mixing well and there is no evidence of convergence problems. Plots for the other two `beta` parameters are similar. As for the rationale for the excellent convergence, the Gibbs sampler in this simple regression example takes advantage of the conditional distributions described in Section 4.4, and all chains immediately move to the true posterior distribution.

We next examine the log-linear model for the data that was introduced in Section 6.3. We provide three sets of initial values as follows:

```
#inits
list(beta=c(0,0), tau=1)
list(beta=c(-20,-10), tau=.1)
list(beta=c(20,10), tau=10)
```

We then run the model with three chains based on these values while monitoring all variables for the first 5000 iterations. Attempting to display the traces immediately leads us to difficulties as *WinBUGS* gives an error message (known as a *trap*) when the `history` button is clicked. Editing the Sample Monitor tool so that only Chains 1 and 2 are retained (using the `chains` __ `to` __ fields) allows the output to be generated without a trap. Figure 6.12) displays the traces for selected parameters from this run.

The plot in Figure 6.12) gives us cause for concern, as if we did not have it already from the trap message. For `beta[2]` we can see that before iteration 1000 one of the chains diverges from the other and develops a

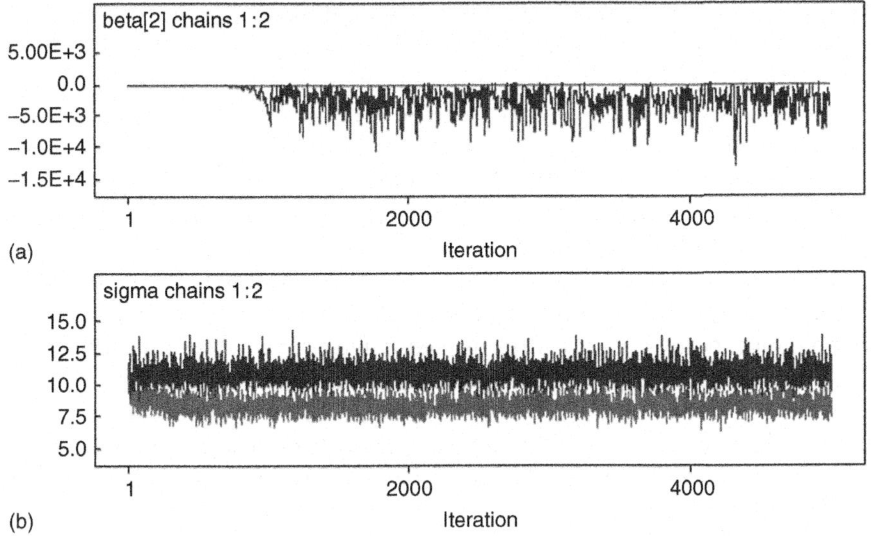

Figure 6.12 (a,b) Multiple-Chain Trace Plot for Selected Log-Linear Model Parameters

much wider standard deviation. For `sigma`, the diverging chain above also has an appreciably higher posterior distribution. What is going on?

This example shows three things. First, it is worthwhile to estimate multiple chains with overdispersed starting values to examine possible convergence issues, even in relatively straightforward models. The single-chain analysis of Section 6.3 did not reveal that this model could potentially run into trouble. The second thing that it shows is that our "conclusions" may be influenced by our starting values if we are not careful, even though our conclusions should not be influenced by them. For example, it could have happened that we only considered the second set of initial values, and then our "conclusions" would have been entirely wrong. The posterior gives a clue here that the second set of initial values is suspect. The "posterior" standard deviation of the beta parameters for chain 2 is very large, like the prior, which seems suspicious given the smaller values in our dataset.

Third, the example shows that some starting values (however well-intentioned) may be poorly chosen or "bad." The latter is the case here. Our log-linear model is equivalent to a model where $y = \exp(\beta_1 + \beta_2 x)$. Consider the case where $x = 10$ in the data. Here, the linear predictor for our second set of starting values would be $\exp(-20 + -10 * 10)$. This is a number extremely close to zero that has over 50 leading zeros after the decimal point. For our third set of starting values we have $\exp(20 + 10 * 10)$. This is a very large number that exceeds 10^{50}. Both of these starting values are so astronomically far away from the posterior distribution that we might expect to run into problems. Also, this model must be estimated by a Metropolis–Hastings algorithm as in Section 4.6. Unlike the Gibbs sampler where the exact full conditional can be obtained and sampled from, here we must choose starting values more wisely because the Metropolis algorithm compares a candidate jump with the existing value. If both the jump and the initial value are extremely far from the posterior, for numerical reasons we might expect progress toward the posterior to be difficult. All computer programs including *WinBUGS* have a finite ability to represent numbers, and will encounter difficulties with extremely large or small numbers. There are many thorny numerical issues that may crop in statistical computing, with many clever solutions (see Thisted, 1988; Altman et al., 2004, for introductions). But to summarize, given finite computer capacities, our starting values should be overdispersed but not unreasonable. It is worth mentioning that *R* also has problems estimating this model using our poorly chosen starting values (see **WinBUGS Code 6.4 Multiple Chains.odc**). For one set of values, *R* complains immediately and refuses to start; for the other, it veers off into a strange portion of the parameter space.

The natural next question to ask is: "what are reasonable starting values?," but we encounter a chicken-and-egg problem in asking it. If we had the posterior distribution already, then it would be much easier to choose reasonable starting values; however, at the outset of the analysis we do not. Still we can often come up with some rough estimates based on educated guesses or approximate analyses. For example, an approximation to our log-linear focus can be obtained by considering $\log y$ and x. We can generate a scatter plot of these two variables in *R* (see **WinBUGS Code 6.4 Multiple Chains.odc** for sample code). This suggests that the intercept is likely positive and in the range 1–4. The slope does not appear pronounced, so slope values should not be too far from zero. We may also do a classical simple linear regression of $\log y$ onto x and use that to guide our starting values. This gives an intercept of approximately 1.8 and a slope of approximately −0.017. In light of this information, we may wish to revise our overdispersed initial values for the `beta` parameters to $(\pm 5, \pm 2)$ and for `tau` to 0.1 and 10. With these initial values, we see that *WinBUGS* converges to the same posterior from multiple starting values (Figure 6.13). Figure 6.13 increases our confidence that *WinBUGS* is finding the true posterior distribution and is sampling from it. Based on the figure, it appears that all chains have reached their stationary distributions by iteration 500. In a fully Bayesian analysis, we do not base the choice of the prior on the data; however, it is acceptable to base the choice of initial values at least in part on data considerations in a fully Bayesian analysis.

Another chicken-and-egg problem arises with regard to the question of how many chains are necessary for a given problem. For better or for worse, it is up to the user to determine whether the actual number

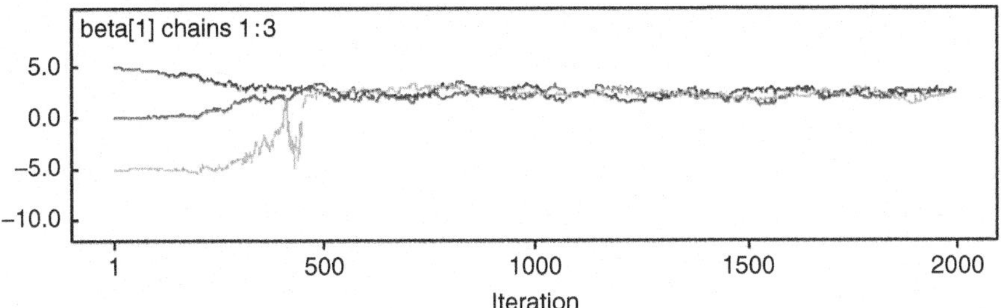

Figure 6.13 Revised Multiple-Chain Trace Plot for `beta[1]` in Log-Linear Model

of chains used for any particular problem is sufficient. As pointed out by Smith and Roberts (1993, pp. 7–8), "however hard you try, someone can always invent a problem for which your intended (finite resource) schedule..." of convergence checks "...will have a high probability of failure." This makes precise recommendations difficult. In general we might expect that, as models become more complex, the advantages of estimating additional chains may be greater. Conversely, fewer chains may possibly be adequate for smaller and simpler models (although there is no guarantee of this). A hybrid approach may be undertaken in which smaller and simpler models are used initially. Convergence may be easier to establish in these models, and the knowledge gleaned from examining the posterior distributions for the simpler models can be used to guide decision making for the more complex extensions of the simpler models. Of course, the more chains that are estimated, the longer it will take the computer to complete a given number of iterations. This means that the user will have to trade off the costs of additional computer and user time against the benefits of more exhaustive convergence checking. One rule of thumb mentioned for selecting initial values for a five-chain run is to "start chain j at $\mu_i + (j-3)\sigma_i, j = 1 \ldots 5$, where μ_i and σ_i are the prior mean and standard deviation of θ_i," where θ_i is the parameter of interest (Kass et al., 1998, p. 96). This might be a very sensible approach for many linear problems but, as we have seen, this would be too overdispersed for *WinBUGS* to run without crashing in our log-linear model of this section.

The Brooks–Gelman–Rubin Diagnostic We have been comparing the traces visually and making an informal assessment of whether the chains have all arrived at the same destination. A more formal approach to making this assessment was provided by Brooks and Gelman (1998), building on ideas first described by Gelman and Rubin (1992). The idea behind this diagnostic is to compare a measure of the variability between the chains to a measure of the variability within the chains. If the ratio of these two measures is large, this indicates that the chains are separated and therefore have not converged. This diagnostic can be a useful summary when the number of chains is fairly large. In such instances, it can become more difficult to identify the individual behavior of the chains graphically as the plots become more densely packed with information. Brooks and Gelman (1998) suggest three rules of thumb be used to assess whether convergence has occurred: check whether the between-chain measure has had sufficient iterations to stabilize; check whether the within-chain measure has had sufficient iterations to stabilize; and check that the ratio has converged to approximately 1.

WinBUGS can calculate the Brooks–Gelman–Rubin diagnostic when multiple chains are being estimated. On the Sample Monitor tool, click on the `bgr diag` button to have *WinBUGS* display a graphic representation of the three measures. The between-chain measure is plotted as the top-most darker gray line, the within-chain measure is plotted as the bottom-most medium gray line, and the ratio is plotted between the previous two lines in light gray. The plotted measures are normalized so that they can be plotted on the same scale in the graphic. The measures are also calculated based on increasing numbers of iterations, so at the left of a given plot only a few of the earliest iterations are used, while at the right the largest number of iterations over the longest interval are used. Figure 6.14 displays the Brooks–Gelman–Rubin plots for the `beta` parameters in a 10,000-iteration run of the log-linear model.

In our example, we see that *WinBUGS* has omitted the first 4000 iterations because it is tuning the Metropolis algorithm (we will discuss the reasons for this in more detail in Section 6.5). It then begins at iteration 4051 after presumably calculating the measures for the first 50 iterations after iteration 4000. We see that the ratio is initially large here, which indicates the chains appear initially dispersed according to the diagnostic. We can deduce, however, from Figure 6.13 that the chains' slow movement causes them to appear initially separated to the diagnostic (observe the "braided" behavior of the three chains in Figure 6.13). Visually, it appears that all three conditions begin to be satisfied around iteration 5000 for both parameters. However, the plot is rather small, and seeing the fine details is difficult. Fortunately, we can ask *WinBUGS* to list out the values

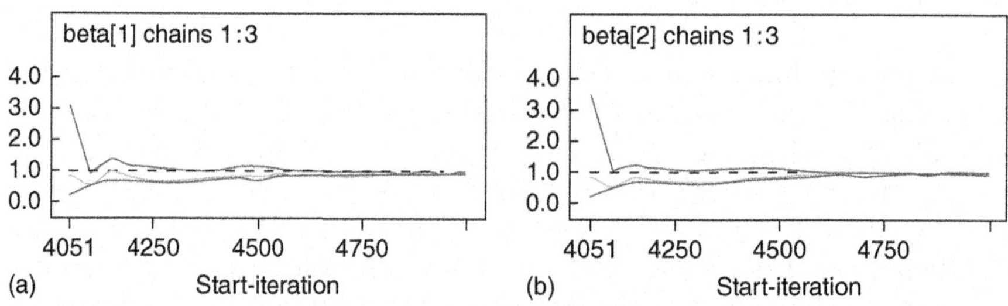

Figure 6.14 (a,b) Brooks–Gelman–Rubin Measures for Log-Linear Model

of the plot to a log file. This is done by double-clicking on a graphic, and then holding down the Control key and doing a left-click on the mouse. The between and within measures appear to be stable after iteration 6000 and the value of the ratio is consistently close to 1 after this point as well. Hence, by the Brooks–Gelman–Rubin diagnostic we might say convergence occurred at iteration 6000 and declare previous iterations as burn-in. We might also say this is a fairly conservative cutoff that is attributable to *WinBUGS* omitting the first 4000 iterations for Metropolis tuning and that convergence had occurred earlier than iteration 6000.

In concluding this discussion, the most important choice about the number of chains is to choose more than one in the first place, at least during early stages of the analysis. We have seen that the decision to perform multiple-chain runs provides a critical convergence check. We would usually also want to report on the use of our multiple-chain runs so as to inform the audience about our actions. The choice of the exact number of chains to run is problem-dependent. Once we are satisfied with the convergence properties of our model and chains, we may want to return to single-chain runs for estimation. The reason for this is that we will need to discard the burn-in period iterations. It may be a more worthwhile use of computer resources to only discard one chain's burn-in iterations rather than numerous chains' burn-in iterations.

6.4.2 Length of Chains

Examining the Monte Carlo error for our parameters is a critical input to deciding how long to run the chain. One reasonably simple way to estimate the Monte Carlo error is the "batch means" method (Roberts, 1996, p. 60; Carlin and Louis, 2000, p. 172) and this method is the one used by *WinBUGS*. The idea is easy to describe. Divide up the chain into numerous small sequences (or batches) and calculate the mean for each batch. Then, see how much these batch means vary around the overall mean that has been estimated from the entire run.

More formally, suppose that the chain has length N and we divide the chain into B batches that have n_B iterations in each batch. Suppose that the batch means, $\overline{\mu}_B$, are approximately independent with mean μ and variance v/n_B. After estimating the overall mean of the entire chain, $\overline{\mu}$, we may estimate the variance among the batch means as

$$\overline{v} = \frac{1}{B(B-1)} \sum_{b=1}^{B} \left(\overline{\mu}_B - \overline{\mu}\right)^2 . \tag{6.4}$$

The Monte Carlo error is the square root of (6.4), and can be described more thoroughly as the standard error of the estimate due to Monte Carlo variability. In deciding how long to run the chain, we would like for the Monte Carlo error to be small relative to the parameter's standard deviation. When this occurs, we would expect that the parameter estimates would not change appreciably with further iterations and thus substantive inferences will not be appreciably affected by Monte Carlo error. One possible rule of thumb for deciding on the number of iterations is that offered by the *WinBUGS* manual: ensure that the Monte Carlo error for each parameter is less than 5% of its standard deviation. Gelman et al. (2003, p. 277) provides a simpler rule. They argue that 100 independent samples "are enough for reasonable posterior summaries," although they indicate more may be needed for quantities such as the posterior probabilities of rare events. However, the Markov chain may be autocorrelated and so we may need many samples to achieve the information that would be provided by 100 independent samples. Hence one useful measure is the *effective sample size* of a given set of samples. This measure takes into consideration the autocorrelation of the chain and attempts to quantify the amount of information in the chain in terms of independent samples.

The effective sample size can be calculated from *WinBUGS'* output using an *R* package called *CODA* (Plummer et al., 2006). To illustrate *CODA*, we return to our log-linear model using a single chain. We burn in the model for 5000 iterations, and then monitor parameters with another 10,000 iterations. We then export the MCMC output to *R* as was described in Section 5.10 (a workflow for loading the *WinBUGS* files and preparing *R* for the analysis appears in **WinBUGS code 6.4 effective sample size.odc**). Assuming that *CODA* has been loaded and the data has been read into *R*, we can use the `effectiveSize` command to calculate the effective sample size. Here, *R* reports that the effective sample size for `beta[1]` is 65.03, `beta[2]` is 74.11, and `sigma` is 9126.86. We see that our estimates for `beta[1]` and `beta[2]` are rather crude despite our reasonably sized run. By Gelman et al.'s 100 independent samples criterion, our 10,000 samples are insufficient for `beta[1]` and `beta[2]` because of the substantial autocorrelation in the chain. We might expect that running the chain for twice as long would cause the effective sample sizes of `beta[1]` and `beta[2]` to be over the 100-samples threshold. Performing the relevant analyses confirms that this is the case, with the new effective sample sizes for `beta[1]` being 138.86, `beta[2]` being 146.40, and `sigma` being 15,374.1.

CODA also offers several other diagnostics to assess whether the length of the chain has been sufficient. The Heidelberger and Welch (1983) diagnostic provides a formal assessment on whether the chain has achieved stationarity by comparing the earlier values of the chain with the later ones. If the stationarity test of the diagnostic fails, then the chain should be run for a greater number of iterations in an attempt to achieve stationarity. If all stationarity tests are passed, the Heidelberger and Welch diagnostic also provides a measure of the relative accuracy of the estimates. The default in *CODA* is for the accuracy tests to indicate whether or not the results can be considered to have two digits of relative accuracy. Accessing these tests is through the `heidel.diag` command. For the single-chain log-linear model iterations used in the previous paragraph, running the tests indicate that all parameters passed the stationarity test. On accuracy, `beta2` failed, which indicates that longer runs would be required in order to achieve two decimals of relative accuracy.

Geweke's (1992) diagnostic also compares earlier and later values of the chain to assess whether the chain has stabilized. The test statistic values of the diagnostic are the standard normal z-scores. If the z-score is suitably extreme, we reject the frequentist null hypothesis of stationarity for that parameter. Here, running the `geweke.diag` command gives the following z-scores for `beta1`, `beta2`, and `sigma`: 0.3913, −0.3566, and −1.2599. These values all lie within the range of −1.96 to +1.96, which would be associated with a 95% standard normal confidence interval; hence by this test there is no need to run the chain further.

The preceding two tests assess stationarity by looking at means of the distributions. Raftery and Lewis (1992) provide a diagnostic that is useful for looking at quantiles of the distributions. The default quantile examined in *CODA* is the 2.5% quantile. Another quantile that may be of interest is the median or the 50% quantile, as it is sometimes used as a measure of central tendency, especially for skewed distributions. The Raftery and Lewis diagnostic also provides an indication of how long the chain must be run in order to estimate the quantile of interest to an accuracy of $\pm r$ with probability p as well as providing an indication of how many additional burn-in iterations may be needed. Finally, the diagnostic provides a summary measure of the overall autocorrelation in a given chain. When this measure I is greater than 5, the chain exhibits high autocorrelation. Using *CODA*'s `raftery.diag` on our data suggests that only a few more iterations would be needed for burn-in, and is likely not an issue. Using the defaults, we find that estimating the 2.5% quantile of `beta1` to a ± 0.005 accuracy with a 95% probability will require a total of 60,234 iterations. For `beta2`, we will need 49,656 iterations. The values of I for both `beta1` ($I = 16.10$) and `beta2` ($I = 13.30$) indicate high autocorrelation. Attempting to use `raftery.diag` to assess the median causes *CODA* to produce a message that a longer run will be needed for such an assessment.

6.5 METROPOLIS–HASTINGS ACCEPTANCE RATES

When using Metropolis sampling, there is a choice to be made of what jumping proposal distribution to use. We have seen in Section 4.6.1 that when the jumps from the proposal distribution are too small, the Markov chain will move but very slowly. This causes very little new information to be learned about the posterior per iteration. If the jumps are too large, the chain will not jump at all for long periods of time, again revealing little information about the posterior per iteration. *WinBUGS*, however, has an automated expert system for trying to ensure that the jumping proposal distribution is neither too broad nor too narrow. For theoretical reasons, we must not base inference on the samples that are generated while *WinBUGS* is in the tuning phase. Fortunately, *WinBUGS* is designed to enforce this restriction, so it will be impossible for you to base inference on the tuning samples. *WinBUGS* reserves the first 4000 iterations for tuning the sampler, and thus accordingly it will not allow you to use anything prior to iteration 4000 for posterior summary statistics. The presence of a check in the `adapting` box on the Update tool indicates that *WinBUGS* will be performing the tuning step (Figure 6.15).

While *WinBUGS* is in this tuning phase, it attempts to ensure that between 20% and 40% of the proposed Metropolis jumps are accepted (Gelman et al., 1995) for reasons discussed in Section 4.6.1. *WinBUGS* can display a graph of the Metropolis acceptance rates for the model being estimated. Figure 6.16 displays moving averages of the maximum, average, and minimum acceptance rates for the log-linear model with the revised initial values used for the three chains. This plot can be generated

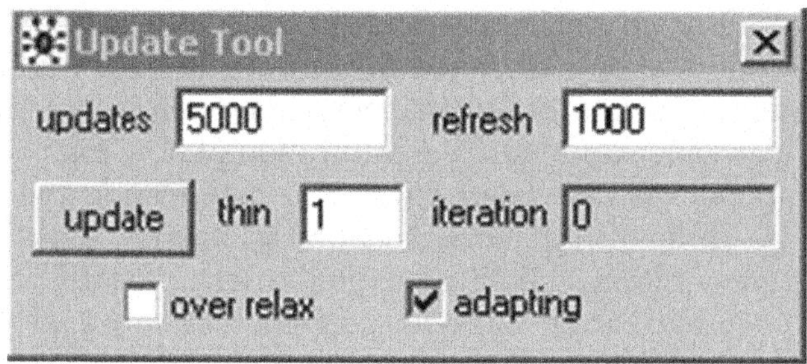

Figure 6.15 Update Tool Indicating a Jumping Proposal Distribution Tuning Phase

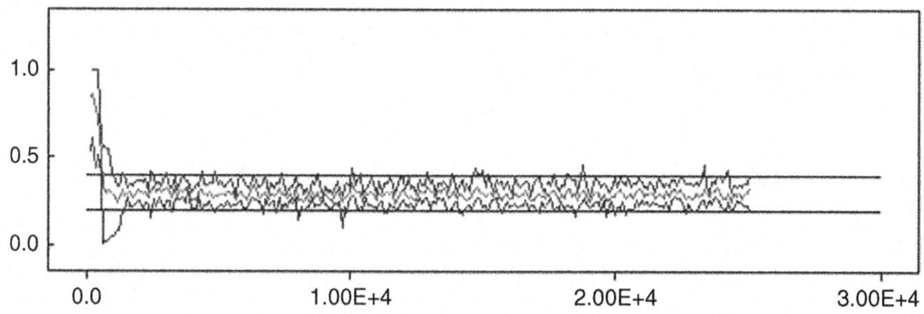

Figure 6.16 Moving Averages of Maximum, Average, and Minimum Metropolis Acceptance Rates for Log-Linear Model

by going to the `Model` menu in *WinBUGS* and selecting `Monitor Met` from the choices. The horizontal bars extending across the plot indicate the range of optimality for acceptances. We see that early values tend to be almost always accepted, which may reflect initially cautious moves on the part of *WinBUGS*. However, within a hundred or so iterations, *WinBUGS* is in the optimal range for jump acceptances.

WinBUGS is typically successful at tuning the Metropolis proposal distributions, but requesting this graphic has a very low opportunity cost and can help rule out one source of slow mixing. If we find that the Metropolis proposal acceptances are outside the optimal range, we should expect to have to run the chain for a larger number of iterations.

6.6 SUMMARY

MCMC is a tool, and like any tool needs to be used appropriately to produce useful results. In this chapter, we have examined how different aspects of our Markov chains need to be reviewed so that our estimates of quantities are both reliable and relatively unaffected by Monte Carlo error. In concluding this chapter, we describe our thought process and recommendations for deciding on the number and length of chains for analyses that are to be published or go to a client. Begin with a series of shorter runs in which the *WinBUGS* code is debugged, convergence checks are performed, multiple chains are investigated, and initial parameter estimates are reviewed for plausibility. One approach for *CODA* is to use this later in the process (roughly in the order it has been described in the book) when the more accessible *WinBUGS* checks such as the trace plot and the Brooks–Gelman–Rubin diagnostics all look satisfactory. For final estimation, consider concluding with a longer run (perhaps 100,000 iterations depending on the number of parameters involved, or alternatively over 10,000 independent samples if autocorrelation is appreciable) and performing a final review of the longer parameter traces. The main reason for concluding with a longer run is to ensure that there are no problems in a lengthier chain that went undiagnosed with short chains. As a bonus, Monte Carlo error will be lower in the final estimates.

Some of the research that appeared during the flowering of MCMC in the 1990s has a minimalist philosophy in terms of length of chains. With ever-increasing computer power, the computational cost of a longer run is very small compared to what it would have been 10 or 20 years ago and there is less incentive to stop with the bare minimum number of iterations required. Technological advances including the likely rise of quantum computing (Hardy, 2013) suggest that computational power will continue to increase, making longer runs even less costly. In this book, we will typically not conclude with a longer run because it is understood that the reader can easily undertake this task himself or herself. For published analyses or analyses that go to clients, however, typically the recipients will not be able to easily undertake this task, so it is much more incumbent on the producer of the analysis to perform this step.

6.7 EXERCISES

1. Reexamine the first model in **WinBUGS Code 5.2 two sample t-test.odc**. Generate traces for `mu1`, `mu2`, `sigma1`, and `sigma2` using 10,000 iterations after a 1000-iteration burn-in. Comment on the appearance of the traces—do they look satisfactory? Why or why not? Zoom in on iterations 5000–6000 using the `beg` and `end` fields. Do the chains appear to be moving fast or slow compared to Figure 6.6?

2. Reestimate the model in **WinBUGS Code 5.5 Proportions 1.odc**. Monitor `pi1`, `pi2`, and `diff` using a 1000-iteration burn-in and 10,000 subsequent iterations. Use the `quantile` tool to create graphs of the running quantiles. Do the graphs give evidence of nonconvergence? Why/why not?

3. Reexamine the first model in **WinBUGS Code 5.6 one-way ANOVA.odc**. Monitor `mu` and `sigma` using a 1000-iteration burn-in and 20,000 subsequent iterations. Generate the autocorrelation function for all of these parameters. Which parameter appears to have the highest degree of autocorrelation? Does the autocorrelation look severe as in Figure 6.7(a)?

4. Using the code in **WinBUGS Code 6.3 re-parameterizing.odc**, fit a log-linear model without mean centering to the Singapore services exports data of Section 5.11. Monitor `beta` and `sigma` using a 5000-iteration burn-in and 20,000 subsequent iterations. Next fit a log-linear model with mean centering to the Singapore data. On a parameter-by-parameter basis, compare the Monte Carlo errors across the two runs.

5. As in the previous exercise, fit a log-linear model without mean centering to the Singapore services exports data. Use the `Correlations` tool to examine the properties of the chain. Produce a scatterplot of `beta[1]` and `beta[2]`. What is the posterior correlation between these two parameters? Use the `print` function to calculate this correlation.

6. As above, fit a log-linear model without mean centering to the Singapore services exports data. Use the `Monitor Met` tool to examine Metropolis acceptance rates for the first 4000 iterations. Plot the graph that the tool creates. At about what iteration do the acceptance rates begin to consistently stay within the desired range?

7. Reexamine the simple linear regression model at the top of **WinBUGS Code 5.8 regression and ANCOVA.odc**. Estimate the model using three chains with three different sets of initial values. For the first set, use `list(beta0 = 100000, beta1 = 100, tau = .001)`. Use `list(beta0 = 0, beta1 = 0, tau = 1)` for the second set. Use `list(beta0 = -100000, beta1 = -100, tau=100)` for the third set. Monitor all iterations of `beta0`, `beta1`, and `sigma`. Run *WinBUGS* for 10,000 iterations. Create the trace plots for all parameters. How many iterations does it take for the chain to converge to the posterior distribution? Why is this number of iterations required for this model?

7

MODEL CHECKING AND MODEL COMPARISON

In the previous chapter, we estimated models with *WinBUGS* and performed convergence checks for the Markov chains. Now that we are producing samples from the posterior distributions, it is time to examine the performance of the model with respect to the data itself. We can examine how well a given model fits the data. We can also compare different models with each other in terms of fit. In this chapter, we will examine ways of checking our models using *WinBUGS*. We will also discuss Bayesian approaches to model comparison.

7.1 GRAPHICAL MODEL CHECKING

We estimate models in an attempt to understand and explain data. Ideally, our model will be a succinct and parsimonious way of describing the relationships occurring within our data. To find out if our model reasonably approximates the relationships in our data, we can examine whether the predictions of the model are consistent with the actual outcomes. We call these activities *model checking* because we are checking the model's predictions against the actual outcomes.

Bayesian Methods for Management and Business: Pragmatic Solutions for Real Problems, First Edition. Eugene D. Hahn.

7.1.1 In Practice: Graphical Fit Plots

In this section, we use the data appearing in McCutchen (1993). McCutchen (1993) examined the determinants of research and development expenditures in the pharmaceutical industry in the 1980s. Twenty pharmaceutical firms were selected from a pool of 36 firms for the study. McCutchen provides each firm's R&D development expenditures in 1980 divided by its 1980 total sales, and we use that in a regression model as a predictor variable (x_1). He also provides the change in R&D expenditures from 1980 to 1985 in millions of dollars for each firm, and we use this as the dependent variable (y). Selected simple linear regression code from the file **WinBUGS Code 7.1 graphical checks.odc** appears below:

```
model {
 for (i in 1:n){
   rndchange[i] ~ dnorm(mu[i],tau)
   mu[i] <- beta[1]+beta[2]*rnd80[i]
   }
#priors
 for (k in 1:2){ beta[k] ~dnorm(0, .00000001) }
...
```

We can examine the fit of this model graphically with the *WinBUGS'* Comparison tool, which is found under the `Inference` menu. Before using the tool, we ensure that `mu` is being monitored during the *WinBUGS* run. Then we enter `mu` in the `node` field of the Tool. Next we type our predictor variable `rnd80` in the `axis` field. Finally, we type our dependent variable `rndchange` in the `other` field. Figure 7.1 shows what our Comparison tool should look like at this stage.

Clicking on the `model fit` button generates the plot in Figure 7.2. We see that this particular set of entries in the tool will plot the predicted values along the y-axis as a function of our predictor variable on the x-axis. It will also superimpose the dependent variable on the plot as black dots. Lines are also drawn on the figure. The dashed outer lines correspond to

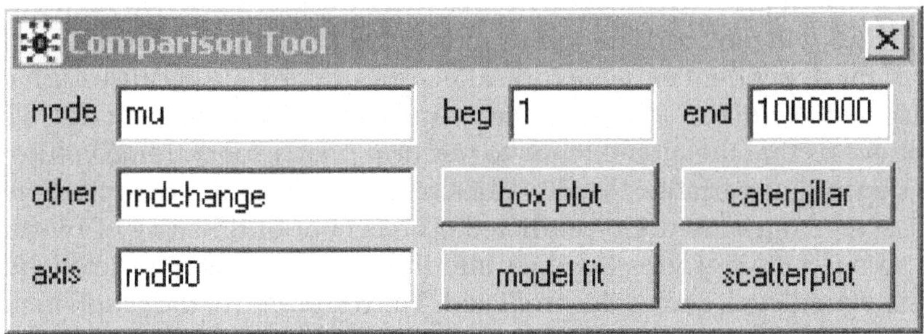

Figure 7.1 Comparison Tool for Pharmaceutical Industry Data

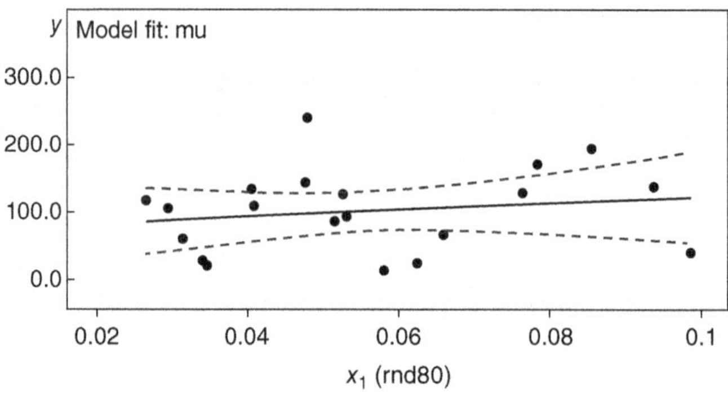

Figure 7.2 Predictor Variable x_1 Fit Plot. Pharmaceutical Data

the 2.5 and 97.5% posterior quantiles of mu, while the solid line is the 50% quantile. *WinBUGS* connects these quantiles at the discrete values of mu with linear segments to produce the lines. Right-clicking on the plot and clicking on `Properties` allows various aspects of the plot to be changed such as text for the axes labels as shown in Figure 7.2. An alternative to the `model fit` plot is the `scatterplot` (bottom right of Figure 7.1).

We can see that mu and the dependent variable `rndchange` are sometimes consistent (i.e., near to each other) but sometimes not. The biggest departure appears at $y_6 = 238$ slightly left of center in Figure 7.1. This value is attributable to Johnson & Johnson (the sixth company in the dataset) which had a very large change in R&D from 1980 to 1985 despite a modest R&D to sales ratio of 4.81% in 1980. There are also a number of other values outside of the 95% posterior interval for mu, indicating that these values are not explained well by the model. Finally, we see the slope of the solid 50% quantile line for mu is shallow and nearly parallel to the x-axis. While this partially reflects internal scaling decision by *WinBUGS*, it is also consistent with the fact that the 95% posterior distribution for `beta[2]` includes the value of zero.

Another way of plotting the data gives us a different way of examining the data. We can make a plot in which we enter `rndchange` in both the `axis` and the node fields. This produces the plot shown in Figure 7.3. *WinBUGS* will plot `rndchange` in sorted order from low to high. The values of the dependent variable (dots) fall on a straight line because we have instructed *WinBUGS* to equate the x- and y-coordinate of each dot. This plot reveals that the model tends to perform poorly for extreme values of the dependent variable. Small values of `rndchange` are overpredicted, while large ones are underpredicted. The benefit of this way of plotting the data is that it groups the data into three groups: overpredicted, adequately predicted, and underpredicted. The resolution of the graph makes any conclusions we might draw somewhat preliminary because the data points can fall almost on top of each other. However, with a careful eye,

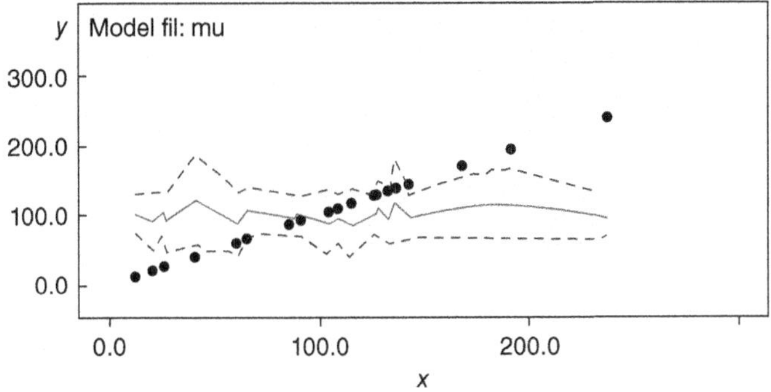

Figure 7.3 Dependent Variable Fit Plot. Pharmaceutical Data

this particular graph shows at left that five to six data points fall below the relevant 95% intervals for mu. At right, at least four data points appear above the relevant 95% intervals. We therefore conclude that at least 10 data points from our sample size of 20 are not predicted well.

Our model is in need of improvement, and we can consider adding a variable to improve the fit. Since Johnson & Johnson is a large company, one possibility is that company size would be a relevant predictor. Detailed information on company size is not available in McCutchen (1993); however, the ratio of cash flow to total sales in 1980 is available. We might expect that companies with larger cash flows relative to sales would have more cash to invest in R&D. We revise our model to include this second predictor variable, x_2, and estimate an additional β parameter for this variable. The 95% credible interval for the cash flow β parameter does not include zero, so this variable improves model fit above and beyond the fit provided by `rnd80`. Figure 7.4 displays fit plots for the revised model. In Figure 7.4(a), we see that plotting mu as a function of `cashflow` leads to most of the points falling within the 95% posterior interval for mu. Figure 7.4(b) shows that fit for extreme values of y has improved somewhat in the revised model. In particular, all but one of the underpredicted values at right are now more adequately predicted. We also appear to be more adequately predicting some of the overpredicted values at left. To further improve the fit, we might examine whether including other variables might lead to better predictions for the extreme values of y. Or, if theory suggests, we might also explore using a quadratic or cubic term in our model.

7.1.2 In Practice: Residual Analysis

Another important tool for model checking is the examination of residuals. The residuals are the observation-level error terms defined as

$$\epsilon_i = y_i - \mu_i.$$

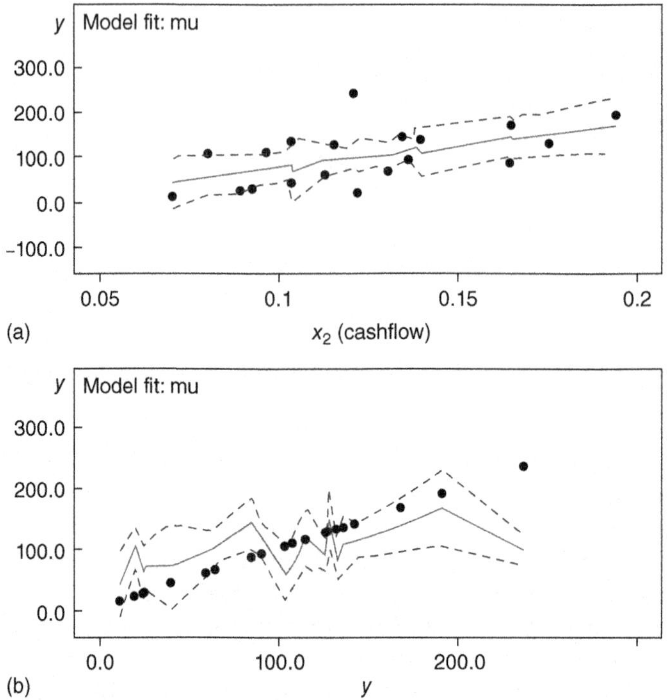

Figure 7.4 Revised Model Fit Plots. Pharmaceutical Data. (a) x_2 Fit Plot. (b) Dependent Variable Fit Plot

We have been visually observing the discrepancies between the data and the model in Figures 7.2 and 7.3, but calculating the residuals allows us to quantify these discrepancies directly. For ease of interpretation, we can also standardize the residuals by dividing them by the standard deviation. The standardized residuals can then be examined with reference to the normal distribution in terms of familiar *z*-scores. Adding the following lines of code in our main `for` loop

```
...
resid[i] <- rndchange[i] - mu[i]
std.resid[i] <- resid[i]/sigma
...
```

instructs *WinBUGS* to calculate the residuals and the standardized residuals (see **WinBUGS Code 7.1b residuals.odc**). We can obtain a graphical plot of these quantities by monitoring them during the course of a run and using the Comparison tool. Typing `std.resid` in the `node` field and clicking the `box plot` button produces the plot in Figure 7.5. We see that the standardized value of ϵ_6 is quite large and its posterior mean is 2.561. We would expect most standardized residuals to fall in the usual

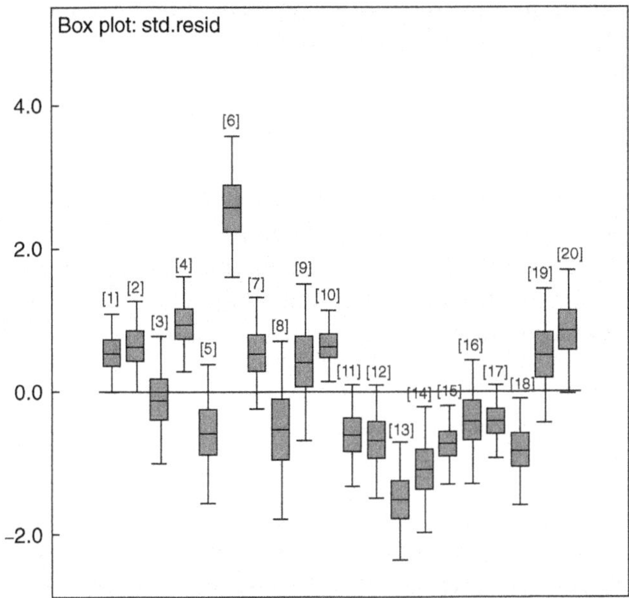

Figure 7.5 Standardized Residuals. Pharmaceutical Data

standard normal 95% interval of $-1.96 < z < +1.96$. Hence we conclude that this residual is an unlikely result to obtain under the standard normal distribution.

7.2 PREDICTIVE DENSITIES AND CHECKING MODEL ASSUMPTIONS

Our observed value for Johnson & Johnson seems inconsistent with its expected value, but how inconsistent is it? The posterior mean of its expected value is $\mu_6 = 99.38$. Reviewing **WinBUGS Code 7.1 graphical checks.odc**, we find that the error standard deviation for the model, σ, is 55.96. We could form a classical normal-based 95% confidence interval for the predicted value of Johnson & Johnson, $\tilde{y}_6$, with standard normal theory, yielding the interval (−10.28, 209.8). This kind of estimation is sometimes known as *plug-in estimation* because we are plugging point estimates into a formula, with the formula here being the usual normal-distribution formula $\mu \pm z\sigma$. However, this approach uses only the point estimates of μ_6 and σ and ignores our uncertainty about them. What we would like to do is find the distribution of the predicted value of the data in such a way that we incorporate our uncertainty about the parameters. This is the posterior predictive distribution that was first discussed in Section 2.8 in the discrete variable context. We have continuous data in our regression analysis here, so we replace the summation in Equation (2.7) with integration to obtain

$$p(\tilde{y}|y) = \int p(\tilde{y}|\theta)p(\theta|y)d\theta, \tag{7.1}$$

which is the posterior predictive distribution for continuous data. We can easily sample from the posterior predictive distribution of our data in *WinBUGS* by adding the following line of code inside the `for` loop that encompasses our functional form and likelihood (see **WinBUGS Code 7.2 posterior predictive.odc**):

```
...
y.pred[i] ~ dnorm(mu[i], tau)
...
```

The newly added code does the following: At each iteration, *WinBUGS* will have estimated τ and will have calculated μ based on its estimates of the β parameters. The new line of code then instructs *WinBUGS* to take a randomly sampled value from a normal distribution having mean μ_i and precision τ at every iteration, and store it in `y.pred[i]`. The values of μ_i and τ change as the model runs, thus capturing the uncertainty in these parameters. So the final distributions of `y.pred[i]` will reflect both the random sampling from a normal distribution as well as the uncertainty in the parameters (μ and τ) themselves as desired. Monitoring `y.pred[6]` gives a 95% posterior predictive credible interval of $(-15.81, 215.8)$, which is wider than the interval obtained by the plug-in approach. Even so, our observed value of $y_6 = 238$ is outside the 95% posterior predictive credible interval, indicating that our value of y_6 is unexpected assuming that the model is true.

7.2.1 The Posterior Predictive p-value

We have seen that y_6 is not in the 95% posterior predictive credible interval for $\tilde{y}_6$. This makes us wonder what would be the chance of having a data value this extreme or more so, assuming that the model were true. The posterior predictive p-value will allow us to find this probability. To implement the posterior predictive p-value in an MCMC run, we can create a binary variable that takes on the value of 1 when an event occurs, and zero otherwise. Then the proportion of 1's in this variable over the MCMC run is our Monte Carlo estimate of the probability of the event. For our model, we can again add a line of code

```
...
p.pred[i] <- step(y.pred[i] - rndchange[i])
...
```

to the same `for` loop right below our `y.pred[i]` code. The `step` function in *WinBUGS* will return the value 1 if the expression in parentheses is greater than zero, and will return zero otherwise. Thus `p.pred` as coded above will be a one-sided or one-tailed posterior predictive p-value. Nominally, it will give the probability that the predictive distribution would generate a value as large as or larger than the data under the current model. To find the probability that the predictive distribution would generate a value as small as or smaller than the data under the current model, we could subtract `p.pred` from 1 (or interchange the variables in parentheses). In the current example, there is only a 1.1% chance that a data value as large as or larger than that of Johnson & Johnson would arise assuming our model was true (see **WinBUGS Code 7.2 posterior predictive.odc**). We might therefore wish to designate y_6 as an outlier.

Outliers usually cause concern when we suspect that they might have an excessive influence on our parameters so that inferences are distorted by their presence. In these cases, one possible strategy is to exclude the outlier. Reviewing Figure 7.4(a) suggests that Johnson & Johnson is unlikely to be changing the slope of the line as estimated by β_3. It appears in the middle of the data and so is probably not twisting the slope of the line upward or downward. It may, however, be impacting the intercept (see Section 7.7 for more details). While this is a matter of opinion, a reflective approach to exclude outliers may be more useful than "automatic" approaches (assuming the outliers are not a result of typographic mistakes or other data quality errors, which definitely should be addressed). Outliers indicate a conflict between the model and the data, but in the absence of certitude in the model we may want to look first at the limitations of our model and possible extensions or expansions of it. More philosophically, outliers can be among the more interesting of our observations and, as anyone who has read business case studies can corroborate, can be sometimes be a source of insight. Truly incommensurable data will often need to be excluded, but our first choice for milder outliers such as y_6 would be to focus on enhancements to the model and theory.

Our posterior predictive p-value for y_6 is fairly small, but in labeling y_6 as an outlier or not it is natural to consider the overall sample size in our analysis. As an analogy from the game of poker, the chances of drawing a three-of-a-kind are fairly small (approximately 2%) but it would not be all that surprising to draw one in the course of 20 rounds of poker. We can therefore consider the probability distribution of observing a value as large as or larger than y_6 in 20 observations. One way to approach this in *WinBUGS* would be to simulate 20 predictive observations from our model. We then create a binary variable that takes on the value 1 if we generate a predictive observation as extreme as or more so than y_6, and zero otherwise. Then we sum up the 20 binary variables to form a posterior predictive outcome distribution for the entire dataset (with the outcome

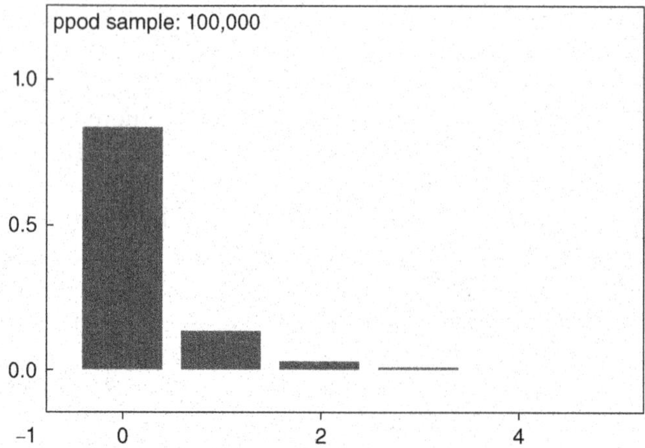

Figure 7.6 Posterior Predictive Outcome Distribution for $\tilde{y} \geq y_6$

here being that the predictive observation being as extreme as or more so than y_6).

Monitoring the variable ppod in **WinBUGS Code 7.2 posterior predictive.odc** shows that we would expect to see between zero and two predictive observations as extreme as y_6 or more so in a sample size of 20 (Figure 7.6). The chance of observing exactly one such predictive outcome in a sample size of 20 is about 13.2%, possibly leading us to conclude that Johnson & Johnson may not automatically be a candidate for deletion.

The posterior predictive p-value can be used for other kinds of model checking. For example, we may want to examine whether our residuals possess skewness. *Skewness* refers to asymmetry in the tails of the distribution such that one tail is longer or heavier than the other. We can measure the amount of skewness in the data using the formula

$$Skewness = \frac{\sqrt{n}\sum_{i=1}^{n}\epsilon_i^3}{\left(\sum_{i=1}^{n}\epsilon_i^2\right)^{\frac{3}{2}}}. \quad (7.2)$$

We may also want to examine the *kurtosis* of our residuals. This refers to the relative heaviness/lightness of the tails in comparison to the central part of the distribution. The kurtosis is often quantified as

$$Kurtosis = \frac{n\sum_{i=1}^{n}\epsilon_i^4}{\left(\sum_{i=1}^{n}\epsilon_i^2\right)^2} - 3. \quad (7.3)$$

Under normal distribution, the values of both *Skewness* and *Kurtosis* should be zero. Positive values of *Skewness* indicate a longer tail to the right, with the opposite holding for negative *Skewness*. For *Kurtosis*, positive values indicate a sharper central peak and heavier tails (more like a witch's hat with a wide brim), while negative values indicate the distribution has less of a defined transition between peak and tails (more like a dome).

Skewness and *Kurtosis* posterior predictive p-values for the pharmaceutical data are calculated in **WinBUGS Code 7.2 posterior predictive.odc**. The posterior mean of the observed *Skewness* is 0.603, while for the predicted *Skewness* it is −0.004. The posterior predictive p-value for *Skewness* as calculated by *WinBUGS* is 0.724. Since this is greater than 0.5, we subtract our value from 1 to obtain the posterior predictive p-value of 0.276 (alternatively, we could interchange the quantities inside our `step` function in the *WinBUGS* code and rerun the model). Since a posterior predictive p-value of 0.276 is fairly large, our model appears to be satisfactory with respect to skewness. The observed *Kurtosis* is 0.409 and the predicted *Kurtosis* is −0.272. *WinBUGS* calculates our test statistic as 0.788, so again we calculate the posterior predictive p-value as 0.212. This fairly large value again gives us no substantive evidence that the residuals are non-normal with regard to kurtosis.

We have created test statistics to examine outliers, skewness, and kurtosis, but the procedure is much more broadly adaptable. In general, we can consider some test statistic T that examines other assumptions or characteristics of our model. It is then a matter of simulating the posterior predictive distribution of T, computing the observed distribution of T, and calculating the posterior predictive p-value of T. For example, Gelman et al. (2003, pp. 163–164) examine whether there is autocorrelation in a sample of binomial trials using the posterior predictive p-value, while Ntzoufras (2009, pp. 383–385) examines autocorrelation and heteroscedasticity in the context of regression.

The posterior predictive p-value and related test statistics are very easy to calculate in an MCMC run and also have an appealingly straightforward interpretation. In addition, they are very broadly adaptable to virtually any kind of test statistic T we might have an interest in. However, a limitation of tests based on posterior predictive p-values is that they can be conservative (Bayarri and Berger, 1998). This means, in practice, that they are less likely to reject the null hypothesis of the appropriateness of the model, and hence to *understate* any issues that may be present. Recall that we are using the data to estimate the parameters of the model, and then we are using the estimated parameters to assess the adequacy of the data. This somewhat circular approach contributes to the conservativeness of the posterior predictive p-value.

One possible way to address this problem is to split the sample into an estimation portion and then a verification portion. We would then estimate the parameters with the estimation portion of the data, and next perform model criticism using the verification portion. This will prevent the circularity; however, in practice it may not always be feasible to split the data as required. Another solution is to find some way of expressing the test statistic in such a way that the model parameters "cancel out," which will also break the circularity and make the test statistic more powerful. However, this solution may require some knowledge of statistical distribution theory to be viable (Lunn et al., 2013, pp. 156–157) and will not be applicable to all possible model characteristics we would like to test. Alternatives to the posterior predictive p-value have been developed (Bayarri and Berger, 2000; Robins et al., 2000), but these are typically less easy to work with. In defense of the posterior predictive p-value, Stern (2000, p. 159) in essence argues that the calibration of the posterior predictive p-value may be less of an issue when we are on guard for its conservativeness.

7.2.2 In Detail: Comparing Posterior Predictive p-Value Test Statistics

Given that the parameters are partly the source of the limitations of the posterior predictive p-value, it may be worthwhile to see whether we can construct test statistics T such that the number of parameters involved is as small as possible. For example, the kurtosis statistic in (7.3) can be written (Cramér, 1946, p. 184) as

$$\frac{\mu_4}{\sigma^4} - 3 \tag{7.4}$$

where μ_4 is the fourth central moment (a deep understanding of which is not essential for the current purposes) and where σ is the standard deviation. Hence one possible approach might be to estimate σ as a parameter and use that in our calculation of the kurtosis test statistic (Spiegelhalter et al., 1996a, p. 45). We can call this test statistic T_2, while we call our original test statistic based on (7.3) T_1. We compare T_1 and T_2 in **WinBUGS Code 7.2 posterior predictive.odc** using the pharmaceutical data. Box plots of the posterior predictive distributions for the two different approaches to calculate kurtosis appear in Figure 7.7. On the left we see the box plot for kurtosis T_1, while on the right we see the box plot for kurtosis T_2 where σ is estimated as a parameter. The latter distribution has much more uncertainty than the former.

Results for the different approaches to calculate kurtosis appear in Table 7.1. The posterior predictive p-value for T_1 is 0.2124, while for T_2 it is considerably closer to the maximum possible value at 0.4002. This suggests that the involvement of σ in the latter approach is contributing to the conservativeness of the latter approach's p-value. We may therefore prefer to calculate the T_1 posterior p-value for a more stringent test of kurtosis in regression residuals.

While this is not critical for the current discussion, advanced readers may notice that the predicted value of T_1 in Table 7.1 is somewhat negative while theory suggests it should be near zero. This indicates the ratio term in (7.3) is smaller than we might expect under normal theory, i.e., it seems

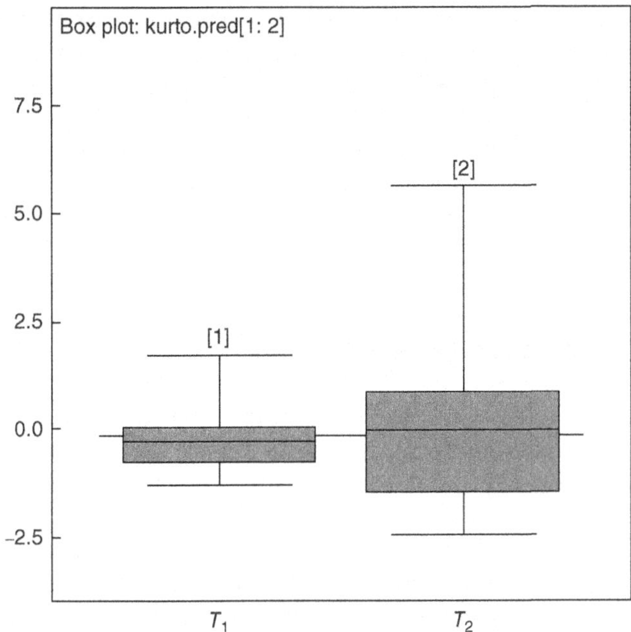

Figure 7.7 Posterior Predictive Distribution Box Plots for Two Different Ways of Calculating Kurtosis

TABLE 7.1 Posterior Means for Kurtosis Statistics

	Kurtosis Using (7.3) (T_1)	Kurtosis Using σ as Parameter (7.4) (T_2)	Adjusted Kurtosis (T_3)
Observed	0.4093	0.7544	0.7776
Predicted	−0.2721	−0.0045	0.0226
Posterior predictive p-value	0.2124	0.4002	0.2124

to be biased downward. Conversely, the predicted T_2 is close to zero. Advanced readers may notice that the denominators of (7.3) and (7.4) can be thought of in terms of variance expressions such that the denominator of (7.3) is expressible in terms of the population variance $\sigma^2_{pop} = \frac{1}{n}\sum_{i=1}^{n}(x_i - \overline{x})^2$, while the method for kurtosis under (7.4) seems to have the properties of the unbiased sample variance $\sigma^2_{sample} = \frac{1}{n-1}\sum_{i=1}^{n}(x_i - \overline{x})^2$. This suggests that we may be experiencing some downward bias in small sample sizes in T_1 which could be addressed by multiplying through by the squared ratio of σ^2_{sample} to σ^2_{pop}, which is $n^2/(n-1)^2$. Since this ratio involves only data quantities, it should not make our posterior predictive p-value more conservative. The third column of Table 7.1 provides results for this adjusted kurtosis measure, called T_3. The posterior mean of its predicted distribution more closely approaches zero like T_2, and the posterior mean of its observed value is close to that of T_2. However, the posterior predictive p-value of T_3 retains the desired lack of conservativeness of T_1. To summarize, our results show that there may be multiple ways of formulating test statistics, and that the posterior predictive p-value may be less conservative when we formulate them in a way to minimize the impact of parameters.

7.3 VARIABLE SELECTION METHODS

Very often, we may want to determine which variables we should include in our functional form. Not all variables will be equally relevant for understanding the data involved in a business scenario, and for reasons of parsimony we may want to see how we can productively simplify our functional form to emphasize the most relevant variables. Usually, the decision is presented as a binary one: we will either include or omit a particular variable from our functional form. Variable selection methods can be used for this purpose.

In classical statistics, there are a number of different methods for variable selection such as examination of R^2 values (unadjusted or adjusted), forward and backward stepwise regression, and the use of likelihood-based measures. Similarly, there are a number of different methods for Bayesian variable selection, reflecting different ideas about how variable selection can or should be performed. As in classical statistics, different Bayesian variable selection methods using different criteria may result in different conclusions about which variables should be selected. Accordingly, the literature on Bayesian variable selection methods contains lively discussions as to which method is appropriate for what situation. In this section, we will examine a few of the more readily applicable Bayesian variable selection methods. We will also summarize the characteristics and emphasis of these methods. One important difference between classical and Bayesian variable selection methods is the role of the prior in Bayesian methods. As we will see, some careful thought needs to be given to the prior in undertaking variable selection.

7.3.1 Kuo and Mallick's Method

Since we have a binary decision about whether to include a variable, and since it is unknown to us whether we should include it or not, we might treat our decision as just another unknown parameter to be estimated using Bayesian inference. We then can simply prepare a binary indicator variable for each explanatory variable and add it appropriately to our model. This binary indicator variable is identical in concept to the group membership dummy variable of Section 5.8.2, except that here it is treated as an unknown to be estimated. Hence we will both estimate our parameters and also examine whether those parameters should be included or not.

Suppose we write our indicator variable as I_k, where k indexes the K variables under consideration. Then instead of our usual regression functional form, we make the minor changes to

$$\mu_i = \beta_0 + I_1\beta_1 x_1 + I_2\beta_2 x_2 + \cdots + I_K\beta_K x_K. \tag{7.5}$$

This looks very much like the usual functional form with an added "interaction term" for each variable. However, these added terms, I_k, will take on the value zero or 1, and hence will either exclude or include the relevant $\beta_k x_k$ depending on whether zero or 1 is assigned to I_k by the Markov chain. This approach to variable selection was proposed by Kuo and Mallick (1998).

Since the I_k variables are random quantities, we must assign priors to them. It is natural to give these variables Bernoulli priors because these are simple binary priors for events like coin flips. We can then assign some probability of variable inclusion. Ntzoufras (2002) suggested a value of 0.5 as a noninformative prior. Setting lower values for the variable inclusion probability will tend to produce simpler functional forms with only the most relevant variables being included, while setting higher values will tend to produce functional forms with more variables being retained. This is roughly similar to stepwise regression where one can also control the algorithm's stringency for variable inclusion/exclusion.

The priors for the indicator variables are independent of the priors for other parameters in the Kuo and Mallick method. So, we could express our model for the regression case with Kuo and Mallick variable selection as follows:

$$\begin{aligned}
y_i &\sim \text{Normal}(\mu_i, \tau) \\
\mu_i &= \beta_0 + I_1\beta_1 x_1 + I_2\beta_2 x_2 + \cdots + I_K\beta_K x_K \\
\beta_1, \ldots, \beta_K &\sim \text{Normal}(0, 0.0000001) \qquad (7.6) \\
\tau &\sim \text{Gamma}(0.001, 0.001) \\
I_1, \ldots, I_K &\sim \text{Bernoulli}(0.5). \qquad (7.7)
\end{aligned}$$

For the indicator variable priors in (7.7), selecting a single variable selection probability such as 0.5 is not the only option. One extension of this model would be to specify that the variable inclusion probability itself is an unknown parameter π. Another would be to specify that each variable had its own separate variable inclusion probability π_k, and that all of these were unknown. We could then specify noninformative priors for them with Uniform(0, 1) distributions or Beta$\left(\frac{1}{2}, \frac{1}{2}\right)$ distributions (Section 2.7). The data would then contribute to our understanding of π_k.

7.3.2 In Practice: Kuo and Mallick Variable Selection

We examine the pharmaceutical data using Kuo and Mallick variable selection in **WinBUGS Code 7.3.1 Kuo Mallick.odc**. Note that when a parameter is excluded ($I_k = 0$), it will not make a contribution to the likelihood. As a result, *WinBUGS* will sample the parameter from its prior. This sampling from the prior can take the parameter very far away from its posterior distribution, which may cause slow mixing. Slow mixing should lead to higher Monte Carlo error, and as a result longer runs may be needed. Aside from the slowness of the method, performance studies (Dellaportas et al., 2002, p. 32; O'Hara and Sillanpää, 2009, p. 93) generally indicate that Kuo and Mallick's method performs similarly to other variable selection methods. To account for any slow mixing, we run the Markov chain for an extended number of iterations (500,000).

Assuming that the intercept always remains in the model, one of the following four functional forms (FF) will be visited:

$$\begin{aligned}
&\text{FF1:} \quad \mu_i = \beta_0 + \beta_1 x_1, \\
&\text{FF2:} \quad \mu_i = \beta_0 + \beta_2 x_2, \\
&\text{FF3:} \quad \mu_i = \beta_0 + \beta_1 x_1 + \beta_2 x_2, \\
&\text{FF4:} \quad \mu_i = \beta_0.
\end{aligned}$$

We can also have *WinBUGS* calculate the posterior probability of each of these four functional forms by adding the appropriate code. For example, suppose we define the variables `ind[1]` and `ind[2]` to correspond to I_1 and I_2, respectively. Then the following code will tell us when `ind[1]` is 1 (i.e., β_1 is included) and when `ind[2]` is zero (i.e., β_2 is excluded).

```
fForm[1] <- equals(ind[1]*(1-ind[2]),1)
```

If FF1 were being sampled at a given iteration, the product term inside the `equals` statement will equal the value 1 appearing after the comma. *WinBUGS* will therefore store the value of 1 in `fForm[1]` at that iteration and zero otherwise. Adding the appropriate code for the remaining three functional forms allows us to find all of their posterior probabilities.

Results of variable selection on the pharmaceutical data using five different sets of priors appear in Table 7.2. We use the default Bernoulli(0.5) prior as well as the variable-repelling Bernoulli(0.2) prior and the

TABLE 7.2 Variables Selection Parameter Posterior Means for Pharmaceutical Data. Kuo and Mallick's Method

	Priors for I_k				
	Bern(0.5)	Bern(0.2)	Bern(0.8)	Bern(π_k)	Bern(π_k)
				Priors for π_k	
Parameter				Unif(0, 1)	Beta$\left(\frac{1}{2}, \frac{1}{2}\right)$
I_1	0.072	0.021	0.221	0.071	0.071
I_2	0.508	0.211	0.803	0.524	0.508
π_1				0.357	0.286
π_2				0.508	0.504
p(FF = 1)	0.040	0.017	0.052	0.039	0.039
p(FF = 2)	0.477	0.208	0.635	0.491	0.475
p(FF = 3)	0.032	0.004	0.169	0.032	0.032
p(FF = 4)	0.452	0.772	0.145	0.438	0.453
Rel. p(FF = 1)	0.077	0.076	0.076	0.073	0.076
Rel. p(FF = 2)	0.923	0.924	0.924	0.927	0.924

Bern: Bernoulli distribution; Unif: uniform distribution.

variable-attracting Bernoulli(0.8) prior. With a fairly small sample size, we may expect some variable selection sensitivity to the prior specification and this is confirmed in Table 7.2. Posterior means for the indicator variables (I_1 and I_2) show the impact of the different priors. The posterior means (inclusion probabilities) are reduced when the Bernoulli selection probability is lower as in the Bernoulli(0.2) prior, and are increased when the Bernoulli selection probability is higher as in the Bernoulli(0.8) prior. If we do not supply a fixed Bernoulli selection probability but instead give each indicator variable probability π_k separate priors, the last two columns of Table 7.2 show that the inclusion probabilities are similar to what is obtained under the Bernoulli(0.5) prior.

Table 7.2 also shows the posterior probabilities for the four possible functional forms, $p(\text{FF} = k)$. Again, these are sensitive to the prior used. Under the Bernoulli(0.5) prior, FF2 has the highest posterior probability at 47.7%. Under the Bernoulli(0.2) prior, which more heavily penalizes functional forms with larger numbers of variables, the intercept-only FF4 has the highest posterior probability at 77.2%. FF2 is again the highest posterior probability model under the priors in the last three columns of Table 7.2. Less sensitivity to the prior is observed if we only compare functional forms with the same size in terms of the number of variables. The last two lines of Table 7.2 contain the relative probabilities of FF1 and FF2 assuming that the larger FF3 and the smaller FF4 were omitted. These relative probabilities are computed as $p(\text{FF} = k)/\left(p(\text{FF} = 1) + p(\text{FF} = 2)\right)$. We see that the relative probabilities are quite consistent across priors, with FF1 having approximately 7.5% posterior probability and FF2 having approximately 92.5% posterior probability. Since the posterior mean of I_k is approximately equal to the sum $p(\text{FF} = k) + p(\text{FF} = 3)$, the sensitivity of I_k to the prior is primarily due to the issue of the size of the functional form in terms of the number of variables.

Finally, Table 7.2 also displays the posterior means of the π_k parameters which were estimated under two of the prior specifications. The posterior standard deviations (results not shown) are quite wide, indicating that these two parameters are not precisely estimated from the data. From a qualitative perspective, however, the posterior densities of π_1 and π_2 look visibly different from each other (see plots in **WinBUGS 7.3.1 Kuo Mallick.odc**). For example, under the Uniform(0, 1) prior, the posterior distribution for π_1 appears triangular with a mode at 0. Conversely, the posterior distribution for π_2 resembles the uniform distribution. Under the Beta($\frac{1}{2}, \frac{1}{2}$) prior, the posterior for π_1 has an even more pronounced mode at 0, while the posterior for π_2 resembles the Beta($\frac{1}{2}, \frac{1}{2}$) distribution. Since the posterior for π_2 tends to resemble the prior, we are not learning a lot about this parameter from the data. However, we do appear to be learning at least something about π_1 from the data.

7.3.3 Gibbs Variable Selection

Variable selection by Kuo and Mallick's method is conceptually simple and easy to program, but may suffer from slow mixing because of the priors in (7.7). Dellaportas et al. (2002) proposed Gibbs variable selection (GVS) to address the slow mixing issue (see also Ntzoufras, 2002). When an indicator variable is 1, GVS uses the prior and the likelihood to sample from the β parameter's posterior as does Kuo and Mallick's method. When an indicator variable is 0, however, GVS samples β from a "pseudo-prior." The pseudo-prior is typically constructed to mimic the β parameter's posterior distribution. As a result, the Markov chain will not make large excursions for β when the indicator variable is 0.

In the GVS approach, the prior for β_k is dependent on I_k. This contrasts with Kuo and Mallick's approach where the priors for β_k and I_k are independent of each other. By introducing this dependence, GVS allows for customizable tuning of the Markov chain. We are free to try any pseudo-prior we like in order to improve the mixing of the chain. This is because the pseudo-prior is formulated such that it cannot make a contribution to the likelihood since it is only activated when the $I_k = 0$. Thus estimation of the β posterior is not affected by the pseudo-prior. In summary, under GVS we sample from the parameter's posterior distribution when the indicator variable is 1, and we sample from the parameter's "mimicked posterior" distribution when the indicator variable is 0. This strategy improves the Markov chain's mixing at the typically minor cost of additional model specification and the necessity of a pilot run to find out all of the needed posterior distributions.

More formally, GVS replaces the β variable priors in (7.6) with the following priors:

$$
\begin{aligned}
y_i &\sim \text{Normal}(\mu_i, \tau) \\
\mu_i &= \beta_0 + I_1\beta_1 x_1 + I_2\beta_2 x_2 + \cdots + I_K\beta_K x_K \\
\beta_1, \ldots, \beta_K &\sim \text{Normal}(\mu_{k_{\text{GVS}}}, \tau_{k_{\text{GVS}}}) \qquad (7.8) \\
\mu_{k_{\text{GVS}}} &= I_k \mu_{k_{Prior}} + (1 - I_k)\mu_{k_{Pseudo}} \qquad (7.9) \\
\tau_{k_{\text{GVS}}} &= I_k \tau_{k_{Prior}} + (1 - I_k)\tau_{k_{Pseudo}} \qquad (7.10) \\
\tau &\sim \text{Gamma}(0.001, 0.001) \\
I_1, \ldots, I_K &\sim \text{Bernoulli}(0.5).
\end{aligned}
$$

We see in (7.8)–(7.10) that the β_k priors depend on I_k. Values of $\mu_{k_{Pseudo}}$ and $\tau_{k_{Pseudo}}$ are taken from a pilot run without the variable selection parameters. The values of $\mu_{k_{Prior}}$ and $\tau_{k_{Prior}}$ are set in accordance with the prior values we wish to use.

7.3.4 In Practice: Gibbs Variable Selection

We examine the pharmaceutical data using GVS in **WinBUGS Code 7.3.2 GVS.odc**. In setting up the code, we will need to refer back to our previously found posterior distributions in **WinBUGS Code 7.1 graphical checks.odc**. The posterior mean estimates appear in this file, and we can calculate the posterior precisions as the reciprocal of the squared posterior standard deviation estimates. We again examine the same set of priors and run the Markov chain for the same number of iterations for purposes of direct comparison. Results appear in Table 7.3. Overall, the results are consistent with those in Table 7.2.

7.3.5 Reversible Jump MCMC

The Metropolis algorithm (Section 4.6) allows us to sample from a distribution by proposing a jump to a new value of a variable and then accepting this jump according to a particular decision rule. However, it applies only when we have a fixed number of variables in the functional form. In variable selection, the number of variables in the functional form (which is often called the *model dimension*) is unknown. We might therefore wonder if there was a way to jump between different model dimensions using an adaptation of the Metropolis algorithm. Green (1995) proposed a way to do this, called *reversible jump MCMC*. To summarize the approach, an additional term is factored in to the Metropolis–Hastings acceptance ratio (4.14) which ensures correct trans-dimensional sampling in the Markov

TABLE 7.3 Variable Selection Parameter Posterior Means for Pharmaceutical Data: GVS Method

	Priors for I_k				
	Bern(0.5)	Bern(0.2)	Bern(0.8)	Bern(π_k)	Bern(π_k)
				Priors for π_k	
Parameter				Unif(0, 1)	Beta$\left(\frac{1}{2}, \frac{1}{2}\right)$
I_1	0.070	0.019	0.222	0.071	0.072
I_2	0.517	0.221	0.814	0.511	0.521
π_1				0.357	0.286
π_2				0.504	0.511
$p(\text{FF} = 1)$	0.039	0.016	0.049	0.039	0.039
$p(\text{FF} = 2)$	0.485	0.218	0.641	0.478	0.489
$p(\text{FF} = 3)$	0.032	0.004	0.173	0.032	0.033
$p(\text{FF} = 4)$	0.444	0.763	0.137	0.451	0.440
Rel. $p(\text{FF} = 1)$	0.074	0.067	0.071	0.075	0.074
Rel. $p(\text{FF} = 2)$	0.923	0.924	0.924	0.927	0.924

Bern: Bernoulli distribution; Unif: uniform distribution.

chain. Additional types of jumps are also used. There are "birth" jumps, where a new variable is added to the functional form, and "death" jumps, where a variable is removed from the functional form. We may also have "replace" jumps where the variables are shuffled but the model dimension stays the same because births and deaths are constrained to be equal (Lunn et al., 2009). Reversible jump MCMC is an example of a method that can be used for model selection (Section 7.4) as well as variable selection, although we will not explore its use for model selection in this book.

In terms of performance, O'Hara and Sillanpää (2009) found that reversible jump runs quickly but mixing may not always be competitive with other methods. Reversible jump MCMC can also be used for other problems besides variable selection. However, applications of reversible jump MCMC typically require a fair degree of programming sophistication for implementation. Fortunately, variable selection problems with certain kinds of models can be very easily conducted in *WinBUGS*. At the time of writing, performing reversible jump in *WinBUGS* requires a download from `http://www.winbugs-development.org.uk/rjmcmc.html`. Following the instructions there will update a copy of *WinBUGS* with the reversible jump interface (Lunn et al., 2006, 2009).

7.3.6 In Practice: Reversible Jump MCMC with *WinBUGS*

Upon successful installation of the *WinBUGS* jump interface, a new menu item will become available on the menu bar entitled `jump`. If this menu is not visible, the local version of *WinBUGS* will need to be updated. At the time of writing, the jump interface is limited to models where closed-form posterior distributions are available for which *WinBUGS* knows how to use Gibbs sampling. Models requiring Metropolis sampling cannot be estimated.

Selected code to perform variable selection on the pharmaceutical data with *WinBUGS'* reversible jump appears below (see also **WinBUGS code 7.3.3 reversible jump.odc**). The jump interface makes use of some new commands that appear in boldface in the listing.

```
model {
 for (i in 1:n){
  rndchange[i] ~ dnorm(mu[i],tau) }
  mu[1:n] <- jump.lin.pred(x[1:n, 1:nVar], k, .00000001)
 #priors/calculated quantities
  k ~ dbin(0.5, nVar)
  tau ~ dgamma(0.001, 0.001)
  ind <- jump.model.id(mu[1:n])
}
```

Much of the code looks very similar; however, we have moved `mu` outside of the `for` loop and reexpressed it with a new command, `jump.lin.pred`. The first argument to `jump.lin.pred` is the predictor variables. These need to all be placed in their own matrix so, we have created a predictor matrix called `x` in *WinBUGS* for this. We need to indicate the dimensions of this matrix, and so we provide the number of rows as `1:n` and the number of columns as `1:nVar`. Both `n` and `nVar` can be supplied in the data portion of the code which appears after the main model code. The next argument, `k`, is a random variable that indicates the model dimension at a given iteration of the chain. The final argument is the prior precision for the β parameters associated with the predictors. The prior means of the β parameters are assumed by *WinBUGS* `jump` to be zero and this cannot be altered. In addition, we cannot access the β parameters directly when using the `jump.lin.pred` command, so separate runs will be needed for posterior inference about these parameters. Since `k` is a random variable, we assign it a prior. The prior above expresses the notion that all functional forms are equally likely.

Finally, we can track which functional form is being visited at a given iteration of the Markov chain by monitoring the `ind` variable. This variable stores an identifying value for each functional form and also allows us to find the posterior probabilities of the functional forms. We could also use our own code to find the posterior probabilities of the functional forms by entering statements such as follows:

```
fForm[1] <- equals(ind,1)
```

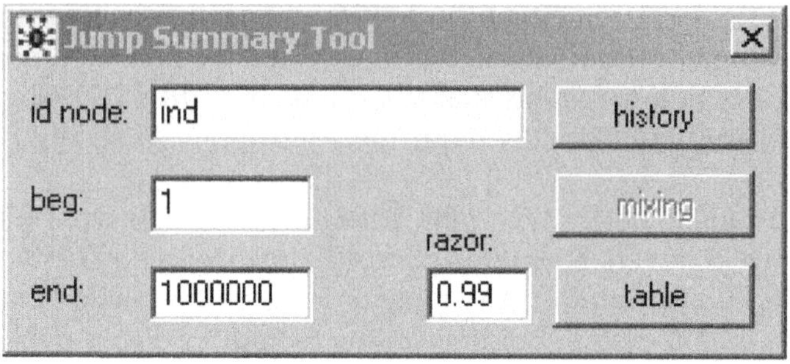

Figure 7.8 Jump Summary Tool

This will set `fForm[1]` to the value of 1 when `ind` takes on the value corresponding to the functional form of interest. We run the Markov chain for a 5000-iteration burn-in and then monitor the parameters `ind` and `k` for 500,000 more iterations. Clicking on the `Summarize...` option in the *WinBUGS* Jump menu causes the Jump Summary tool to appear (Figure 7.8).

We enter the jump id variable `ind` in the dialog box and click `table`. *WinBUGS* will then generate results for the posterior variable selection probabilities based on the reversible jump Markov chain. The output appears in Table 7.4. The first column, entitled model structure, indicates which predictor variables are in or out of the functional form by using 1's and 0's. For example, model structure 01 means that the first predictor variable was out and the second was in. This particular functional form (FF2) had a posterior probability of 0.488 under reversible jump estimation. This probability is similar to other values of $p(\text{FF} = 2)$ under noninformative priors using Kuo and Mallick's method (Table 7.2) and GVS (Table 7.3). In the second portion of the output, we can also see the marginal posterior probability of different variables. This can be compared with $p(\text{FF} = 1) + p(\text{FF} = 3)$ and $p(\text{FF} = 2) + p(\text{FF} = 3)$ from

TABLE 7.4 Reversible Jump Variable Selection Results for Pharmaceutical Data

model structure	posterior prob.	cumulative prob.
01	0.48838	0.48838
00	0.43976	0.92814
10	0.03808	0.96622
11	0.03378	1.0
variable no.	marginal prob.	
1	0.07186	
2	0.52216	

Figure 7.9 Reversible Jump Trace Plot. Pharmaceutical Data

previous tables. Clicking on the `history` button in the Jump Summary tool creates a trace plot of which variables were selected at a given iteration. This can be used to see whether the reversible jump chain might have become stuck at some point. The trace plot for the current analysis (Figure 7.9) does not seem to have unreasonable patterns. Finally, standard density plots for `ind` and `k` are available in the Sample Monitor tool.

7.4 BAYES FACTORS AND *BAYESIAN INFORMATION CRITERION*

We are not limited to variable selection methods in the Bayesian approach. We can also consider model selection methods. Model selection methods allow us to compare different likelihoods and priors with a given dataset in addition to different variables. For example, we could compare a linear regression model with a normal likelihood and certain variables to a log-linear regression model having exactly the same variables. Model selection methods are therefore more general than variable selection methods, but in most applications model selection methods can be used for variable selection problems. This section and subsequent ones in this chapter examine a number of approaches to model selection decisions.

The classic way to perform Bayesian model selection is by the use of Bayes factors. Bayes factors naturally arise from Bayes' theorem in (1.3). We need to add a few more notations to distinguish between different models for the model selection context. Suppose we are looking at Model 1, written as M_1. Then rewriting (1.3) to condition on M_1, we have

$$p(\theta|y, M_1) = \frac{p(y|\theta, M_1)p(\theta|M_1)}{\int p(y|\theta, M_1)p(\theta|M_1)d\theta}. \tag{7.11}$$

The expression in (7.11) tells us that our conclusions are dependent on our model M_1, which makes sense. If, for example, we forgot to include an important variable in M_1, our inferences might be different from if we had remembered to include the important variable. Next, suppose we have Model 2, or M_2. The posterior distributions under M_2 would be found by using Bayes' theorem again, giving an expression very much like (7.11) except with M_2 replacing M_1. In general, we can denote the model number as k and discuss M_k as one of the candidate models.

To compare M_1 and M_2, we could calculate the posterior probabilities of the two models, given the data. We have already found these kinds of probabilities in Tables 7.2 and 7.3 and that they are useful for comparing functional forms. In notation, we would like to find $p(M_1|y)$ and $p(M_2|y)$. We would also like to compare these two probabilities to allow us to find the odds of one model versus the other. Application of Bayes' theorem leads to

$$\frac{p(M_1|y)}{p(M_2|y)} = \frac{p(y|M_1)}{p(y|M_2)} \times \frac{p(M_1)}{p(M_2)}. \tag{7.12}$$

The term $p(y|M_k)$ is called the *marginal likelihood* of Model k. This term *averages over* the uncertainty in the parameter values θ_k by integration. The formula for the marginal likelihood is

$$p(y|M_k) = \int p(y|\theta_k, M_k)p(\theta_k|M_k)d\theta_k \tag{7.13}$$

with θ_k being the parameters for M_k. The marginal likelihood can also be thought of as the prior predictive probability of the data under M_k (Kass and Raftery, 1995), and can be contrasted with the posterior predictive quantities of Section 7.2. The marginal likelihood is also sometimes called the *normalizing constant* (as in Section 2.5) because it arises as the normalizing constant of Bayes' theorem (1.3).

The ratio of the two marginal likelihoods is called the *Bayes factor*. Equation (7.12) can therefore be broken up as follows:

$$\textit{Posterior Odds} = \textit{Bayes Factor} \times \textit{Prior Odds}.$$

This shows that the Bayes factor can be thought of as the change from prior to posterior odds. In the event of equally likely prior probabilities for the two models, the prior odds will equal 1 and the Bayes factor will equal the posterior odds. Therefore, it is common practice to report the Bayes factor in the event that a consumer of the information would like to utilize his/her own prior odds to find his/her own posterior odds.

The Bayes factor can also be thought of as the ratio of the likelihoods that have averaged over (i.e., integrated over) the parameter uncertainty. Suppose, instead, that we ignored the parameter uncertainty in θ and only compared the likelihood of M_1 exactly at representative parameter values for M_1, namely $\hat{\theta}_1$, to the likelihood of M_2 exactly at representative parameter values for M_2, namely $\hat{\theta}_2$. Then the Bayes factor reduces to the simpler likelihood ratio from classical statistics

$$LR = \frac{p(y|\hat{\theta}_1)}{p(y|\hat{\theta}_2)}. \tag{7.14}$$

We briefly review the classical likelihood ratio test because it is useful to compare it with the Bayesian approach. The classical likelihood ratio test applies when we have two functional forms and the first is a subset of the second. If one functional form is a subset of the other, we refer to the first one as being *nested* in the second. The maximized likelihood of the more complex functional form, $p(y|\hat{\theta}_2)$, will be at least as large as the maximized likelihood of the subset, $p(y|\hat{\theta}_1)$, because the more complex functional form will have more parameters and more potential explanatory power than the subset. Therefore, in classical testing we need to account for the difference in the number of parameters in the two functional forms. We write the number of parameters for Model 1 as d_1 and the number of parameters for Model 2 as d_2 (which again is greater than d_1). Then the likelihood ratio test is performed by comparing $-2\ln LR$ to a to a chi-squared distribution having $d_2 - d_1$ degrees of freedom. If the calculated value of $-2\ln LR$ is large even after accounting for the differences in degrees of freedom, we conclude that the two functional forms are not equivalent and we select the more complex one. Note also that we can write the test value as a difference instead of as a ratio using the rules for logarithms, namely

$$LRTest = -2\ln p(y|\hat{\theta}_1) - -2\ln p(y|\hat{\theta}_2). \tag{7.15}$$

Returning to Bayesian approaches, perhaps the most natural way of interpreting the Bayes factor is to convert it to the posterior model probabilities. Under equal prior odds, this is accomplished by finding

$$p(M_k|y) = \frac{p(y|M_k)}{\left(p(y|M_1) + p(y|M_2)\right)}. \tag{7.16}$$

Alternatively, Jeffreys (1961, p. 432) provided a qualitative interpretation of the Bayes factor based on log-scale increments as in Table 7.5. Values of the Bayes factor that are smaller than 1 (or, equivalently, negative $\log_{10}$ Bayes factors) provide support for M_2 over M_1. To use Table 7.5 for such values we simply take the reciprocal (or eliminate the negative sign for the $\log_{10}$ Bayes factors) and use that value for interpretation of the extent to which M_2 is favored.

TABLE 7.5 Bayes Factors and Jeffreys' Interpretation

Bayes Factor	$\log_{10}$ *Bayes Factor*	Interpretation
1–3.2	0–1/2	Evidence not worth more than a bare mention
3.2–10	1/2–1	Substantial
10–32	1–1.5	Strong
32–100	1.5–2	Very strong
>100	>2	Decisive

The Bayes factor is conceptually simple and arises immediately from elementary rules of probability. However, both it and the marginal likelihood can also be very sensitive to the prior specification, which complicates its usage. Bernardo and Smith (1994, ch. 6) describe that Bayes factors are most appropriate in the (somewhat rare) circumstance where one of the models under consideration is believed to be "the true model" but that we do not know which of the models is the true one. If we do not believe there is a true model, or that we have not specified the true model among the models we are examining, then Bayes factors as a decision criterion do not possess their most optimal characteristics. Furthermore, computing the Bayes factor or the marginal likelihood can be somewhat tricky in practice except in special cases. As such, we do not attempt to extensively cover computational details in this book. We do present one approach in the following section.

7.4.1 In Practice: Calculating the Marginal Likelihood for a Simple Proportion

One way of calculating the marginal likelihood is by using the definition from (7.13). Since MCMC makes calculus more convenient, in principle we can always find the marginal likelihood by using MCMC with the definition. Let us work through (7.13) to see how we might do this. The rightmost part ($d\theta$) tells us that we will be (weighted-) averaging over our parameters θ. The second rightmost part, $p(\theta|M_1)$, tells us that we can get the values of θ by simulating from our priors that we have chosen for M_1. *WinBUGS* will happily do this for us. The first term on the right-hand side is the likelihood. So we will insert our simulated values of θ in our likelihood and monitor this term in *WinBUGS*.

We have already manually calculated the marginal likelihood for a proportion in Section 2.5. There we found that the normalizing constant for that model and set of data was 1/6. A program for performing the same calculation using MCMC appears in **WinBUGS Code 7.4.1 marginal likelihood.odc**, and is as follows:

```
model
{
 pi ~ dunif(0,1)
 #monitor the likelihood function
 like <- exp(logfact(n) - logfact(y) - logfact(n-y))
           * pow(pi,y) * pow(1-pi, n-y)
}
#data
list(y=4,n=5)
```

We simulate from the prior distribution for the proportion, `pi`. We enter the formula for the likelihood function as `like`. The leading factorial term in the likelihood has to be implemented with the `logfact` function from *WinBUGS*. The denominator factorial terms are subtracted in accordance with the rules of logarithms and then the result is exponentiated. Alternatively, since the factorial term reduces to a constant $120/25 = 5$ as described in Section 2.5, we could have just typed in the calculated value of "5." Monitoring the `like` parameter for 100,000 iterations gives an estimate of 0.1667, as we would expect given its true value.

In practice, the method we have just described is not used very often. Instead, much ingenuity has been dedicated to finding alternative ways of calculating the marginal likelihood. One reason for this is that if we have large datasets or noninformative priors (or both), the values of the likelihood can become extremely small. This may lead to a software program being unable to represent the likelihood accurately using a given numerical precision. As a result, we will not be able to use the arithmetic mean for our MCMC-based estimate. The log likelihood is often used to deal with small values of the likelihood but the log likelihood does not appear in our definition (7.13) so it cannot be used. One possible workaround for this kind of scenario would be to simulate the values of θ using a program like *WinBUGS* and then exporting the simulations to a program such as *Mathematica* by Wolfram Research where the likelihoods could be calculated and then averaged. *Mathematica* is capable of arbitrary precision calculations and so should have no problem with representing very small numbers.

7.4.2 Bayesian Information Criterion

Since the marginal likelihood can be challenging to compute, we might be content with an approximation. Perhaps the most well-known approximation to the marginal likelihood is the Bayesian information criterion (*BIC*, Schwarz, 1978). *BIC* uses the maximum-likelihood values $\hat{\theta}$ and involves a simple extension to the likelihood ratio (7.14). We begin by having d_k as the dimension of Model k (i.e., the number of variables in Model k) and n as the number of observations in the data set. Then, the

BIC is an approximation to the Bayes factor of M_1 versus M_2, namely

$$BIC_{12} = \frac{p(y|\hat{\theta}_1)}{p(y|\hat{\theta}_2)} n^{(d_2-d_1)/2}. \tag{7.17}$$

The new term (after the first likelihood ratio term) involves the number of observations and the difference in model dimensions. This term will penalize the more complex model. If the more complex model has only a trivially better fit, the penalty will cause the resulting value to become larger, which means that the less complex model will be preferred (see the first column of Table 7.5). Of course, if the more complex model has a substantially better fit, the penalty can be overcome and the more complex model will be preferred. We see that this penalty gets larger and larger as n increases. Both *BIC* and Bayes factors tend to penalize complex models more than the standard methods from classical statistics. Thus, in using *BIC* we will often find that it favors a different functional form (and a less complex one) than one favored by standard classical methods. Despite this fact, *BIC* is a reasonably common frequentist model selection criterion and especially popular when model parsimony is desired.

With some rewriting, the *BIC* also gives us an approximation to the marginal likelihood as

$$p(y|M_k) \approx p(y|\hat{\theta}_k) n^{-d_k/2}. \tag{7.18}$$

Using this approximation, we can find the *BIC* posterior probability of Model k using (7.16). Schwarz (1978) developed the approximation by considering very large sample sizes, while Kass and Wasserman (1995) showed that the prior implied by *BIC* is equivalent to the information provided by one observation from the data.

A common alternative definition of (7.17) for testing involves taking the logarithm of this formula and multiplying it by −2. This gives

$$BICTest = -2 \ln p(y|\hat{\theta}_1) - -2 \ln p(y|\hat{\theta}_2) + (d_1 - d_2) \ln n, \tag{7.19}$$

which might remind you of the formula for the likelihood ratio test (7.15). The penalty term for model complexity in (7.19) is then $(d_1 - d_2) \ln n$, showing that both the sample size and the difference in the number of variables have an additional impact on the value of *BICTest* above and beyond that of the likelihood.

The *BIC* has a number of attractive features, including its widespread use and its ease of computation. Since the *BIC* was derived based on considering very large sample sizes (known as *asymptotic considerations*), the sometimes distracting effects of priors are eliminated. The necessary maximum-likelihood estimates will not be available in Bayesian analyses. However, one could consider substituting the likelihood evaluated at the posterior means for an "approximate approximation." Some of the limitations of *BIC* include the fact that our priors will typically not be the same as the implied *BIC* priors, and our sample sizes will not usually be very large. The *BIC* also cannot be described as a pure Bayesian approach compared to the Bayes factor or other approaches discussed previously in this chapter. This is because it involves making additional assumptions beyond the simplicity of Bayes' theorem which are motivated mostly by convenience. Finally, the dimension of the model may not always be well defined as in the random-effects models which appear in Chapter 8. Despite these considerations, *BIC* may be worthwhile to consider. The *BIC* version of the marginal likelihood (7.18) can also function as an easy computational check if one is computing the Bayes factor or the marginal likelihood by a more involved method.

***In Practice: Computing* BIC *in* WinBUGS** In this example, we use *WinBUGS* to obtain the *BIC* version of the marginal likelihood as in (7.18). The *BIC* gives a point-value result and makes use of point values such as the posterior means of the parameters. Hence, there is no need to simulate *BIC* to form a distribution. Instead, we can compute the *BIC* in *WinBUGS* from point values using the `Node info` function of *WinBUGS*. The code to calculate the marginal likelihood for the example in Section 7.4.1 using the *BIC* approach can be found below as well as in **WinBUGS Code 7.4.2 BIC.odc**. We will need the posterior mean for π in order to find *BIC*. This was calculated in Section 2.6, or we can have *WinBUGS* calculate it. *WinBUGS* gives the posterior mean as 0.7142 in **WinBUGS Code 7.4.2 BIC.odc**. The value of the posterior mean is then supplied as data in the following code to find the *BIC* approximation to the marginal likelihood.

```
model {
  like <- exp(logfact(n) - logfact(y) - logfact(n-y))
            * pow(pi.post,y) * pow(1-pi.post, n-y)
  BIC <- like*pow(n,-0.5)
 }
list(y=4,n=5,pi.post=0.7142)
```

For clarity and ease of debugging, we calculate the first term in (7.18) in a variable called `like`. The variable `BIC` is the product of `like` and `n` to the $-1/2$ power as required by the second term of (7.18). In running *WinBUGS*, we do not proceed past the compile step of the process. This is because we do not need to perform simulations and so we do not continue with the loading of initial values. Instead, we go to the `Info` menu and select `Node info`. We then type into the dialog box the terms we want calculated and click the `values` button. The output will appear in the *WinBUGS* log file. Here, typing `BIC` in `Node info` gives an estimate of the marginal likelihood as 0.1663. This compares reasonably with the result of Section 7.4.1.

7.5 DEVIANCE INFORMATION CRITERION

Our final model comparison method involves the deviance information criterion (*DIC*, Spiegelhalter et al., 2002). The motivations behind *DIC* can be explained by comparing it to a model comparison method from classical statistics called *AIC* (*Akaike's information criterion*, Akaike, 1973).

7.5.1 *AIC* and Classical Non-nested Model Selection

We have seen that in classical statistics we may perform variable selection using the likelihood ratio test by calculating (7.15) and comparing it with the relevant chi-squared distribution. However, this test is limited to variable selection where we have two different functional forms and one functional form is nested within the other. Akaike (1973) developed a classical model comparison criterion that is applicable to non-nested functional forms as well as to the comparison of entirely different likelihood functions. To find the *AIC* for Model k, we calculate

$$AIC_k = -2 \ln p(y|\hat{\theta}_k) + 2d_k \tag{7.20}$$

using the maximum-likelihood values for Model k, $\hat{\theta}_k$. The rightmost term of (7.20) shows that *AIC*'s penalty for model complexity does not involve the sample size, in contrast to *BIC*. It only involves 2 times the number of parameters. In general, *AIC* will tend to favor more complex models than *BIC* because of its more modest penalty for complexity.

The model with the smallest *AIC* value is considered to be the most appropriate model. The difference in *AIC* values is used to compare models informally. Burnham and Anderson (1998, p. 128) provide the following interpretations based on Monte Carlo studies. A difference in *AIC* of 2 or less provides no credible evidence as to the superiority of the model with the smaller *AIC* value. A difference in *AIC* of 2–4 provides weak evidence, while a difference of 4–7 provides strong evidence. Finally, a difference in *AIC* of 10 or more provides very strong evidence that the model with the larger *AIC* value is not the best model. The motivation of *AIC* is predictive, and *AIC* attempts to find the model with the best short-term out-of-sample predictive ability.

The other term in *AIC* besides the penalty term is −2 times the log likelihood. This term is called the *deviance*, given by

$$deviance = -2 \ln p(y|\theta). \tag{7.21}$$

Note that the deviance can be calculated for any values of θ, not just the maximum-likelihood values; hence the use of the more general θ in (7.21). We have seen a deviance term in a number of equations in this chapter, and it is found in many other statistical contexts as well (Hastie, 1987). This is because the deviance quantifies how well θ fits the data (or more accurately, its *lack* of fit). The deviance is related conceptually (and computationally) to the more familiar mean-squared error in simple normal models.

7.5.2 *DIC*: A Bayesian Version of *AIC*

Akaike's work was highly influential and has generated a very large number of related information criteria designed for different purposes or to account for various aspects of *AIC* (Burnham and Anderson, 1998; Ando, 2010). One might consider using the posterior means with the deviance so as to develop a Bayesian version of *AIC*, and this notion features prominently in *DIC*. *DIC* is calculated as follows: We first find the posterior mean of the deviance calculated over the MCMC run, which is called $\overline{D}$. Next we take the posterior means of all the parameters θ and calculate the deviance with these posterior means substituted into the deviance formula, i.e., $-2\ln p(y|\overline{\theta})$. The result is called $\hat{D}$. We will also need the number of parameters for Model k to use in a Bayesian version of (7.20). However, we have mentioned previously that for some models (such as random-effects models) the number of parameters is not well defined and that this poses a problem for *BIC* and *AIC*. Also, informative priors can constrain the parameters so that the parameters cannot represent the data as well as they could without such restrictions. Spiegelhalter et al. (2002) showed that one can estimate the *effective* number of parameters for Model k as

$$\delta_k = \overline{D}_k - \hat{D}_k.$$

Alternatively, dividing the variance of the deviance over the MCMC run by 2 can be used to estimate δ_k (Gelman et al., 2003, p. 182). Having found all the needed terms, Spiegelhalter et al. (2002) showed that *DIC* for Model k can be calculated as

$$\begin{aligned} DIC_k &= \overline{D}_k + \delta_k \\ &= \hat{D}_k + 2\delta_k. \end{aligned} \tag{7.22}$$

The second of these two expressions especially parallels (7.20) although either can be used. Model comparison proceeds by finding *DIC* for all models of interest. The model with the smallest value of *DIC* has the most support according to the criterion. Like *AIC*, *DIC* is a less formal model comparison measure than the Bayes factor or the likelihood ratio test. Spiegelhalter et al. (2002, p. 613) suggest that the guidelines for interpreting *AIC* can be applied to *DIC*. Differences in *DIC* of 2 or less are considered modest, while differences of 3 or greater provide more substantial grounds for preferring the model with the smallest *DIC*. Examples of *DIC* being used in a business context include Berg et al. (2004), who compared models for S&P 100 returns using it.

Since there is considerable overlap between the developers of *DIC* and those of *WinBUGS*, *WinBUGS* has been programmed to provide *DIC* with a few mouse clicks and no extra code. It is therefore the easiest to implement of the Bayesian model comparison methods discussed so far. However, a few considerations are worth bearing in mind when using *DIC*. First, as we have seen, it relies heavily on posterior means, so it should be used with caution whenever the posterior mean is unlikely to be a useful summary of a parameter. Parameters with more than one mode or with extreme skewness may not be appropriate. Spiegelhalter et al. (2002, pp. 596–600) examine an alternative calculation of δ_k using the posterior medians (which are more resistant to skewness) and find that the performance is generally consistent; however, they note that it is possible to encounter issues (pp. 612–613). *WinBUGS* currently does not support direct calculation of *DIC* based on the posterior medians, but it could be done manually. To do this, you can enter the word `deviance` in the *WinBUGS* Sample Monitor tool after the burn-in iterations have finished running, along with the other needed parameters. *WinBUGS* has been programmed to be able to automatically determine the deviance associated with a given model, and entering `deviance` will allow you to see the MCMC estimates of the deviance during the run. Next, run the model for the desired number of iterations and request output summary statistics for all of the parameters including `deviance`. The standard *WinBUGS* output statistics include the posterior median, giving the median version of $\overline{D}_k$. The posterior medians of the other parameters will also appear in the standard output. The posterior medians of the parameters would then need to be substituted into −2 times the log-likelihood function to give the median version of $\hat{D}_k$. This calculation could be done in a separate *WinBUGS* program along the lines of the *WinBUGS BIC* calculation of Section 7.4.2, or could be done using some other program. Finally, find δ_k using (7.22). Note that the mean will not be an appropriate summary statistic for certain models with discrete variables, and so *WinBUGS* will not calculate *DIC* for some models. If this is the case, the *DIC* option will be unavailable (grayed out) in *WinBUGS*.

A more serious issue is that we may be able to write the likelihood in different, but equivalent, ways and this may cause large changes in *DIC* for what is essentially two versions of the same model (see Section 7.5.4).

In particular, it may be possible to write a model where parameters are formally integrated out, or are not formally integrated out but are integrated out using MCMC. These two ways of writing the same model will typically produce different values of *DIC*. This problem can be avoided by only using one type of likelihood function for all models under consideration. This issue is something to be aware of when using *DIC*; however, in the more usual case, in business we are concerned with variable selection issues where this problem would be less likely to occur. Similar complexities arise in the use of *DIC* for missing data models (Celeux et al., 2006, however, see Chen, 2006 for an example where *DIC* performs well) and count models (Millar, 2009). A more philosophical criticism of *DIC* is that it is not Bayesian and that it would not tend to favor the kind of models that would be favored by Bayes factors (Dawid, 2002). Akaike (1983) actually showed that *AIC* does lead to the same kind of conclusions as Bayes factors calculated under certain very informative priors. However, these priors are not common in practice and in any case it is clear that *DIC* draws inspiration more from *AIC* than from Bayes factors. *DIC*, like *AIC*, will tend to favor more complex models than would Bayes factors. A number of other concerns about (as well as praise for) *DIC* have been raised in the discussion following Spiegelhalter et al. (2002), but if the goal is to more informally select a model with the best out-of-sample predictive performance, *DIC* is arguably one of the more used Bayesian criteria for this task. To summarize, *DIC* is a simple but still relatively new way of trying to come to grips with a very complex problem: model selection among a wide array of complex models. Instead of only focusing on *DIC* results alone, we find it useful to consider them along with the results from other tests and diagnostics when trying to assess the adequacy of a model.

7.5.3 In Practice: *DIC* for Variable Selection

We now reexamine the pharmaceutical data and use *DIC* to select among the four functional forms previously considered in Section 7.3.1. Code for this analysis appears in **WinBUGS Code 7.5.3 DIC variable selection.odc**. Since *WinBUGS* calculates *DIC* automatically, we only need to write the code for the functional form of interest. After performing the burn-in iterations, we access the *DIC* tool by going to the `Inference` menu and selecting the `DIC` entry. We click `set` button on the *DIC* tool after the burn-in iterations have been completed so that *WinBUGS* will exclude the burn-in from *DIC* computations (Figure 7.10). If for some reason we have clicked this button by mistake, we can also `clear` it using the tool. For illustrative purposes, we also bring up the Sample Monitor tool and enter the word `deviance` for our first functional form. We

Figure 7.10 *DIC* Tool

then run *WinBUGS* for the desired number of iterations (here 20,000). When these have completed, we click `DIC` on the *DIC* tool. *WinBUGS* reports *DIC* as 225.7 for this model. The reported value of `Dbar` equals the posterior mean of `deviance`, $\overline{D}$, as described previously in Section 7.5.2. Both values are 222.6 here.

Repeating the process for the remaining functional forms yields the results appearing in Table 7.6. FF2 has the lowest *DIC*, so it is preferred. This is consistent with the results from Section 7.3 under the noninformative variable selection priors. Values of δ_k are slightly larger than the true number of variables in each functional form but these differences may be partially attributable to Monte Carlo error.

The third column of Table 7.6 presents the difference between the *DIC* value for a given functional form versus that of the minimum *DIC* value of FF2. We can use this difference for interpretation. There is some evidence that FF1 and FF4 can be removed from consideration according to the *DIC* differences. However, the difference in *DIC* for Model 3 is fairly small, suggesting that it cannot be easily excluded as a possible best model. This recommendation contrasts with the results from Section 7.3, where FF3 received little support except under the "variable-attracting" Bern(0.8) prior. This also contrasts with conclusions we might reach by examining 95% credible intervals. The 95% credible interval for β_1 in FF3 is wide (−1521, 1103) and nearly centered on zero (see **WinBUGS Code 7.5.3 DIC variable selection.odc**).

Another interesting way of comparing the results of Table 7.6 to, say, Table 7.2 is to calculate *DIC* weights mentioned by Spiegelhalter et al. (2002, p. 617). To do this, we calculate $\exp(-DIC_k/2)$ for each of the four

TABLE 7.6 ***DIC* and *BIC* Results for Pharmaceutical Data**

Functional Form	*DIC*	δ_k	Difference from Min. *DIC*	*DIC* Weights	*BIC* Probabilities
FF 1	225.74	3.13	6.56	0.026	0.027
FF 2	219.19	3.14	0	0.684	0.720
FF 3	221.35	4.21	2.16	0.232	0.160
FF 4	224.11	2.05	4.92	0.058	0.093

functional forms. We then divide each of these values by the sum of all of the values. The results are treated as a "weight of evidence," and an argument presented by Burnham and Anderson (1998, pp. 124–125) suggests that these weights are the *AIC/DIC* equivalent of posterior probabilities. In this book, we will not make such a claim but instead we only present these numbers comparatively with respect to the actual posterior probabilities calculated in Table 7.2 and elsewhere. We see that the *DIC* weight for FF3 is 23.2% in Table 7.6 versus the posterior probability of 3.2% for FF3 under the Bern(0.5) prior in Table 7.2. FF4 is also an interesting case to look at because it has no variables in the model at all, retaining only an intercept. The *DIC* weight for FF4 is 5.8% in Table 7.6 versus the posterior probability of 45.2% for FF4 under the Bern(0.5) prior in Table 7.2. This corroborates our previous discussion that *DIC* tends to favor more complex models while Bayesian methods favor simpler models. In fact, Chen et al. (2008) find that *DIC* favors complex models even more heavily than *AIC* itself.

For completeness sake, the last column of Table 7.6 includes the posterior probabilities arising from the use of *BIC* marginal likelihood formula. Details on the calculation can be found in Section 7.7. We see that *BIC* in this example also selects FF2 as preferred. A *BIC* Bayes factor comparing FF2 and FF3 can be calculated by taking the ratio of their posterior probabilities. This gives a 4.49/1 Bayes factor for FF2 versus FF3, which is substantial evidence on the Jeffreys' scale. A similar calculation for *DIC* (which does not have the same theoretical justification, but is done for comparative purposes) shows that *DIC* does not give substantial evidence favoring FF2. *BIC* also favors the smaller FF4 considerably more than does *DIC*. Comparison of the *BIC* results with those of Table 7.2 shows that *BIC* in this example seems to occupy a middle ground between *DIC* and the formal Bayesian approaches considered earlier in Section 7.3. Discrepancies between the two are not entirely surprising given the approximate nature of *BIC* and that it is based on a different set of priors than those used in the models we estimated in Section 7.3.

7.5.4 In Practice: Likelihood Transformations and *DIC*

We adapt an example from Spiegelhalter et al. (2002, pp. 608–609) to illustrate how equivalent likelihoods may lead to different values of *DIC*. Spiegelhalter et al. (2002) show that a likelihood based on the *t*-distribution can also be equivalently written in terms of a random-effects model based on the normal distribution. While we have not discussed random-effects models yet (Chapter 8), to summarize, this model will "average out" the individual random effects in estimating the model and in doing so will give the *t*-distribution. We reexamine data from Section 6.3 for this analysis. Code appears in **WinBUGS Code 7.5.4 DIC Transformations.odc** and results appear in Table 7.7.

TABLE 7.7 ***DIC*** **for Two Equivalent Models**

Likelihood	β_1	β_2	*DIC*
t-Distribution	4.763	−0.0118	618.4
Normal random-effects	4.763	−0.0122	566.2

We see that we are getting the same values for β_1 and β_2 under both models. However, the *DIC* for the normal random-effects model is decisively smaller. Spiegelhalter et al. (2002) indicate that this is to be expected because the two different models have differences in focus (a topic which is out of scope for this book but the interested reader can see their article for details). Spiegelhalter et al. (2002) would argue that the normal random-effects model with its smaller *DIC* would have better out-of-sample predictive performance than the t-distribution model.

As a practical matter, the last two sections have shown that interpreting differences in *DIC* across models is easiest when the likelihood is the same across all models as in Section 7.5.3. By contrast, comparing a random-effects model with a non-random-effects model (as above) can become more involved because of the focus issue.

7.6 SUMMARY

Comparing functional forms and models is very important to business analytics. We need to be able to discard less appropriate models and retain more appropriate ones to guide decision making. In Sections 7.1 and 7.2, we described ways of pre-screening models in terms of being a reasonable representation of the data. Once we have a collection of models that appears reasonable, often we would then want to take the best of these for subsequent usage. This was the topic of Sections 7.3–7.5. This chapter would have been much shorter if the canonical approach to Bayesian model comparison, Bayes factors, were easy to obtain in a wide variety of instances. Since this is not the case, multiple options exist. For a more formal Bayesian approach, the methods of Section 7.3 would be the most suitable. *BIC* can be used as a large-sample approximation to a formal approach, while *DIC* has a more informal, practical orientation in terms of finding the model with the best predictive performance.

7.7 EXERCISES

1. Reanalyze the pharmaceutical data of Section 7.2 after editing the data to remove the sixth observation of all variables (Johnson &

Johnson) in **WinBUGS Code 7.2 posterior predictive.odc**. What, if anything, changes in the parameter estimate results? What is your interpretation?

2. Is the standard deviation of the 2011 credit flows data in Section 5.6.3 significantly different from other years? Modify the code in **WinBUGS Code 5.6 one-way ANOVA.odc** so as to test this possibility.

3. Reanalyze the first ANOVA model in **WinBUGS Code 5.6 one-way ANOVA.odc** and place a uniform prior from 0 to 100 on `sigma`. Add the backtransformation to `tau` as described in Section 5.4. How do the results change?

4. Find the *BIC* posterior probabilities for Table 7.6. To find the first term in (7.18), calculate the likelihood using the posterior means of our parameters. A short-cut method for finding this in the current example is to refer back to the *DIC* numbers in **WinBUGS Code 7.5.3 DIC variable selection.odc**. Find the $\hat{D}_k$ values from the *WinBUGS* file and then calculate the first term as $\exp(-\hat{D}_k/2)$. The second term involving the sample size and the number of parameters is then calculated and multiplied with the first. Once this has been done for all of the functional forms, divide the results by their sum to get the posterior probabilities.

5. Find the *BIC* posterior probabilities for Table 7.6 by writing out the normal likelihood manually instead of using $\hat{D}_k$. Once you have these values, finding the second term and subsequent steps are as in the previous exercise.

6. Reexamine the second model in **WinBUGS Code 5.8 regression and ANCOVA.odc** using reversible jump. Use the two predictor variables, `rsrch` and `country`, in this parallel slopes ANCOVA model. Which of the two predictors is most preferred by reversible jump? What is the posterior probability of this functional form?

7. Reexamine the final model in **WinBUGS Code 5.8 regression and ANCOVA.odc** using reversible jump. Use the two predictor variables, `rsrch` and `country`, and also manually create the interaction variable in a program such as Excel for `jump` to analyze. The interaction variable can be created as the product of `rsrch` and `country`. Supply this interaction variable as data with the other two predictors in a matrix for `jump`. Which functional form is most preferred by reversible jump? How does this compare with the conclusions in **WinBUGS Code 5.8 regression and ANCOVA.odc** based on credible intervals?

8. Modify the final model in **WinBUGS Code 5.8 regression and ANCOVA.odc** to perform variable selection using Kuo and

Mallick's method. Use Bernoulli(0.5) priors for all indicator variables, I_k. What are the posterior probabilities of the variables being included?

9. Estimate *DIC* for each of the three models in **WinBUGS Code 5.8 regression and ANCOVA.odc**. Which of the three functional forms is best supported by *DIC*? Next, use Burnham and Anderson's scale from Section 7.5.1 to describe the differences between functional forms. Describe how these conclusions compare with those in **WinBUGS Code 5.8 regression and ANCOVA.odc** based on credible intervals.

8

HIERARCHICAL MODELS

The models we have considered so far in this book have made an important assumption: that each observation arises in a consistent, independent manner with respect to all other observations after accounting for its predicted value. The data may not always have this property. Consider salaries at a firm. A worker's salary from year to year does not usually jump up and down in an independent manner. Instead, if you know the salary at a given time, you may have a fairly good idea where it has been in the previous six months and where it would be six months in the future. It would probably be easier to guess where this worker's salary would be in six months than it would be to guess the salary of some other randomly selected worker.

In this chapter, we will examine models for data that have a structure where certain data points are believed to be more similar than others. This data is often called *panel data*, *repeated measures data*, or *longitudinal data*. Models for this kind of data also have various names that have arisen in different disciplines. These names include hierarchical models, multilevel models, random-effects models, variance-components models, and mixed models. No matter what the name is, hierarchical models have much to offer to those wishing to understand business problems and are a natural extension of conventional linear models. Many business datasets arise as panels, and so it is important to have models for this type of data.

Bayesian Methods for Management and Business: Pragmatic Solutions for Real Problems, First Edition. Eugene D. Hahn.

8.1 FUNDAMENTALS OF HIERARCHICAL MODELS

As in our salaries example above, we often have reason to believe that our data is *clustered* or *nested* in a particular structure. In business we may find the following:

- Marketing research reveals that consumers in different states/provinces have different attitudes toward corporate behavior or different receptiveness to a particular advertising campaign, but that commonalities and correlations exist across states as well.
- Certain firms in an industry have an excellent track record over time with regard to a particular metric (or set of metrics) while other firms underperform (Hansen et al., 2004; Short et al., 2007).
- Technology adoption rates tend to be faster for some individuals (Cenfetelli and Schwarz, 2011) or in some industries than others.
- Both firm characteristics and country-level factors across firms are relevant in understanding corporate disclosure in international accounting (Dong and Stettler, 2011) or in foreign direct investment (Arregle et al., 2009). In the same vein, firms from different Asian countries may be more or less effective in terms of quality management (Kull and Wacker, 2010).
- Certain employees or certain MBA students are more proactive in applying for advancement opportunities over time than others.

The first example shows that hierarchical models are relevant where we may expect some similarities in responses due to correlations and shared characteristics across groups or markets. The remaining examples above emphasize that hierarchical models are useful for examining scenarios where change, growth, or evolution is an important aspect of the business reality. Since both these kinds of business scenarios occur so regularly, hierarchical models have much to offer in understanding topics of business and management interest.

We begin with a graphical representation of different kinds of models. Figure 8.1 shows a single-level model with three groups. We model the means of each group with θ_i where i indexes the groups. Each group has three data points $y_{i,j}$, where j indexes the data points. Part of the model

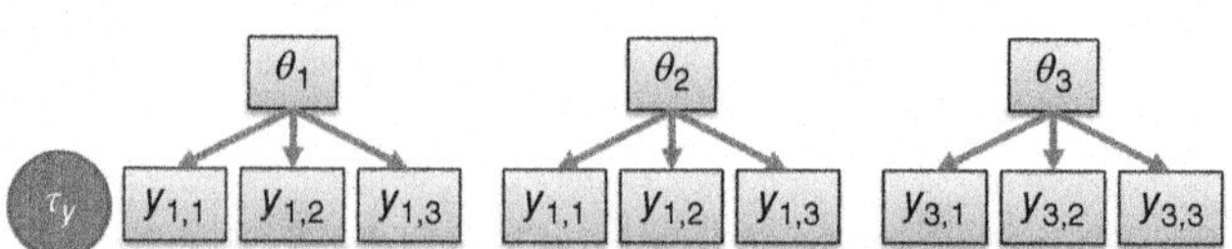

Figure 8.1 Single-Level Model

involves the precision of the data points τ_y, where by our usual convention $\tau_y = 1/\sigma_y^2$. We can see that each group's mean is independent of the remaining means. In the current model, these mean parameters would be called *fixed-effects* parameters. Fixed-effects parameters arise when we have a (usually small) set of parameters that completely define a population. For example, we can completely define the population of companies by splitting companies into the two groups: "companies with less than 100 full-time employees," and "companies with 100 or more full-time employees." The model depicted in Figure 8.1 is exactly the same as our one-way ANOVA (analysis of variance) model with homogeneous variance discussed in Section 5.6.1 where we considered U.S. consumer credit flows over multiple years.

Suppose we did not have fixed-effects parameters but instead randomly sampled our groups from a larger population. In this situation, we would call our parameters *random effects*. We would expect that the overall population parameters (such as national competitiveness) would influence our group means (i.e., our random effects at the state or province level). Since we have a random sample, we could also use our group means to estimate these overall population parameters. For example, we could use state-level or province-level results to learn about overall national outcomes.

We now write the population mean of θ as μ_θ and the population precision of θ as τ_θ. This gives a model as appearing in Figure 8.2. There are now two sources of variation that occur at different levels: the lower level data precision τ_y, and the higher group-level precision in the population, τ_θ. The group-level means are no longer independent of each other. For example, suppose in our random sampling from the population we found that θ_1, the state-level result in State 1, was very large. This would naturally lead us to believe that the value of μ_θ was larger at the national level. It would also lead us to believe that higher values might be possible in the remaining values of θ, i.e., in other states.

Because the θ parameters are related to each other through μ_θ and τ_θ, a hierarchical model permits something called *borrowing strength*. Suppose that, for example, we could only obtain a small sample size to estimate θ_3. This would mean θ_3 cannot be precisely estimated. We may still be able to

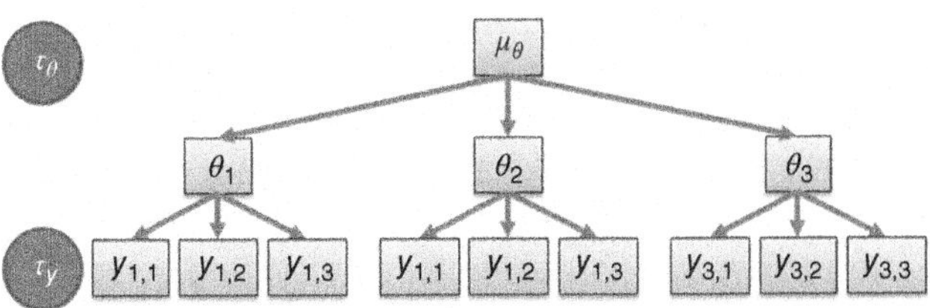

Figure 8.2 Two-Level Hierarchical Model

learn something about θ_3 given that it is a sample from a population about which we are learning from the other groups. As a result, we may be able to reduce our uncertainty about θ_3 through the information from the other groups. In a hierarchical model, random-effects parameters with smaller sample sizes or greater uncertainty will contribute less to our understanding of the overall population parameters, while better estimated parameters will contribute more. Also, these less well estimated random effects will be more influenced by the population parameters. This is often called *shrinkage* because the random effects will be pulled or shrunk toward the population values more than the better estimated random effects. This property makes sense in practical contexts such as business and management. If we have good information about something, we will use it. If not, we will want to recruit other meaningful information from similar cases to help us understand something that is not well estimated. The hierarchical model also lets us think about (and more importantly, model) different sources of variance, recognizing that different sources of variance arise at different levels of the dataset.

We can begin to refine some of these concepts by returning to our dataset from Section 5.6. In this dataset, we considered U.S. consumer credit flows over a five-year period. Our one-way ANOVA results (Figure 5.16) were based on the assumption that the credit flows in each year were independent of each other. This is a simplification of the true situation, which we now expand upon with a hierarchical model.

Before we begin specifying this model, we offer the following caution: **we must be especially mindful about the specification of priors for higher level variance/precision terms**. In particular, research by Liu and Hodges (2003) and Gelman (2006) shows that our usual "noninformative" gamma prior can be (very) informative at higher levels. This is particularly true when the number of random-effects terms is small as in our example.

Detailed research by Liu and Hodges (2003) shows that this can happen in other situations as well, such as when the number of observations per random-effects term is small. Concern about the prior for higher level variances/precisions can be traced back to at least Stone and Springer (1965); however, the problem was not always well appreciated in the earlier part of the MCMC era. Liu and Hodges (2003) documented that the higher level variances can easily have bimodal posterior distributions while Gelman (2006) showed that the usual gamma prior can lead to the overshrinking of random effects, and may even lead to an undefined posterior distribution for very small J.

Gelman instead recommended a uniform prior on the higher level standard deviations or, otherwise, a member of the folded t-distribution family. The uniform prior is easier to conceptualize but may have a slightly disagreeable tendency to lead to overestimate σ_b (Gelman, 2006, p. 521). The t-distribution can be tuned in a more complex manner but may require more expert judgement about what the "right" settings are to achieve the "right" level of random-effects dispersion. Advanced expert judgement may also be needed for other more complex priors (Gustafson et al., 2006). Therefore, we use the uniform distribution here.

It turns out that there is a simple way to counterbalance the tendency of the uniform prior on σ_b to lead to overestimates of variance components. By placing a uniform prior on a slightly lower power of σ_b, that is, a uniform prior on σ_b^w with w slightly less than 1, the effect will be to pull back the values for σ_b toward zero. The smaller the value of w in this root-uniform distribution, the greater will be the effect.

Having discussed the priors, we formulate a hierarchical model for the data consistent with the diagram in Figure 8.2 as follows:

$$\begin{aligned} y_{i,j} &\sim \text{Normal}(\mu_j, \tau_y) \\ \mu_j &= \alpha + b_j \\ \alpha &\sim \text{Normal}(0, 0.000001) \\ \tau_y &\sim \text{Gamma}(0.001, 0.001) \\ b_j &\sim \text{Normal}(0, \tau_b) \\ \sigma_b &\sim \text{Uniform}(0, U). \end{aligned} \tag{8.1}$$

The first-level likelihood for $y_{i,j}$ and the functional form for μ_j look very similar to the effects-coded ANOVA of Section 5.6.2. We have an overall intercept α and "group-mean" terms b_j at the first level of the model. The difference from the effects-coded ANOVA is that here the former group means b_j are assumed to be drawn from a common or shared normal distribution that has mean zero and standard deviation σ_b. This source of variation is at the second level of the model. Because these intercept terms are now random, this model is known as a *random-intercepts* model. Also, since the normal prior is for the terms at the second level of the model, this second-level prior is sometimes called a *hyperprior*.

To finish the specification, we have a gamma prior on τ_y and, importantly, a uniform(0, U) prior on σ_b. The choice of the upper bound $U > 0$ for the uniform distribution on σ_b will require some thought. Setting U too low will arbitrarily force the posterior distribution of σ_b to be too small and cause inferences to be inappropriately precise. Setting U to a large value will not constrain σ_b, but when the sample sizes are small, the overestimation (and decreased precision) described by Gelman (2006) may occur. In the absence of prior knowledge, it is therefore better to err on the side of U being too large. Another possibility is to model a subset of the data to gain some insight into the distribution of σ_b. If you then discard the subset, the insight can be used to form an informative prior for

σ_b for the remaining data. Once we have settled on a prior for σ_b, we then calculate τ_b in the usual manner, namely $\tau_b = 1/\sigma_b^2$.

8.1.1 In Detail: Hierarchical Model Error Terms

It is worthwhile to take a closer look at the error terms in a hierarchical model. With our functional form in (8.1), the first-level error terms are

$$\epsilon_{i,j} = y_{i,j} - (\mu_j + b_j).$$

The first-level error terms $\epsilon_{i,j}$ have variance σ_y^2. The b_j terms are equivalent to the error terms at the second level. The total variance of $y_{i,j}$ after accounting for our fixed term α is the sum of the variances of the two sets of error terms: the first-level error terms $\epsilon_{i,j}$, and the second-level random error terms b_j. So the total variance of $y_{i,j}$ after accounting for α is $\sigma_b^2 + \sigma_y^2$. It is useful to examine the covariance of observation $y_{i,j}$ with some other observation y_{i^*,j^*}. Doing so reveals

$$\begin{aligned} \text{cov}(y_{i,j}, y_{i^*,j^*}) &= \text{cov}(\alpha + b_j + \epsilon_{i,j}, \quad \alpha + b_{j^*} + \epsilon_{i^*,j^*}) \\ &= \text{cov}(b_j + \epsilon_{i,j}, \quad b_{j^*} + \epsilon_{i^*,j^*}). \end{aligned}$$

To make matters more concrete, consider our credit flow dataset and let i index the credit flow types and let j index the years. Because α is a constant, it does not contribute to the covariance and drops out of the formula. When the observations come from different years so that $j \neq j^*$, the covariance is zero because of our assumption of error-term independence in our model specification. However, within a given year j (such that $j = j^*$), the covariance among the i credit flow types is

$$\text{cov}(b_j, b_{j^*}) = \sigma_b^2.$$

Interestingly, this means that our hierarchical model provides a way for accounting for dependence within a given year. We can examine the ratio of the covariance within the years to the total variance. This gives

$$\rho_{\text{icc}} = \frac{\sigma_b^2}{\sigma_b^2 + \sigma_y^2},$$

which is known as the *intra-class correlation*. The intra-class correlation is a measure of the dependence across the i units (across credit flow types) within the j units (within years). We can also say that ρ_{icc} measures the between-years proportion of variance.

Suppose instead we wanted to estimate a model where there was dependency across years within a given credit flow type. We might want to use such a model as a simple way to account for dependence across time. Then by following the same logic above, we can write a model where the yearly random-effects terms b_j would instead be credit flow random-effects terms b_i. The change of subscripts in the resulting model would give us a simple way of accounting for dependency across years (j) for a given credit flow type (i).

We can also consider another way of writing the model in (8.1). The functional form of this model is $\mu_j = \alpha + b_j$. The b_j terms have a mean of zero in this functional form. But, if the b_j terms have a mean of zero, what would the mean of $(\alpha + b_j)$ be? Since α is a constant, the mean of $(\alpha + b_j)$ is α. We can therefore change our model specification and "push" the first-level α parameter up into the second level. This gives

$$\begin{aligned} y_{i,j} &\sim \text{Normal}(\mu_j, \tau_y) \\ \mu_j &= \mu_j \\ \tau_y &\sim \text{Gamma}(0.001, 0.001) \\ \mu_j &\sim \text{Normal}(\alpha, \tau_b) \\ \alpha &\sim \text{Normal}(0, 0.000001) \\ \sigma_b &\sim \text{Uniform}(0, 125). \end{aligned} \tag{8.2}$$

In terms of Figure 8.2, α has been pushed up to the second level and is equivalent to μ_θ in the figure. Now we see more clearly that α is the population mean of the random effects. Note that we could have omitted the functional form $\mu_j = \mu_j$ in writing out this model, but have included it both for comparative purposes and for completeness.

The practice of specifying lower level terms higher in the model is known as *hierarchical centering* (Gelfand et al., 1995). When possible, hierarchical centering is something to be on the lookout for because it typically helps to improve MCMC performance. This means that we will need fewer iterations to reach a given level of Monte Carlo error. It is also worth looking out for the one possible problem with hierarchical centering, which is that we may over-parameterize if we are not mindful. Recall in Section 6.1 we tried to estimate a single mean with two parameters. Neither parameter was identifiable and the Markov chain could not converge. Here, if we tried to estimate an intercept α at the first level and also a population mean for the intercepts at the second level, we would experience the same problem (Section 8.7).

8.1.2 In Practice: The One-Way Random-Effects ANOVA Model

We can estimate random-effects ANOVA models using the code in **WinBUGS Code 8.1 one-way random effects ANOVA.odc**. Here, we assume that the 5 years for which we have data are a sample from a broader population of years. This model also provides a simple way of handling the fact that in each year market events may cause different types of credit flows to behave similarly. This is a conceptual change over the earlier analysis, which assumed that different credit flow types behaved independently of each other in a given year. Key portions of the code appear below. Of emphasis is the prior mean of `b[j]` being set to zero to ensure identifiability, and also the uniform prior on `sigma.b`.

```
model
 {
  for (i in 1:holders) {
      for (j in 1:years) {
         y[i,j] ~ dnorm(mu[j], tau.y) } #likelihood
      }
 #functional form for mu + prior for b
      for (j in 1:years) { mu[j] <- alpha + b[j]
                              b[j] ~ dnorm(|0|, tau.b) }
 #prior for tau.y, sigma.b and alpha
  alpha ~ dnorm(0,0.000001)
  tau.y ~ dgamma(0.001, 0.001)
  sigma.b ~ dunif(0,125)
  ...
 }
```

Selected results are presented in Table 8.1. As expected, we observe some shrinkage in the values of μ in this analysis versus the one-way independence ANOVA results of Table 5.12. The shrinkage contributes to our estimates of changes from year to year being smaller than they were

TABLE 8.1 One-Way Random Effects ANOVA Results for Consumer Credit Flows

Parameter	Mean	Std. Dev.	95% Credible Interval
μ_1	17.73	15.00	(−12.37, 46.42)
μ_2	−1.019	12.77	(−25.78, 24.54)
μ_3	−30.67	14.12	(−58.11, −3.035)
μ_4	−26.15	13.58	(−52.96, 0.132)
μ_5	−5.611	12.57	(−30.02, 19.53)
$\mu_1 - \mu_2$	18.75	18.06	(−15.14, 55.43)
$\mu_1 - \mu_3$	48.40	22.50	(1.588, 90.63)
$\mu_1 - \mu_4$	43.89	21.66	(0.441, 85.00)
$\mu_1 - \mu_5$	23.34	18.58	(−11.11, 60.97)
σ_b	33.39	20.88	(5.342, 91.26)
σ_y	31.66	5.41	(23.14, 44.33)
ρ_{icc}	0.465	0.247	(0.021, 0.903)

previously. Year 1 remains higher than Years 3 and 4 according to the 95% credible intervals for $\mu_1 - \mu_3$ and $\mu_1 - \mu_4$. Unfortunately, with the small sample size we cannot make precise statements about ρ_{icc}, as is shown by its broad credible interval.

8.1.3 In Practice: Hierarchical Centering

In the previous model, the functional form for the `mu[j]` parameters was divided into a fixed intercept `alpha` and random intercepts `b[j]`. Hierarchical centering can be performed in this model by eliminating this division and then relocating `alpha` to the second level of the hierarchy as the mean of the random effects, which are now `mu[j]`. Hierarchical centering requires only a minor modification of the above program listing as below (see also **WinBUGS Code 8.1 one-way random effects ANOVA.odc**).

```
...
  #priors for mu's
      for (j in 1:years) { mu[j] ~ dnorm(alpha, tau.alpha) }
  #prior for tau.y, sigma.alpha and alpha
   alpha ~ dnorm(0,0.000001)
...
```

Improvements with respect to Monte Carlo error for selected parameters can be seen in Table 8.2. Five thousand iterations of burn-in and 50,000 iterations for estimation were used for both sets of results. The centered parameter α has a nearly 79% reduction in Monte Carlo error. The σ_α parameter (which was σ_b in the noncentered model) also had a sizable reduction in Monte Carlo error.

Our previous discussion indicated that the centered and noncentered models were essentially equivalent, but we could also check whether the model fit has been affected by the relocation of `alpha` to a higher level. A natural tool for informally examining this question is the deviance information criterion (*DIC*). *WinBUGS* reports a *DIC* of 248.73 for the noncentered model and 248.66 for the centered one. Hence, the models are

TABLE 8.2 Monte Carlo Error Estimates for Hierarchically Centered Versus Noncentered Models

Parameter	Centered MC Error	Noncentered MC Error	MC Error % Reduction
α	0.103	0.484	78.8
μ_1	0.138	0.158	13.1
σ_α, σ_b	0.218	0.367	40.6
σ_y	0.035	0.038	5.4

not appreciably different in terms of fit (as we would expect). In looking at the *DIC* results, we can review the effective number of parameters in the model, δ_k. In hierarchical models, the effective number of parameters can be considerably less than the specified number of parameters because of the dependence among random effects. Both models have eight specified parameters (`alpha`, `tau.y`, `sigma.b`, and five random intercepts). Here, *DIC* indicates that δ_k is approximately 5.7 in both models examined so far. The dependence across observations in a given year has led to the effective number of parameters being smaller than the 8 specified.

8.1.4 In Practice: Examining Alternative Priors for Variance Components

The choice of the prior for σ_b can be influential, especially when the number of groups is small. Here we compare the uniform prior above with alternative priors for σ_b. One possibility is to extend the range of the uniform distribution. We examine doubling it to 250 in Table 8.3. We also consider the root-uniform distribution discussed earlier in this section. Finally, a gamma prior (which is not recommended) is shown for reference purposes. In order to reduce the influence of Monte Carlo error, the estimates below are based on 995,000 iterations of the sampler after 5000 iterations of burn-in.

Because of the skewness of the posterior distribution of σ_b, the most pronounced differences occur at the upper limit of the 95% credible interval. Doubling the range of the uniform prior from 125 to 250 has increased the upper limit of the credible interval from about 90 to over 105 in addition to increasing all other quantities. Using the root-uniform distribution with w of either 1.05 or 1.1 and upper bound of 125 has the effect of shifting the quantities downward as compared to a uniform distribution with an upper bound of 125. The same effect occurs when the upper bound of both distributions is 250. Finally, we see that (the not-recommended) Gamma(0.001, 0.001) produces posterior quantities that are markedly different from the other priors.

TABLE 8.3 Posterior Estimates of σ_b under Different Priors

Prior	Mean	Std. Dev.	95% Credible Interval
Uniform(0, 125)	33.77	20.59	(6.088, 90.18)
Uniform(0, 250)	35.56	25.81	(6.106, 105.1)
Root-Uniform(1.05, 125)	33.07	20.34	(5.102, 88.52)
Root-Uniform(1.1, 125)	32.63	20.21	(4.869, 87.81)
Root-Uniform(1.05, 250)	34.44	24.67	(5.261, 100.2)
Gamma(0.001, 0.001)	17.73	17.89	(0.053, 60.29)

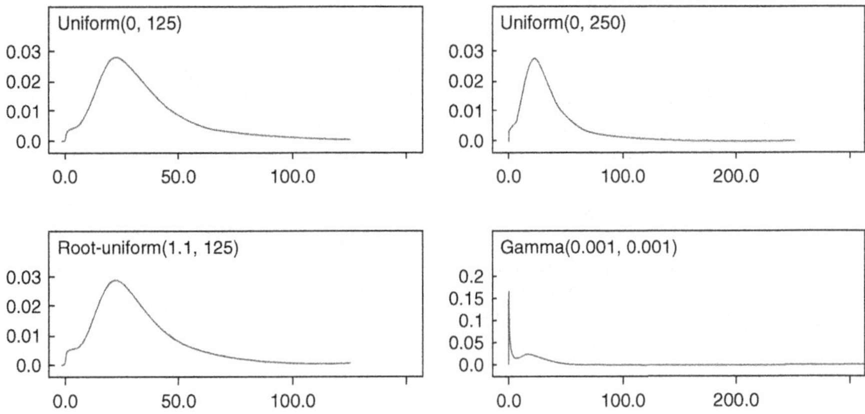

Figure 8.3 Posterior Densities for σ_b under Different Priors

Figure 8.3 displays the density plots for σ_b using four of the priors in Table 8.3. The density arising from the gamma prior stands out from the others. There is a substantial amount of density concentrated in the 0–1 range, a secondary mode of density in the 10–40 range, and a long upper tail. This gives the bimodal posterior distribution mentioned by Liu and Hodges (2003). Since the Gamma(0.001, 0.001) prior has a large spike of density near zero and a long upper tail, the posterior density amounts to a superimposed image of the prior distribution on top of the density favored by the likelihood. This is especially visualizable here because the data information about σ_b is very weak. Inferences about σ_b would have to be made cautiously, with particular caution needed for the credible interval. Fortunately, the posterior distributions of `mudiff` do not change very much under the different priors (excluding the gamma prior), and in business those parameters are likely to concern us more.

8.1.5 In Practice: Longitudinal Modeling

When we have longitudinal data that consists of repeated measurements on the same entities over time, we would expect that there would be some correlation between the measurements over time. Laird and Ware (1982) provided an early discussion of how random-effects models can be used for this kind of data, and today one of the more common uses for random-intercepts models is to account for serial dependence in longitudinal data. Thus, another way we could expand upon the one-way independence ANOVA model (5.1) would be to account for serial dependence. Suppose we are willing to keep the years as fixed effects as in the independence ANOVA but allow for longitudinal dependence in the *credit types*. We include random intercepts `b[i]` for the credit types and assign them a hierarchical prior. The yearly fixed effects receive independent priors as in (5.1).

```
model
  {
  for (i in 1:holders) {
    for (j in 1:years) {
       y[i,j] ~ dnorm(mu[i,j], tau.y) #likelihood
       mu[i,j] <- alpha[j] + b[i] }
    }
 #priors for alpha's and b's
   for (i in 1:holders) { b[i] ~ dnorm(0, tau.b) }
   for (j in 1:years) { alpha[j] ~ dnorm(0, 0.000001) }
 #prior for tau.y & sigma.b
   tau.y ~ dgamma(0.001, 0.001)
   sigma.b ~ dunif(0,125)
 ...
```

The results for the yearly fixed-effects α from this model are very similar to those obtained under the independence model in Table 5.12. There is a slight increase in the posterior standard deviations for α, indicating that we have only lost a little precision as a result of serial dependence. A look at the results for ρ_{icc} shows that its posterior mean is small at 0.146. However, the 95% credible interval for ρ_{icc} is again very wide because of the small sample size.

Comparing the values of ρ_{icc} estimated so far suggests that there is more dependence within years ($\rho_{\text{icc}} = 0.465$) than there is within credit types ($\rho_{\text{icc}} = 0.146$). However, these conclusions are at best tentative given the wide credible intervals. The *DIC* can be used to examine model fit. The *DIC* for this model is 250.6, which gives a poorer fit than previous models. The *DIC* output gives the number of effective parameters as 7.89. Since the five fixed-effects α_j and the first-level precision τ_y would be expected to contribute about one each to the number of effective parameters, the remaining five random effects plus σ_b collectively contribute slightly less than 2 effective parameters to *DIC*.

8.2 THE RANDOM COEFFICIENTS MODEL

In addition to random intercepts, we can also consider those models where the slope terms are random. Figure 8.4 displays an illustration of datasets where random slopes could be useful. The ovals in the figure display the approximate locations of the group-level data, which are indicated with small dots. Within the ovals, the line indicates the regression line for that particular group. The figure also displays a dashed line which represents the regression line that would be estimated if the groups were ignored. In Figure 8.4(a), the dashed line indicates that ignoring the grouping information would cause the slope to be estimated as zero when, within groups, the slope is clearly positive. Conversely, Figure 8.4(b) shows that ignoring the grouping information would lead

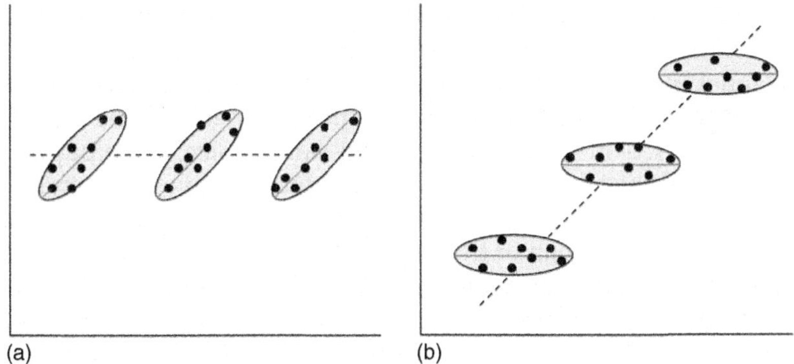

Figure 8.4 (a,b) Slopes Within and Across Groups

us to conclude that the slope is positive when it is actually zero within groups.

A closer look at Figure 8.4 reveals that the within-group slopes are all identical. In this case, the parallel slopes ANCOVA (analysis of covariance) model discussed in Section 5.8.2 could be applicable. Another possibility is that the groups (and their slopes) are independent of each other. If so, the ANCOVA model for unequal slopes could be used. A third possibility would be that the groups have similarities to each other, perhaps by all being in the same industry or by being influenced by common market forces. This might lead to data such as that appearing in Figure 8.5. Suppose we have a random sampling of groups from a greater population, and that we were to estimate the regression line for each group. Then the within-group slopes should not be entirely independent from group to group, but instead may be related. Visually we see in Figure 8.5 that the slopes are not all exactly equal, nor do they appear to be completely independent. A hierarchical model with random slopes could be used for

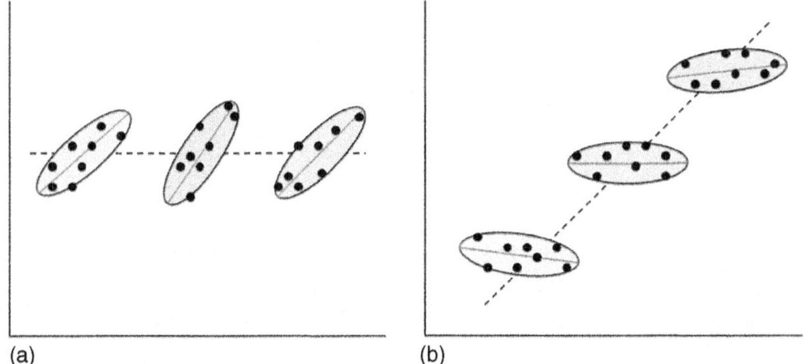

Figure 8.5 (a,b) Random Slopes Hierarchical Model

this kind of data. The random slopes model is also known as the *random coefficients* model (Raudenbush and Bryk, 2002).

The random coefficients model possesses the useful hierarchical model properties discussed in the previous Section compared to the ANCOVA models discussed in Section 5.8.2. If the sample size per group varies among groups, the random coefficients model will have the useful property of shrinkage. This will cause the more poorly estimated slopes to be closer to the overall mean of the slopes. If the number of groups is large and the sample size per group is small, an ANCOVA model may produce a very large number of poorly estimated parameters. In this situation, it may become harder to see the forest for the (possibly blurry) trees. The random coefficients model by contrast will borrow strength across all the groups to provide a better estimate of what the population slope value is given the data. Finally, the random coefficients model conceptually is more of a middle ground between our two ANCOVA extremes of exactly equal slopes across groups and completely independent slopes across groups. We therefore may find that a hierarchical model fits better with our conceptual/theoretical framework for understanding how a market segment or industry operates. We can always use the model comparison methods of Chapter 7 to test whether one or the other of the two ANCOVA extremes is a more appropriate model for the data than would be the random coefficients model.

Suppose the groups are indexed by i and are drawn from a population. Also suppose that the observations for group i are indexed by j. Then the specification for the random coefficients model is as follows:

$$\begin{aligned}
y_{i,j} &\sim \text{Normal}(\mu_{i,j}, \tau_y) \\
\mu_{i,j} &= \alpha_i + \beta_i(x_{i,j} - \overline{x}) \\
\alpha_i &\sim \text{Normal}(\mu_\alpha, \tau_\alpha) \\
\beta_i &\sim \text{Normal}(\mu_\beta, \tau_\beta) \\
\tau_y &\sim \text{Gamma}(0.001, 0.001) \\
\mu_\alpha &\sim \text{Normal}(0, 0.00000001) \\
\mu_\beta &\sim \text{Normal}(0, 0.00000001) \\
\sigma_\alpha &\sim \text{Uniform}(0, U_\alpha) \\
\sigma_\beta &\sim \text{Uniform}(0, U_\beta).
\end{aligned} \tag{8.3}$$

The functional form shows that we are estimating a regression model for each and every group i. These regression slopes are correlated because of their relationships at the second level. The intercepts are also correlated because of their relationships at the second level. However, the intercepts and the slopes are not specified to be correlated to each other in (8.3). We see this because the α_i parameters are influenced by μ_α and τ_α but not by any of the parameters having to do with β. The converse is also true.

In the above functional form, we have subtracted out the overall mean $\overline{x}$ from the values of the predictor variable $x_{i,j}$ because performing this subtraction usually reduces convergence issues. The intercepts α_i can now be interpreted as the expected value for an observation that has a predictor variable equal to the overall mean $\overline{x}$. Another option would be to center the predictors $x_{i,j}$ around the group means $\overline{x}_j$. This would change the interpretation of the random intercepts to being estimates of the individual group means $\overline{x}_j$.

This model has also been hierarchically centered. We can contrast this model with a hierarchically uncentered version since the uncentered version is more common in classical statistics. Uncentering gives

$$\begin{aligned}
&\ldots\\
\mu_{i,j} &= \mu_\alpha + \mu_\beta(x_{i,j} - \overline{x}) + \alpha_i + \beta_i x_{i,j}(x_{i,j} - \overline{x})\\
\alpha_i &\sim \text{Normal}(0, \tau_\alpha)\\
\beta_i &\sim \text{Normal}(0, \tau_\beta)\\
&\ldots
\end{aligned}$$

with the rest of the specification as before. We see that the functional form of the hierarchically centered version is easier to write and simpler to follow. However, the uncentered version does show more explicitly that μ_β can be thought of as a slope term that predicts the data, which might not be as apparent when it is discussed as the population mean of the slope terms. It is worth emphasizing that the β_i terms have a *different* interpretation in the hierarchically uncentered model. They are now the group-specific "slope departures" from (or deviations around) μ_β. If we would like to use the hierarchically uncentered model and also find the actual (undifferenced) value of the slope for group i, it would be necessary to find $\mu_\beta + \beta_i$ for each group. We can instruct *WinBUGS* to do this calculation while sampling from the hierarchically uncentered model if needed. This would give us the posterior distributions of the undifferenced slopes just as in Section 5.8.3 where we implemented this idea previously. By contrast, the hierarchically centered model gives us the more useful undifferenced β_i coefficients naturally.

To summarize, the hierarchically uncentered random coefficients model is of some academic interest because it gives us a different way of thinking about our model. However, the centered version would be preferred by users of MCMC for performance reasons, and the β_i coefficients it generates have a natural interpretation as slopes instead of the less natural interpretation of "slope departures."

8.2.1 In Practice: Structuring Data for Hierarchical Models

Our data comes from a compensation study of the managers of charitable organizations in Maryland in the 1990s originally examined by

Baber et al. (2002). Charitable organizations' tax filings to the U.S. Internal Revenue Service (Form 990) provided the raw data. Baber et al. (2002) analyze 664 publicly available observations from 331 charities. We examine only a subset of these and consider 374 observations from 96 charities. The subset was taken as a more manageable dataset for illustrative purposes. Our focus is on the relationship between the charity's revenue and the charity CEO's salary.

In some applications of hierarchical models, we will have an equal number of observations per group. This gives a so-called rectangular dataset where the number of columns is the same for all groups. We have used the `.Dim` statement in *WinBUGS* to load an array of rectangular data. However, many times in a hierarchical modeling context we may have a varying number of observations per group. This leads to what is known as *nonrectangular, imbalanced,* or *ragged data* for analysis. This kind of data is often the easiest to process in *WinBUGS* when it is supplied as columns (Section 5.7.3). A sample of the dataset appears below, where we can see that, within each organization (`org`), the indicator number for observations (`obs`) can have a maximum value of 4, 5, or 3, causing raggedness. The data also contains a unique identifier number for each row (`recno`) as well as the revenue and salary information.

```
recno[] org[]   obs[]   revenue[]   salary[]
1       1       1       2060953     123922
2       1       2       1814222     127339
3       1       3       1186939     133707
4       1       4       1600476     141447
...
18      5       1       12304124    185531
19      5       2       17595960    189102
20      5       3       16257081    196454
21      5       4       18630427    232234
22      5       5       18819655    248124
23      6       1       31153694    173333
24      6       2       33848677    188750
25      6       3       37830044    202010
...
```

We would like to have the CEO's salary as $y_{i,j}$, where i denotes the organization and j denotes a given observation on the organization. The following code accomplishes this task:

```
for (k in 1:374) {
     y[org[recno[k]],obs[recno[k]]] <- salary[k] }
```

WinBUGS processes each of the 374 rows of the data by accessing the `org` and `obs` identifiers for that record (`org[recno[k]]` and `obs[recno[k]]`, respectively), and then writing the value of salary for that record to the appropriate `y[i,j]`. A similar code can be inserted in the loop for `x[i,j]`. When doing any such data reorganization in *WinBUGS*, it is highly recommended that the created data is checked for accuracy. This can be done after the *WinBUGS* compile step. Clicking on the *WinBUGS* `Info` menu and selecting `Node info` will display a dialog box where `x` and `y` can be entered. Clicking `values` sends the contents of `x` and `y` to the log file where they can be reviewed for accuracy.

8.2.2 In Practice: The Random Coefficients Model

Our *WinBUGS* code for the random coefficients model follows the specification in (8.3). However, we need to make sure that *WinBUGS* handles the raggedness of the data when calculating the likelihood and estimating the functional form. The observation-level index j runs from 1 to its maximum value J. Since J varies from organization to organization, we must supply this as data to *WinBUGS*. We place this in a variable called `maxobs`. *WinBUGS* also needs to be informed of the total number of organizations I, and this is given as `totorg`. With this information, we can write the likelihood and the functional form for the model as below:

```
model
  {
  for (i in 1:totorg) {
    for (j in 1:maxobs[i]) {
        y[i,j] ~ dnorm(mu[i,j], tau.y)
        mu[i,j] <- alpha[i] + b1[i]*(x[i,j] - mean.x) }
    }
 ...
```

The functional form is centered around the overall mean of the data. The data has also been transformed so that `y[i,j]` is in thousands of dollars in CEO salary and `x[i,j]` is in millions of dollars of charity revenue. Since there are 96 charities, the data should supply a sizable amount of information about σ_α and σ_β. From our earlier discussion, the random intercepts will be estimates of CEO salaries in thousands of dollars given that a charity's revenue is equal to the overall mean. Since we might suspect that salary will have a standard deviation in the tens or possibly hundreds of thousands, we use a uniform prior on (0, 1000) for σ_α as a diffuse prior that broadly encompasses these possibilities. Forming a prior for the standard deviation of the slopes is not as intuitive. However, if this standard deviation is very wide, either the charities will be paying CEOs with little regard to revenue, or the CEOs will be agreeing to salaries with little regard to revenue. Neither of these situations seems particularly likely, so we would expect the standard deviation of σ_β to be smaller than that of σ_α. We therefore select a uniform prior on (0, 100) for σ_β. A Gamma(0.0001, 0.0001) prior was used for τ_y.

Full code and data for the model appears in **WinBUGS Code 8.2 random coefficients.odc**. Key results from our model are in Table 8.4. The expected CEO salary for a charity with revenue equal to the overall mean is $\mu_\alpha = \$137{,}200$. As charity revenue increases by \$1 million dollars, CEO salary increases on average by $\mu_\beta = \$1615$. The dependence among salaries within groups over time, ρ_{icc}, is high at 0.917.

We see that the standard deviation of the random intercepts, σ_α, is fairly large with an estimated value of 45.82. Because the intercepts represent the expected CEO salary if each charity's revenue was equal to the average revenue, σ_α indicates there is substantial heterogeneity among baseline CEO compensation after statistically adjusting for differences in charity revenue. Figure 8.6 displays the intercepts using box plots with associated 95% credible intervals. In some cases, CEOs would be very well compensated if the charity's revenue were equal to the mean, while in others the compensation is considerably less. Figure 8.7 displays the random slopes. Visually we see more homogeneity here. This suggests that once a baseline salary has been established, changes in CEO compensation over time tend to be more similar and tied to the charity's revenue changes. Still

TABLE 8.4 Random Coefficients Model: Compensation Data

Parameter	Mean	Std. Dev.	95% Credible Interval
μ_α	137.2	5.970	(125.5, 148.8)
μ_β	1.615	0.244	(1.144, 2.108)
σ_α	45.82	4.238	(38.17, 91.26)
σ_β	1.299	0.205	(0.928, 1.735)
σ_y	13.67	0.633	(12.49, 14.96)
ρ_{icc}	0.917	0.016	(0.882, 0.944)

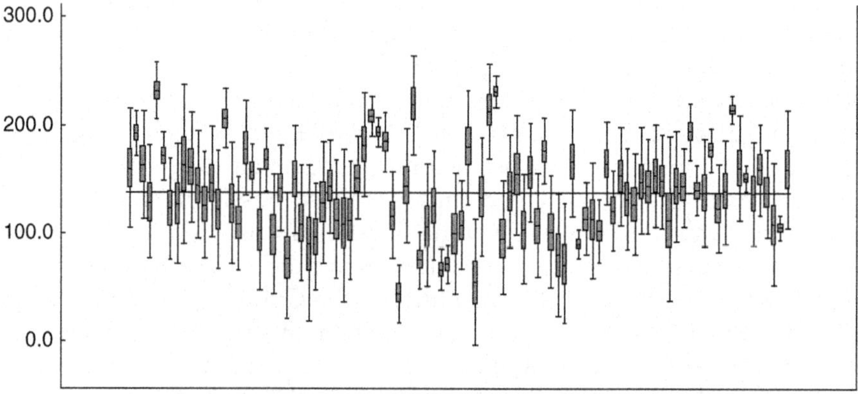

Figure 8.6 Box Plots of Random Intercepts: CEO Compensation

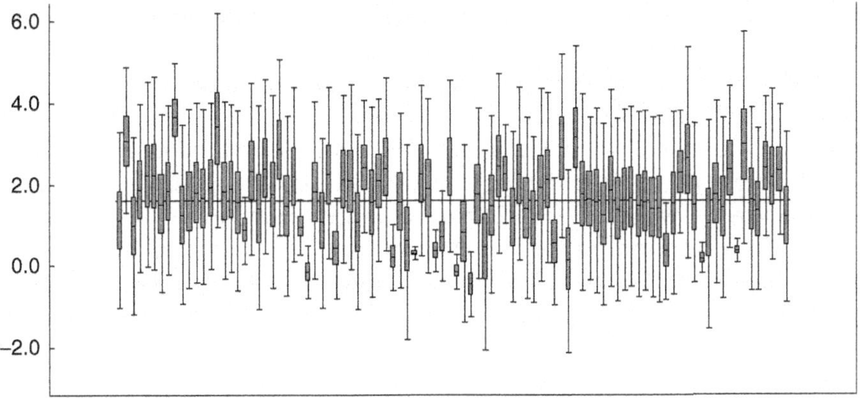

Figure 8.7 Box Plots of Random Slopes: CEO Compensation

there are differences. Close inspection of the box plots reveal that a few charities have random slopes at or just above the zero point, indicating a decreased relevance of changes in revenue to changes in salary.

It is interesting to contrast the predicted values from our random coefficients model to the predicted values that we might get if we performed 96 independent simple linear regressions with one regression per group. The sample sizes per group are small, and so the α and β coefficients are not reliably estimated. Figure 8.8 shows the data and predicted values for Charity 1 using the random coefficients model as well as a simple linear regression model for that group. The predicted value equation is the usual

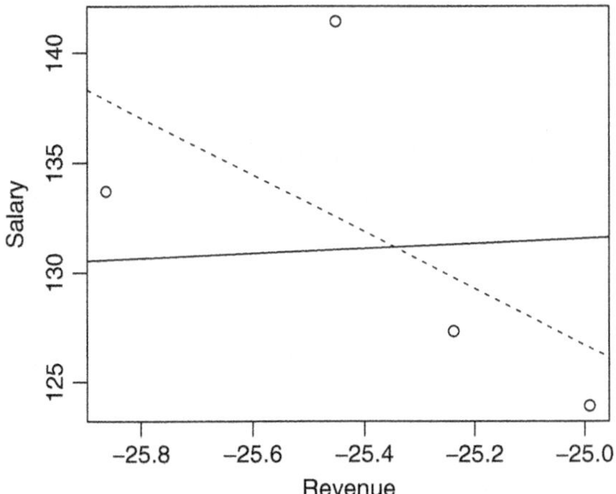

Solid line: random coeffcients model; Dashed line: simple linear regression using data for this group only.

Figure 8.8 Predicted Values for Random Coefficients Model Versus Group-Specific Simple Linear Regression Model: Charity 1 CEO Compensation

$\hat{y} = \alpha + \beta x$, where the value of x has been centered as described above and the values of α and β are estimated from the models.

Under the random coefficients model, we have the intercept and slope as 159.7 and 1.126 for Charity 1. The predicted values for this model appear as a solid line in Figure 8.8. We can also estimate the classical simple linear regression line for Charity 1's data. This gives the intercept and slope as −196.3 and −12.92 (*WinBUGS* gives similar results, see **WinBUGS Code 8.2 random coefficients.odc**) and is represented by a dashed line in the figure. The two lines are strikingly different and lead to different conclusions. The simple linear regression model follows the four data points more closely. The overall trend by the dashed line is that higher revenue leads to lower salaries at this charity (the negative values on the x-axis are due to centering around the revenue mean). Despite the negative sign, the slope is not significantly different from zero according to its classical p-value ($p > 0.35$). By contrast, if we regard this charity as a member of the broader charity population, then the random coefficients model shrinks the weak estimates of α and β back toward the population values. The results that have been shrunk back seem more plausible, as it would be counterintuitive for a CEO's salary to continually decrease when revenues increase and to continually increase when revenues decrease.

8.2.3 In Practice: Changing Random Coefficients to Be Non-random

We next explore what happens if we remove our random slopes and instead estimate the fixed slope over all observations. This will estimate the equivalent of the dashed line in Figure 8.5. Because we found ρ_{icc} to be large and significant, there is substantial clustering within charities. Hence we would expect β to change substantially. This change will probably be for the worse because the dashed line in Figure 8.5 is typically not as useful. We nevertheless proceed to see what would happen if we failed to model the slopes as random.

We modify our model specification in (8.3) to estimate a single β with a vague normal prior after removing the hierarchical parameters μ_β and σ_β, giving

$$\begin{aligned}
y_{i,j} &\sim \text{Normal}(\mu_{i,j}, \tau_y) \\
\mu_{i,j} &= \alpha_i + \beta(x_{i,j} - \overline{x}) \\
\alpha_i &\sim \text{Normal}(\mu_\alpha, \tau_\alpha) \\
\beta &\sim \text{Normal}(0, 0.00000001) \\
\tau_y &\sim \text{Gamma}(0.001, 0.001) \\
\mu_\alpha &\sim \text{Normal}(0, 0.00000001) \\
\sigma_\alpha &\sim \text{Uniform}(0, U_\alpha).
\end{aligned} \tag{8.4}$$

The posterior mean of β is then 0.416 with 95% credible interval (0.291, 0.540). The slope over all observations is much lower than the population mean of the group-specific slopes, similar to the diagram on Figure 8.5(a). The estimate of σ_α increases noticeably to 57.73 with credible interval (50.01, 66.94). As a result, box plots for the random intercepts in this model (analogous to Figure 8.6) show noticeably greater variability (see **WinBUGS Code 8.2 random coefficients.odc**).

DIC was expressly designed as a tool to compare hierarchical models and so we examine *DIC* values here. The *DIC* is 3208.1 for the model of (8.4), while it is 3128.8 for the random slopes model of (8.3). The large difference indicates that the random-slopes model provides a substantially better fit. We can therefore discard our current model based on its less useful fixed effect β as well as poorer *DIC*.

8.2.4 In Practice: Multiple-Predictor Random Coefficients Models

In addition to CEO salary being related to revenue, we might expect that salaries increase over time. We can add multiple predictor variables to our functional form and then add the appropriate hierarchical priors to examine this possibility. Our functional form becomes

$$\mu_{i,j} = \alpha_i + \beta_{1i}(x_{i,j} - \overline{x}_1) + \beta_{2i}(x_{i,j} - \overline{x}_2)$$

and we add a normal prior for the β_{2i} parameters that has mean μ_{β_2} and precision τ_{β_2}. For the current data, the year 1994 was taken as the approximate mean of x_2 and centering was done around this value.

Results from estimating this model appear in Table 8.5. CEO salaries increased by about \$5600 per year on average as indicated by the μ_{β_2} parameter. Adding year as an x_2 variable led to a decline in the effect of revenue as measured by μ_{β_1}. However, revenue still remained a relevant predictor of salary with a 95% credible interval well away from zero.

TABLE 8.5 Multiple-Predictor Random Coefficients Model: Compensation Data

Parameter	Mean	Std. Dev.	95% Credible Interval
μ_α	131.6	5.931	(119.8, 143.2)
μ_{β_1}	1.117	0.232	(0.680, 1.593)
μ_{β_2}	5.233	0.843	(3.587, 6.908)
σ_α	45.48	4.342	(37.65, 54.68)
σ_{β_1}	1.089	0.202	(0.707, 1.507)
σ_{β_2}	5.642	0.923	(3.858, 7.490)
σ_y	10.67	0.577	(9.60, 11.86)
ρ_{icc}	0.947	0.011	(0.922, 0.965)

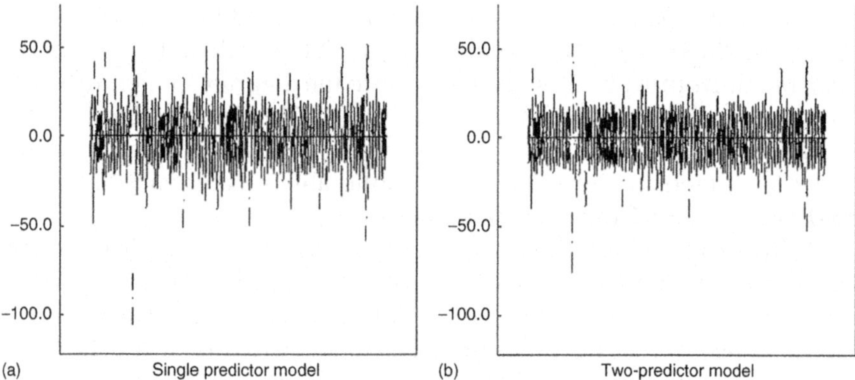

Figure 8.9 (a,b) Box Plots of Residuals for Random Coefficients Models

One reason for considering the two-predictor model involves a plot of the residuals appearing in **WinBUGS Code 8.2 random coefficients.odc**. In Figure 8.9(a), we see the box plots of the 396 residuals for the single-predictor random coefficients model. These residuals show greater variability around zero and some observations are clearly not being explained well by the model. The two-predictor model was considered in an attempt to provide a more comprehensive account of CEO salary predictors. After fitting the two-predictor model, the residuals do show a noticeable improvement as displayed on Figure 8.9(b). They have moved closer to the zero line, and the dispersion around the line at any given point is more consistent than before. While some outlying points remain, other outliers appear to have been better predicted than before. The ability of the model to predict the data can also be assessed through *DIC*. The *DIC* is 2988.9 for the current model, which is a considerable improvement over the single-predictor *DIC* of 3128.8. The evidence suggests that the two-predictor model is preferable.

8.3 HIERARCHICAL MODELS FOR VARIANCE TERMS

A hierarchical structure may also be placed on variance parameters. One use of a hierarchical model on variances is to provide a middle ground between the extremes of the homogeneous variance model and the heterogeneous variance model in ANOVA. In Section 5.6.1, our conclusions regarding the credit flows data were somewhat unsatisfactory as a result of the homogeneity of variances assumption. The considerable year-to-year variation in the raw data (Figure 5.14(a)) was not easy to reconcile with the homogeneous variance aspect of the model. In contrast, our heterogeneous variances ANOVA model fared much better in this regard (Figure 5.17(a)). However, the downside of this model is that it treats every year as entirely separate. Because of this, there is no learning about parameters that carries over from year to year. Thus the independently estimated variances were fairly large and we were not as able to precisely estimate them. The two models are both more extreme than the likely underlying reality, leading us to consider how we might specify a hierarchical model for the variances.

We have been using a normal prior for all of our second-level parameters considered so far. However, this is not recommended for variance parameters because they must be positive. If *WinBUGS* samples a negative precision, it will stop and display an `undefined real result` error. However, under the right (or perhaps wrong!) circumstances, *WinBUGS* will happily allow standard deviation terms to become negative. This can happen if the model is parameterized in such a way that the negative sign is converted to a positive sign in estimating τ (see **WinBUGS Code 8.3 hierarchical variances.odc** for details). This is not an error on the part of *WinBUGS* but instead would be a conceptual error translated into *WinBUGS'* programming code. One way to make the parameters positive is to use a normal prior for the logarithm of the standard deviation parameters. Then we exponentiate these terms to make them greater than 0.

To get a sense for assigning values to our prior, we can try taking a large number of simulations from a normal distribution we are interested in, exponentiating the simulations, and examining the result. Figure 8.10 displays the histograms of these simulations for four different normal distributions. The *R* code to produce plots like these appears in **WinBUGS Code 8.3 hierarchical variances.odc**. The top row indicates that,

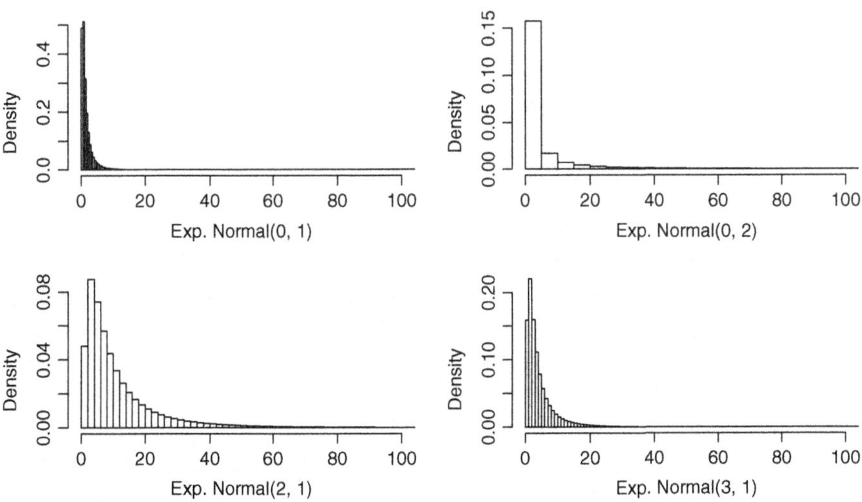

Figure 8.10 Simulating from Normal Distributions and Exponentiating the Result

as we make the standard deviation of the normal distribution larger (from 1 to 2 when the mean remains at zero), the simulations are less likely to cluster near zero and more likely to become very large. The second row shows that, as the mean of the normal distribution increases (from 2 to 3), the simulations again are less likely to cluster near zero and more likely to become very large. All of these distributions have skewness and a long, thin upper tail. Since changing the mean seems to have a more modest effect, we might specify a prior where the mean is unknown but the standard deviation is fixed to a reasonably small value. Large values of the exponentiated normal distribution's mean and standard deviation do not seem plausible as priors for the credit flows data because we have little reason to believe that our data will support values that are extremely large.

Instead of our simulation approach, it turns out that there is a distribution which we could use called the *log-normal distribution*. Although the name is possibly confusing, this distribution is exactly what we have in Figure 8.10. Namely, if we took the logarithm of the values in Figure 8.10, they would be transformed back to their original normal shape. The log-normal distribution is available in *WinBUGS* and can be called by using `dlnorm` as the distribution. However, we will stick with the more familiar normal distribution and then transform it as above.

Adding hierarchically structured variances to our fixed-effects ANOVA model can be done as follows:

$$
\begin{aligned}
y_{i,j} &\sim \text{Normal}(\mu_j, \tau_j) \\
\tau_j &= 1/\sigma_j^2, \\
\sigma_j &= \exp s_j \\
\mu_j &\sim \text{Normal}(0, 0.00000001) \\
s_j &\sim \text{Normal}(\mu_s, \tau_s) \\
\mu_s &\sim \text{Normal}(0, \tau_\mu).
\end{aligned}
\tag{8.5}
$$

Here, we give the s_j terms a normal prior with mean μ_s and precision τ_s. We set τ_s to 0.25 here in order to allow the s_j terms to move somewhat around μ_s but to also to try to weakly constrain them from becoming extremely large. To give μ_s itself the flexibility to move, we give it a normal prior with mean zero and precision τ_μ of 0.25. Since selecting values for τ_s, μ_s, and τ_μ is not especially intuitive, prior sensitivity analysis would be recommended for this kind of model. We could also place a prior on τ_s if desired. Finally, we can monitor $\exp \mu_s$ to investigate its distribution. This parameter can be thought of as an estimate of the population mean for the standard deviations after transforming from the logarithmic scale. Selected *WinBUGS* code for this portion of the model is as follows:

```
...
#priors and functional form for sigmas
   for (j in 1:years) {
      sigma[j] <- exp(s[j])
      s[j] ~ dnorm(mu.s, .25) }
   mu.s ~ dnorm(0,.25)
#other calculated parameters
   exp.mu.s <- exp(mu.s)
```

Another possibility would be to use the typical gamma distribution to create the hierarchical structure for our precision variables. We can specify that the precisions come from a common gamma distribution with parameters a and b. A model using this approach appears below:

$$\begin{aligned} y_{i,j} &\sim \text{Normal}(\mu_j, \tau_j) \\ \mu_j &\sim \text{Normal}(0, 0.00000001) \\ \tau_j &\sim \text{Gamma}(a, b) \\ a &\sim \text{Gamma}(\cdot, \cdot) \\ b &\sim \text{Gamma}(\cdot, \cdot). \end{aligned} \tag{8.6}$$

In thinking about assigning (hyper-)priors to a and b, Table A.1 shows that the mean of the gamma distribution is a/b and the variance is a/b^2. Since the usual way of expressing noninformativeness is to take a and b as fixed, small values, we can choose values for our a and b gamma hyperpriors that tend to produce small values. We might want to observe such models with more care given the caution against small parameter gamma priors (Liu and Hodges, 2003; Gelman, 2006) as described earlier in this chapter. Still, an important difference here is that we have not given fixed, small values to a and b but instead have given them hyperpriors, which allows for variability in a and b. This should allow flexibility in a and b, which may have a counteracting effect on the problem. Alternatively we could assign hyperpriors to a and b that tend to produce moderate or large values for them. Large values for a and b will be highly informative for our τ_j, and so we would expect this to constrain the posteriors with our small dataset. A selected *WinBUGS* code to implement this portion of the gamma hierarchical model is as follows:

```
...
  for (j in 1:years) {
        tau[j] ~ dgamma(a,b) }
  a ~ dgamma(.3,1)
  b ~ dgamma(.1,1)
```

In the above code, the mean of the distribution for a is 0.3 while the mean of the distribution for b is 0.1. The variances for both of these distributions are small. Overall these priors will tend to produce small values for τ_j with occasional moderate values being possible. We refer to this as Hyperprior 1. For Hyperprior 2, we use the Gamma(3, 10) distribution for a and the Gamma(1, 10) distribution for b. These distributions will favor small values of a and b with little chance of more moderate values occurring. Hence the model with these hyperpriors is the one to watch more closely. Finally, we use strongly informative Gamma(10, 1) distributions for both a and b in Hyperprior 3. These distributions will favor small values for σ_j. We compare the results of estimating the hierarchical model for variances under the four specifications below.

The one model of possible concern was the gamma model with Hyperprior 2. However, the densities of σ_j all appeared to be reasonable for this model and did not have the undesirable concentrated spike near zero as seen in the density at the bottom right of Figure 8.3. The means for the μ_j are similar across all models. The means for σ_j and the standard deviations for the μ_j and σ_j parameters are also similar across models, with the exception of the gamma under Hyperprior 3. The gamma under Hyperprior 3 produces smaller values for σ_j and the standard deviations of μ_j as expected. We estimated the model using Hyperprior 3 for comparative purposes, but we might be less inclined to base our main conclusions on it because of its informativeness.

The values of σ_j in Table 8.6 can be contrasted with the σ_j estimated in the unequal variances ANOVA model of Table 5.14. Our hierarchical models create a small amount of shrinkage in the means of σ_j here as opposed to in the unequal variances ANOVA. A more noticeable difference is that the posterior standard deviations of σ_j are smaller under the hierarchical model, indicating that the σ_j parameters are more precisely estimated. Therefore, the hierarchical model here allows us to eliminate the debatable homogeneous variance assumption of classic ANOVA but gives more precision to our estimates than does the unequal variances ANOVA. This, in turn, allows us to improve our estimates of the more important μ_j parameters and their differences.

The log-normal model for the variances has the μ_s parameter which we can exponentiate to estimate the population mean of the σ_j. This is estimated to be 22.43. Posterior summaries of a and b for the gamma models are also included in Table 8.6. We see a has a small posterior mean, even in the third gamma model where the informative prior would encourage a larger a. The posterior means of b are more clearly influenced by the choice of hyperprior values.

8.4 FUNCTIONAL FORMS AT MULTIPLE HIERARCHICAL LEVELS

One interesting aspect of a hierarchical model is that we can place functional forms at multiple levels. For example, we might have first-level

TABLE 8.6 Parameter Posterior Means and Standard Deviations in Hierarchical Variance Models: Credit Flows Data

	Log-Normal		Gamma, Hyperprior 1		Gamma, Hyperprior 2		Gamma, Hyperprior 3	
Parameter	Mean	SD	Mean	SD	Mean	SD	Mean	SD
μ_1	27.05	15.16	27.01	14.62	26.88	14.77	26.97	12.10
μ_2	1.809	23.99	1.803	24.19	1.673	23.99	1.814	20.11
μ_3	−38.10	20.99	−37.95	21.35	−37.93	21.18	−38.02	17.67
μ_4	−32.12	16.50	−32.01	16.61	−32.09	16.43	−32.11	13.58
μ_5	−4.498	5.891	−4.54	5.569	−4.470	5.422	−4.492	4.690
σ_1	30.21	14.65	29.68	14.32	29.68	14.26	25.33	10.04
σ_2	48.78	22.11	48.91	23.52	48.81	23.32	41.70	16.59
σ_3	42.95	19.41	43.26	21.08	43.11	20.46	36.81	14.83
σ_4	33.59	15.35	33.41	16.67	33.16	15.95	28.40	11.15
σ_5	11.75	5.838	11.18	5.420	11.01	5.186	9.738	3.911
$\exp \mu_s$	22.43	22.42						
a			0.146	0.071	0.156	0.056	0.712	0.185
b			0.812	0.951	0.178	0.137	13.31	3.732

slopes which explain how x predicts y in various groups. We might want to examine if there is some overarching relationship that would help us predict whether the slope would be positive or negative in a new group. Perhaps this new group is a business we want to acquire. If we have the relevant data about this business, we can predict the slope of the relationship in advance of acquiring the business. We can also answer general questions such as "can the slope of the relationship between x and y be predicted by z or not" or "what variables best predict the slope of the relationship between x and y."

Another reason we may be interested in this kind of model is that we may have explanatory data that only exists at a higher level. For example, we might expect that companies with an employee stock ownership plan have more motivated employees because the employees' wealth is partially linked to stock performance (Shipper et al., 2013). Here, company characteristics could have an impact on employee actions and outcomes. The possibility of creating functional forms to explain outcomes at multiple differing hierarchical levels opens up many new possibilities for generating business insights. Relationships that exist at macro levels may have impacts at micro levels, and vice versa. We can set up functional forms that simultaneously capture these relationships that exist at different levels for an integrated perspective. This contrasts favorably with a piecemeal approach that would treat the levels as separate and unrelated.

The Bayesian approach makes this kind of multilevel model easy to develop. Writing out multilevel functional forms is very much the same as writing them out for the single-level regression model. We will need to add additional priors to handle our new parameters, but the hard work of finding of the posterior distributions is handled by MCMC and *WinBUGS*. We can focus our attention on other issues such as what functional forms would best represent the business context, what the appropriate priors for parameters are, and whether convergence has occurred for the parameters.

In fact, we have already been specifying multilevel functional forms. As mentioned at the begining of Section 2.1, for very simple functional forms such as $\pi = \pi$ or $\mu = \mu$ we often omit writing out the functional form explicitly but they are still there. We have already estimated second-level means such as μ_α and can now consider estimating the second-level *conditional* mean. Since the conditional mean is another name for the predicted or expected value in a regression analysis, this implies that we can predict the second-level means with a regression-style functional form. This kind of model has been called the *intercepts- and slopes-as-outcomes model* by Raudenbush and Bryk (2002) because we predict the first-level intercepts and slopes with the second-level data.

We can write out a version of this model with one predictor at each level as follows:

$$
\begin{aligned}
y_{i,j} &\sim \text{Normal}(\mu_{i,j}, \tau_y) \\
\mu_{i,j} &= \alpha_i + \beta_i(x_{i,j} - \overline{x}) \\
\alpha_i &\sim \text{Normal}(\mu_{\alpha_i}, \tau_\alpha) \\
\beta_i &\sim \text{Normal}(\mu_{\beta_i}, \tau_\beta) \\
\tau_y &\sim \text{Gamma}(0.001, 0.001) \\
\mu_{\alpha_i} &= \alpha_{21} + \beta_{21} * (x_{2i} - \overline{x}_2) \\
\mu_{\beta_i} &= \alpha_{22} + \beta_{22} * (x_{2i} - \overline{x}_2) \\
\alpha_{21} &\sim \text{Normal}(0, 0.00000001) \\
\beta_{21} &\sim \text{Normal}(0, 0.00000001) \\
\alpha_{22} &\sim \text{Normal}(0, 0.00000001) \\
\beta_{22} &\sim \text{Normal}(0, 0.00000001) \\
\sigma_\alpha &\sim \text{Uniform}(0, U_\alpha) \\
\sigma_\beta &\sim \text{Uniform}(0, U_\beta).
\end{aligned}
\tag{8.7}
$$

We have added second-level intercept parameters $\alpha_{2\cdot}$ and second-level slope parameters $\beta_{2\cdot}$ to the model. We have also added second-level functional forms for μ_{α_i} and μ_{β_i}. This makes the expected values of the first-level α_i and β_i to be conditional on the values of x_{2i} as well as the values of the second-level intercept and slope parameters just described.

Our basic model of (8.7) can easily be extended to represent other sets of relationships we might think are relevant. For example, we can consider having multiple predictors at either the first or second levels. We could also remove the second-level functional form for either α_i or β_i if desired. These variations would involve changing the functional forms and adding or deleting prior distributions as needed. There is no reason either why we have to stop at two levels for this model or for any of the models we have considered in this chapter. We may have three or more levels depending on the hierarchical structures in a particular dataset.

8.4.1 In Practice: Second-Level Functional Forms

Salaries for individuals often increase over time. Here we examine the CEO salary data and consider the effect of time on salary. Since we have multiple consecutive years from each charity, time is likely to be an important explanatory variable. Of course, some salaries increase more over time than do others. We might expect salary increases to be related to the profitability of a charity. Since CEO salaries are part of a charity's overall expenses, we define profit as revenue minus program expenses, where program expenses refer to money spent on the programs that the charity sponsors for its beneficiaries. Looking at profit as total revenue minus total expenses (including the CEO's salary) leads to very similar conclusions because the CEO's salary is typically small relative to program expenses.

We will specify that time predicts salary at the first level of the hierarchy. The relationship between time and CEO salary for a given charity will thus be quantified by this first-level slope. We expect random variations in the first-level slopes across charities. At the second level, we will see whether the first-level slopes can be predicted by the charity's profit. Our measure of the charity's profit is total revenue of the charity across all available time periods minus total program expenses across all available time periods. Profit has been measured in thousands of dollars in our dataset. We might expect that the greater the charity's profit, the faster the CEO salaries increase over time.

We can also predict the intercepts in addition to predicting the slopes. Because we have mean-centered the data at the first level, the intercept indicates what the CEO's salary is expected to be at the average value of time for the data. The average of time for this data is approximately 2.5, so we have set the mean time equal to 2.5 in our *WinBUGS* code to be described below. The intercepts will therefore estimate what each CEO's salary would be at year 2.5 in the dataset. We will assume that the intercepts are random and examine whether these first-level intercepts can be predicted by the charity's profit at the second level.

The interpretational difference between the two functional forms is as follows: Since the intercept estimates the CEO salary at a consistent point in time (at the 2.5-year mark), the model for the intercepts helps us to understand how we might predict whether a salary is high or low at a given time compared to other CEO salaries. It is a model of *salary magnitude* at a fixed, consistent point in time. In contrast, the model for the slopes helps us to understand how we might predict whether salaries will increase more rapidly or slowly over time. It is a model of *growth rates* or *change rates* in each CEO/charity pairing.

The code for estimating this model appears in **WinBUGS Code 8.4 second level modeling.odc**. Selected portions of the code appear below.

```
...
#second level model
for (i in 1:totorg) {
   alpha[i] ~ dnorm(mu.alpha[i],tau.alpha)
   mu.alpha[i] <- alpha2[1] + b2[1]*(profit[i] - mean.
        profit)
```

```
    alpha.resid[i] <- alpha[i] - mu.alpha[i]
    b1[i] ~ dnorm(mu.b1[i],tau.b1)
    mu.b1[i] <- alpha2[2] + b2[2]*(profit[i] - mean.profit)
    b1.resid[i] <- b1[i] - mu.b1[i]
    }
 #priors
    alpha2[1] ~ dnorm(0,0.00000001)
    alpha2[2] ~ dnorm(0,0.00000001)
    b2[1]     ~ dnorm(0,0.00000001)
    b2[2]     ~ dnorm(0,0.00000001)
    ...
```

The main change from our previous hierarchical models is the addition of the functional forms at the second level. A new variable called `profit` has been added to the data listing. We use `profit` to predict the `alpha` parameters through their conditional means `mu.alpha`. The second-level intercepts (`alpha2`) and slopes (`b2`) are given noninformative priors. We can examine the fit of the second-level functional forms by creating variables `alpha.resid` and `b1.resid` as indicated in the code. Plotting these second-level residuals can give us insight into what areas of improvement there might be at the second level.

8.4.2 In Practice: Interpreting Second-Level Coefficients

Running the model in **WinBUGS Code 8.4 second level modeling.odc** for 50,000 iterations after 5000 burn-in iterations produced the results shown in Table 8.7. The parameters α_{21} and β_{21} appeared in the functional form for the intercepts α_i. We now discuss how to interpret them.

The α_i intercepts themselves are the predicted CEO salaries at the 2.5-year approximate average of the time periods. This means that α_{21} is the predicted value of a CEO salary at the 2.5-year time period for a charity with a profit equal to the mean profit among charities. Phrased differently, it is the expected salary at the mean time under the condition

TABLE 8.7 Second-Level Coefficients: Compensation Data

Parameter	Mean	Std. Dev.	95% Credible Interval
α_{21}	122.6	6.000	(110.8, 134.3)
α_{22}	6.927	0.871	(5.234, 8.655)
β_{21}	0.002012	0.000559	(0.0009133, 0.003103)
β_{22}	0.000162	0.0000871	(−0.00000778, 0.000335)
ρ_{icc}	0.966	0.006152	(0.9523, 0.9763)
σ_α	58.11	4.374	(50.32, 67.44)
σ_β	6.565	0.8625	(4.941, 8.328)
σ_y	10.87	0.5798	(9.797, 12.08)

of mean charity profit. The value of α_{21} is 122.6 in Table 8.7. It may come as no surprise that α_{21} is very close to the overall mean of CEO salaries in the data, which is 121.25 thousand dollars.

β_{21} indicates how much a CEO's salary at time equal to 2.5 years is being influenced by overall charity's profit. The value of β_{21} in Table 8.7 (0.002012) shows that there is a small impact of the overall charity's profit. For each thousand dollars of the charity's profit above (or below) the mean profit, CEO salaries would be expected to increase (or decrease) by \$0.002 thousand dollars (i.e., \$2) at the 2.5-year mark. The 95% credible interval for β_{21} excludes zero, so even though the effect of profit is small in magnitude, it appears to be positive.

The β_i slopes are the predicted yearly increases in CEO salaries. So α_{22} is the predicted yearly increase or *raise* of a CEO salary at the 2.5-year time period for a charity with a profit equal to the mean profit among charities. The value of α_{22} is 6.927 in Table 8.7, which means the average annual raise in CEO salary is \$6927. β_{22} indicates how much the raise at time equal to 2.5 years is being influenced by the overall charity's profit. The credible interval for β_{22} in the table includes zero, so we would conclude that the overall charity's profit does not seem to increase or decrease the CEO raises.

Speaking more broadly, when interpreting the β_{22} coefficient we need to remember that it measures "the rate of change in a measure of change." This can be hard to conceptualize, so it is worthwhile to try to think of a more managerially relevant metric to describe what is happening. For example, a rise in salary is a measure of change. Other measures of change that may be managerially relevant include common metrics such as interest rates or depreciation rates. Those with a background in finance and options theory can also think about metrics for derivatives.

The value of *DIC* for this model is 2999.82. The current model fit is better than with the baseline random coefficients model (8.3) ($DIC = 3128.8$). However, the multiple-predictor random coefficients model we estimated in Section 8.2.4 has a better fit than the current one ($DIC = 2988.9$). This suggests that adding more predictors to our current second-level functional forms might improve the fit.

Box plots of the second-level residuals help us to pinpoint what needs to be improved. In Figure 8.11(a), we see the second-level residuals for the α_i parameters. Fit here seems quite poor, so we could definitely use a better second-level functional form (Section 8.7). In terms of a managerial topic, we see that there is a lot of diversity in absolute CEO salaries that we currently cannot explain very well. In Figure 8.11(b), we see the second-level residuals for the β_i parameters. In terms of a managerial topic, we have done a better job in explaining yearly CEO raises although we should also consider ways to obtain additional improvements.

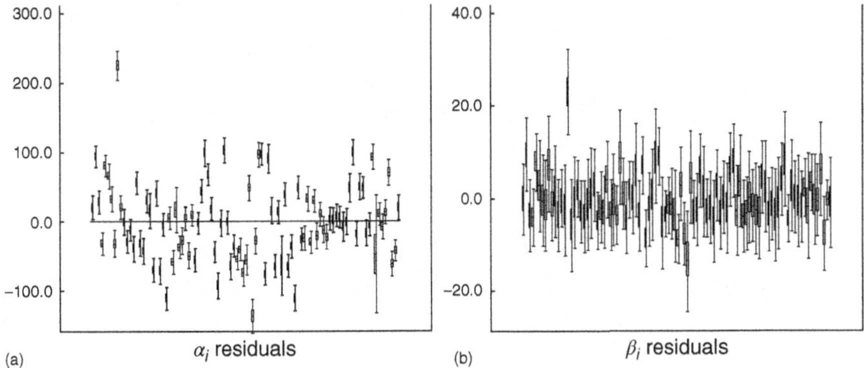

Figure 8.11 (a,b) Box Plots of Second-Level Residuals

8.5 IN DETAIL: MODELING COVARYING HIERARCHICAL TERMS

In Section 8.2, we discussed the random coefficients model (8.3) and examined it in the context of the CEO compensation dataset. We found we were able to improve the fit by expanding on the basic model. In this section, we introduce another method that may improve model fit. The α and β coefficients are defined to be independent of each other in the random coefficients model (8.3). But is this assumption warranted? Figure 8.12 displays a scatter plot of α and β in the salary data when we estimated model (8.3). The scatter plot, which can be created in *WinBUGS* by using the `scatter` option on the `Comparison Tool`, shows that α and β seem to be related. Larger values of one parameter tend to be associated with larger values of the other.

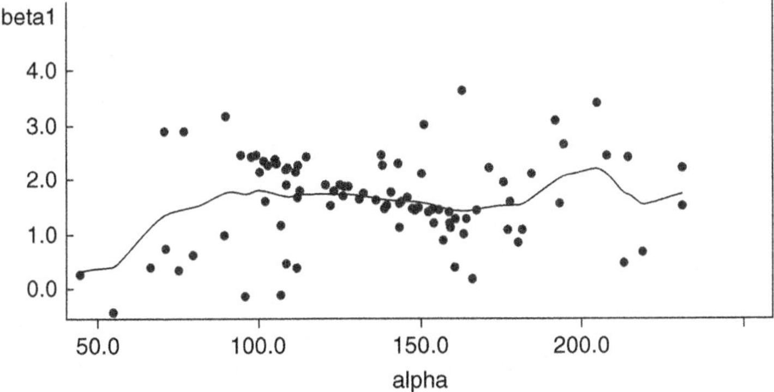

Figure 8.12 Scatter Plot of α and β Parameters in Random Coefficients Model (8.3) for CEO Data

We may find that allowing parameters to covary further improves the fit of the model, and we'll explore that idea in this section. We have previously assumed that the α_i parameters follow a (univariate) normal distribution, and that the β_i parameters follow their own separate normal distribution. If we assume that the α_i and β_i are correlated, then we might suppose they have a bivariate normal distribution. A formal treatment of the bivariate normal distribution has the potential to become more demanding mathematically (e.g., using concepts from matrix algebra) but here we focus on the essentials.

Just as in our random coefficients model (8.3), the random intercepts and random slopes have population means μ_α and μ_β, respectively. They also have population variances σ_α^2 and σ_β^2. However, now the intercepts and slopes can covary and we can quantify this with a covariance parameter $\sigma_{\alpha\beta}$. We introduce some notation in order to communicate more succinctly as well as to prepare ourselves for programming *WinBUGS* later. To describe the bivariate means as collection or as a group, we use a bold symbol, namely $\boldsymbol{\mu}$ where $\boldsymbol{\mu} = (\mu_\alpha, \mu_\beta)$. The covariance matrix is written as $\boldsymbol{\Sigma}$. For reference purposes, in the bivariate case we can write out the covariance matrix explicitly as

$$\boldsymbol{\Sigma} = \begin{pmatrix} \sigma_\alpha^2 & \sigma_{\alpha\beta} \\ \sigma_{\alpha\beta} & \sigma_\beta^2 \end{pmatrix}.$$

We see this matrix is *symmetric*, i.e., the covariance $\sigma_{\alpha\beta}$ appears above and below the diagonal line formed by the variances.

Since *WinBUGS* is set up to work with precisions instead of variances, we need a symbol for the precision matrix. We use $\boldsymbol{\Sigma}^{-1}$ for the precision matrix. Now we can write out our model using a notation such as follows:

$$(\alpha_i, \beta_i) \sim \text{Bivariate Normal}\left(\boldsymbol{\mu}, \boldsymbol{\Sigma}^{-1}\right).$$

This indicates that the α_i and the β_i parameters have a bivariate normal distribution with the given set of means and the given covariance matrix. We may also be interested in finding the correlation. We can calculate the correlation using the standard formula

$$\rho_{\alpha\beta} = \frac{\sigma_{\alpha\beta}}{\sigma_\alpha \sigma_\beta}.$$

8.5.1 Specifying Priors for the Bivariate Normal

We now look at prior distributions. The conventional, broad normal priors can be used as noninformative priors for our bivariate means μ_α and

μ_β. Priors for $\mathbf{\Sigma}$ require some more thought. We know that the correlation ρ cannot go outside the range −1 to +1, so we have to be careful about how we sample $\sigma_{\alpha\beta}$. Perhaps the easiest way to sample from $\mathbf{\Sigma}$ is through a step-by-step approach. In Step 1, we place our usual uniform priors on σ_α and σ_β. We then take the sampled values of σ_α and σ_β, and create the needed σ_α^2 and σ_β^2 parameters. For Step 2, we insert σ_α^2 and σ_β^2 into $\mathbf{\Sigma}$.

In Step 3, we can place a prior on $\rho_{\alpha\beta}$ that follows the −1 to +1 restriction on the range. A natural choice might be a uniform distribution for $\rho_{\alpha\beta}$ in the range −1 to +1 to represent noninformativeness. Or, for a more informative prior, we could use a Beta distribution that has been transformed to the range −1 to +1. Then we take our sampled values and calculate

$$\sigma_{\alpha\beta} = \rho_{\alpha\beta}\sigma_\alpha\sigma_\beta,$$

which is found from a rearrangement of the formula for the correlation above. This will produce a valid value for $\sigma_{\alpha\beta}$, which we can then insert into $\mathbf{\Sigma}$ as Step 4.

There is a more general name for our valid covariance matrix where all the variances are positive and the covariance is set up so that the correlation is between −1 and +1. This kind of matrix is called a *positive-definite* matrix. We need to make sure that we are giving *WinBUGS* positive-definite matrices because *WinBUGS* will crash if it does not receive one, or tries to sample from a non-positive-definite matrix. Typical problems when operating *WinBUGS* can involve not programming the constraints properly, or forgetting that the matrix must be symmetric.

Now that the priors have been decided, we can write out the full model. Our random coefficients model with covarying coefficients is

$$\begin{aligned}
y_{i,j} &\sim \text{Normal}(\mu_{i,j}, \tau_y) \\
\mu_{i,j} &= \alpha_i + \beta_i(x_{i,j} - \overline{x}) \\
(\alpha_i, \beta_i) &\sim \text{Bivariate Normal}(\boldsymbol{\mu}, \mathbf{\Sigma}^{-1}) \\
\tau_y &\sim \text{Gamma}(0.001, 0.001) \\
\mu_\alpha &\sim \text{Normal}(0, 0.00000001) \\
\mu_\beta &\sim \text{Normal}(0, 0.00000001) \\
\rho_{\alpha\beta} &\sim \text{Uniform}(-1, 1) \\
\sigma_\alpha &\sim \text{Uniform}(0, U_\alpha) \\
\sigma_\beta &\sim \text{Uniform}(0, U_\beta)
\end{aligned} \tag{8.8}$$

where $\boldsymbol{\mu}$, $\mathbf{\Sigma}$, and $\rho_{\alpha\beta}$ are defined as above.

8.5.2 In Practice: The Covarying Random Coefficients Model

We reexamine the compensation dataset using the model in (8.8). The full code listing appears in **WinBUGS Code 8.5.1 covarying random coefficients.odc**. A good portion of the code involves minor changes from the first model in the previously discussed **WinBUGS Code 8.2 random coefficients.odc**. New code blocks and changes are described below.

We have relabeled the `alpha[i]` and `b1[i]` parameters in the previous functional form to the more general term `coef`. The `coef` parameters are now in an array where the first dimension `i` indexes the charities as before and the second dimension indexes the coefficients themselves. This allows us to place a bivariate normal prior on `coef` in the priors Section of the code. We see the bivariate normal prior is assigned using `dmnorm`. *WinBUGS* can also accommodate a trivariate or higher normal distribution using `dmnorm`. We supply the means for the bivariate normal distribution as `mu.coef[]` and the precision matrix as `invSigma[ , ]`. The dimensions of the precision matrix `invSigma[ , ]` are intentionally left blank in this portion of the code.

```
...
 #Functional form
  mu[i,j] <- coef[i,1] + coef[i,2]*(x[i,j] - mean.x)
...
 #Priors
 for (i in 1:totorg) {
   coef[i, 1:2] ~ dmnorm(mu.coef[], invSigma[ , ] ) }
 mu.coef[1] ~ dnorm(0,0.00000001)
 mu.coef[2] ~ dnorm(0,0.00000001)
 sigma.alpha ~ dunif(0,1000)
 sigma.beta ~ dunif(0,100)
 rho ~ dunif(-1, 1)
...
 #Other calculated parameters
 sigma.alpha.beta <- rho*sigma.alpha*sigma.beta
 Sigma[1,1] <- pow(sigma.alpha, 2)
 Sigma[1,2] <- sigma.alpha.beta
 Sigma[2,1] <- sigma.alpha.beta
 Sigma[2,2] <- pow(sigma.beta, 2)
 invSigma[1:2 , 1:2] <- inverse(Sigma[ , ])
...
```

The normal priors for `mu.coef[]` are identical to the former priors for `mu.alpha` and `mu.b1`, and the priors for `sigma.alpha` and `sigma.beta` are also the same as before. The one truly new parameter is `rho` and it receives a uniform prior across its range as a noninformative prior. Next we construct the covariance matrix from our standard deviations and `rho`. Finally, we generate our precision matrix. Here we tell *WinBUGS* the dimensions of `invSigma` on the left-hand side of the statement. On the right-hand side, we instruct *WinBUGS* to compute the matrix inverse of `Sigma` as needed. This calculation slows down *WinBUGS* considerably.

Table 8.8 displays the results for the covarying random coefficients model with respect to the CEO compensation data. The `mu.coef[]` parameters have been labeled to match with their counterparts from Table 8.4 so that the two tables can be compared more easily. The new parameter for our bivariate model is $\rho_{\alpha\beta}$ in Table 8.8. Note that it should not be compared with the intraclass correlation ρ_{icc} from Table 8.4, because the latter represents a different kind of correlation. We see that $\rho_{\alpha\beta}$ indicates the α_i and β_i parameters have a strong positive correlation. We can discuss this finding in more detail as follows:

The slopes and intercepts are shrunk back toward their population values in random coefficient models. In the nonvarying model (8.3), the coefficients were shrunk back toward their means. In the covarying model (8.8), *WinBUGS* was able to improve fit by shrinking some intercepts and slopes one way and others in another way. In particular, fit was better when higher intercepts were allowed to be associated with higher slopes and when lower intercepts were associated with lower slopes. So, instead of shrinking the slopes and coefficients only back to *point values* (the population means), *WinBUGS* is now shrinking them back to a *line*. This can be seen clearly by plotting the values of α_i and β_i in our current model as in Figure 8.13. The values of α_i and β_i appear more closely around a straight line than before.

We also find that μ_α and μ_β are slightly larger in Table 8.8 as compared to Table 8.4. The slight decrease in σ_y is a promising sign because it suggests that the first-level residuals seem to be explained slightly better as well. *DIC* can be used to examine the model fit issue more concretely. *DIC* under the noncovarying random coefficients model is 3128.8, while under the covarying random coefficients model it is 3112.0. Thus we conclude from the credible interval for $\rho_{\alpha\beta}$ and *DIC* that the covarying random coefficients model is an improvement over the noncovarying random coefficients model.

TABLE 8.8 Covarying Random Coefficients Model: Compensation Data

Parameter	Mean	Std. Dev.	95% Credible Interval
$\rho_{\alpha\beta}$	0.693	0.101	(0.466, 0.862)
μ_α	142.8	8.049	(127.5, 159.2)
μ_β	2.044	0.338	(1.426, 2.754)
σ_α	56.87	6.216	(45.71, 70.00)
σ_β	1.681	0.326	(1.114, 2.389)
σ_y	13.37	0.618	(12.22, 14.64)

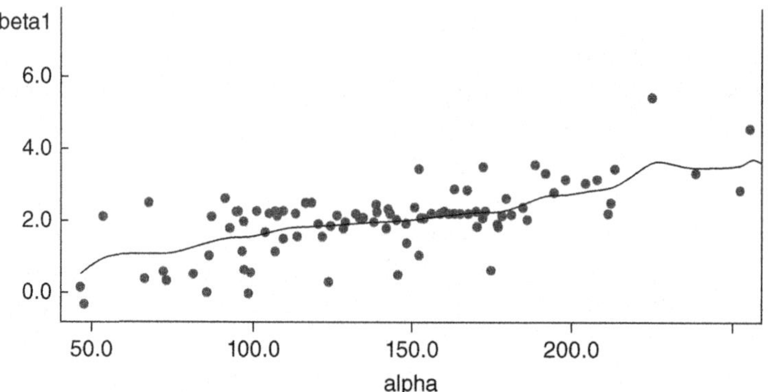

Figure 8.13 Scatter Plot of α and β Parameters in Random Coefficients Model (8.8) for CEO Data

8.5.3 In Practice: Case Studies in the Covarying Random Coefficients Model

It is good news that $\rho_{\alpha\beta}$ improves model fit; however, it does more than this. Importantly, it explains a key feature of the data that helps generate new insights about compensation strategies. This feature of the data might have gone unnoticed without the help of our covarying random coefficients model. Below we explore in more detail what implications there are from covarying random coefficients by looking at two of the organizations in the dataset in more detail.

Charities 15 and 51 provide two case studies that help explain why $\rho_{\alpha\beta}$ improves model fit. The raw data points and predicted value lines appear in Figure 8.14. Three different predicted value lines appear in the figure. The dashed line is from a simple linear regression which takes into consideration only the four data points of the selected charity. The solid

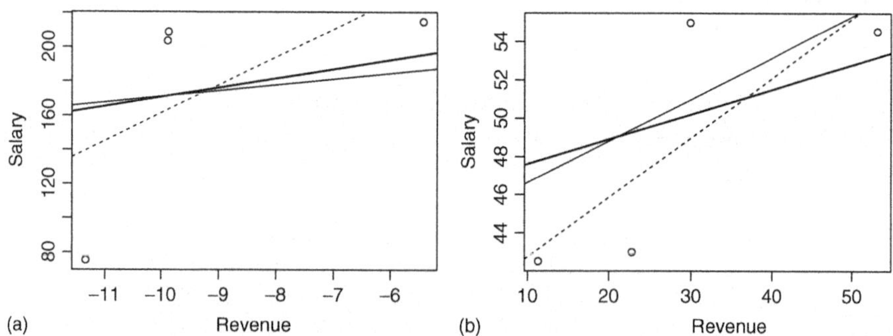

Figure 8.14 (a,b) Predicted Values under Three Models: CEO Compensation

line indicates the predicted values from the noncovarying random coefficients model (8.3). The heavy solid line indicates the predicted values from the covarying random coefficients model (8.8). The x-axis indicates the charity's annual revenue after the mean revenue (approximately \$27 million) has been subtracted away.

On the left of the figure, we see that Charity 15 is below average in terms of revenue. While the CEO's initial salary may seem nonexcessive at \$75,000, the CEO was well-compensated given the charity's more modest \$18 million revenue average. This can be seen from the intercepts. The simple linear regression intercept was approximately \$324,000, which is the predicted salary of this CEO if Charity 15 had revenue equal to the average revenue of the charities considered. We may be somewhat skeptical of the simple regression intercept because of the small sample size. The noncovarying random coefficients model shrinks this back to \$205,000, while the covarying random coefficients model permits a higher value at \$225,000. All of these intercepts are comparatively high, suggesting that, if all charities had the same revenue, this CEO would be more highly compensated than most other CEOs.

Figure 8.14 also shows that the CEO's salary more than doubled when the charity's revenue rose by about (a comparatively small) \$1.5 million. The slope under simple regression was 16.35, indicating that for each additional million in revenue, this CEO's salary was expected to increase by \$16,350. The noncovarying random coefficients model shrinks this slope back to 3.41 while the covarying model allows the slope to be larger at 5.44.

At the opposite end of the spectrum, Charity 40 had a sizable \$56 million average revenue but the CEO was paid a modest \$49,000. The CEO received only about \$8000 in salary increases despite the charity's revenues doubling over the study period from \$380 million to \$800 million. The simple regression intercept indicates that this CEO would be expected to make \$39,600 if the charity's revenue were equal to the average. This can be compared with the previously mentioned expected salary of \$324,000 for CEO 15 if Charity 15's revenue were equal to the average. The noncovarying random coefficients model shrinks this back to \$44,500, while the covarying random coefficients model permits a slightly higher value at \$46,200. The covarying model permits the shallowest slope of 0.129, indicating that for each additional million in revenue, the CEO's salary was expected to increase by \$129 (a factor of 100 smaller than that of Charity 15's CEO). The slope under the noncovarying model was 0.217, while that under simple regression was 0.311.

In summary, $\rho_{\alpha\beta}$ captures the fact that there seems to be notable differences in pay-for-performance across charities. Some CEOs are making more than what would be expected, and these same CEOs' salaries also increase at accelerating rates. Our case studies indicate that these differences may be related to whether a charity is smaller or larger in terms of revenue. Additional models could be developed to explore this observation in more detail.

8.6 SUMMARY

In the era of large data, hierarchical models are especially useful and are worthwhile to understand. Hierarchical models help us to better model data patterns at the individual level, and then help us to aggregate the individual-level patterns into more comprehensible "big picture" results. Fortunately for us, the principles for hierarchical models are very similar to those for single-level models. We may specify the means or other functional forms at the different levels as we see fit. Although the resulting models are inevitably more complex, we can directly rely on our knowledge of single-level models to develop and understand our multilevel hierarchical models. As we have seen, ignoring the hierarchical structure of our data may lead to incorrect conclusions. Hence, the ability of hierarchical models to address this important type of business data in a reasonably straightforward way makes it an essential tool for advanced business insight.

8.7 EXERCISES

1. What are some of the scenarios in which hierarchical models are used?
2. Give two examples of multilevel business datasets that you have worked with or with which you are familiar.
3. Using the credit flows data, respecify the model of (8.2) so that it has a first-level intercept AND a population mean. Estimate the model in *WinBUGS* and comment on the behavior of the Markov chain for these parameters.
4. Remove the random intercepts and their standard deviation (and precision) from the longitudinal random-intercepts model of Section 8.1.5 to obtain the fixed-effects ANOVA model. Find the *DIC* of the fixed-effects ANOVA model and compare it with the *DIC* of longitudinal random-intercepts model of Section 8.1.5. Comment on the value addition of the random effects in this dataset.
5. Reexamine the first model in **WinBUGS Code 8.2 random coefficients.odc** using a Gamma(0.001, 0.001) prior for `tau.y` instead of the currently used Gamma(0.0001, 0.0001) prior. Estimate the parameters appearing in Table 8.4. Use 5000 iterations of burn-in and 50,000 iterations. Compare your results with those of Table 8.4. How much do the parameter values change under this different prior?

6. Find *DIC* for the models of Table 8.6 using the source code which appears in **WinBUGS Code 8.3.1 hierarchical variances.odc**. Which (if any) model is preferred by *DIC*? What (if anything) does this tell you about the influence of the prior in Bayesian model comparisons using *DIC* in this dataset?

7. Modify **WinBUGS Code 8.4 second level modeling.odc** to add `x[i,j]` (charity revenue) as an additional covariate at the first level. Append the following code to the first-level functional form: `+ b12[i]*(x[i,j] - mean.x)`. Give the `b12[i]` parameters a mean of `mu.b12` and a precision of `tau.b12`. Assign the same priors to `mu.b12` and `tau.b12` that are used for `b2[1]` and `tau.b1`. What is the posterior mean and 95% credible interval for `mu.b12`?

8. Use the same model as developed in the previous question. Monitor the first-level residuals in `resid`. Compare the values of `resid` in this model as compared to the original model of **WinBUGS Code 8.4 second level modeling.odc** using the `box plot` function in *WinBUGS*. How (if at all) have the residuals improved?

8.8 NOTATION INTRODUCED IN THIS CHAPTER

Notation	Meaning	Example	Section Where Introduced
μ_θ	A second-level (or higher) mean of θ parameters	μ_θ	8.1
τ_θ	A second-level (or higher) precision of θ parameters	τ_θ	8.1
σ_θ	A second-level (or higher) standard deviation of θ parameters	σ_θ	8.1
ρ_{icc}	The intra-class correlation	ρ_{icc}	8.1.1
$\boldsymbol{\mu}$	A collection, vector, or set of parameter means	$\boldsymbol{\mu} = (\mu_\alpha, \mu_\beta)$	8.5

(continued)

Notation	Meaning	Example	Section Where Introduced
$\mathbf{\Sigma}$	A covariance matrix	$\mathbf{\Sigma} = \begin{pmatrix} \sigma_\alpha^2 & \sigma_{\alpha\beta} \\ \sigma_{\alpha\beta} & \sigma_\beta^2 \end{pmatrix}$	8.5
$\mathbf{\Sigma}^{-1}$	A precision matrix	$\mathbf{\Sigma}^{-1}$	8.5
$\sigma_{\alpha\beta}$	The covariance between α_i and β_i terms	$\sigma_{\alpha\beta}$	8.5
$\rho_{\alpha\beta}$	The correlation between α_i and β_i terms	$\rho_{\alpha\beta}$	8.5

9

GENERALIZED LINEAR MODELS

Most of the models that have been discussed so far in this book have assumed that the data has a normal distribution. However, there is much variety in business data and a normal distribution is not always the most appropriate one for a particular analysis. In this chapter, we explore models for non-normal data. We will see that we can greatly expand our analytical toolbox with some basic changes to our *WinBUGS* code.

9.1 FUNDAMENTALS OF GENERALIZED LINEAR MODELS

The normal distribution plays a central role in statistics because of its attractive properties and its applicability. Historically, normal linear models were the first to receive extensive development, and these models (including ANOVA, simple/multiple regression, and ANCOVA) remain important cornerstones of statistical modeling. The true distribution of a given dataset is almost always unknown; however, the normal distribution often makes for a good approximation. Still, many kinds of data do not have a normal distribution even approximately. In Section 2.1, we discussed how coin flips could be modeled with the binomial likelihood function, while in Section 3.7 we discussed modeling count data with the Poisson distribution. A coin flip taking on the vales of heads or tails clearly does not have a normal distribution,

Bayesian Methods for Management and Business: Pragmatic Solutions for Real Problems,
First Edition. Eugene D. Hahn.

and nonnegative integer-valued counts are not well approximated by a normal distribution unless the sample size and the magnitude of the data are both large.

While our models of Sections 2.1 and 3.7 had appropriate likelihood functions, they were very simple and had only a basic predictive capability. What we would like to do is to write out a more detailed functional form for our non-normal outcome data so that we can see whether certain predictors have a relationship with our non-normal outcome data. Conceptual insights and computational advances in the field of statistics have led to the area that is now called *generalized linear models* (Nelder and Wedderburn, 1972; McCullagh and Nelder, 1989), which accomplish what we would like to do. Generalized linear models bring together a non-normal likelihood for our outcome data with a linear functional form that permits many predictors.

Fortunately for us in the MCMC era, Bayesian estimation of these models can be performed with the same familiar methods we have already been using. There are only two main things that are different in generalized linear models. First, we will be using different likelihood functions than *WinBUGS'* normal likelihood function `dnorm`. These different likelihood functions correspond to different distributions, which of course have different properties. We will have to have a basic understanding of some of these distributions' properties in order to use them effectively. The second thing that is different is that we now may have to think about something called the *link function*.

The Link Function Instead of diving right into a discussion of the link function, we will first talk about the *inverse link function* because we have already been using it. The inverse link function is, simply put, a transformation of our functional form. Since our linear functional forms can produce both positive and negative values, we might use a transformation to ensure that negative numbers are not passed to a distribution for counts such as the Poisson distribution. For example, instead of writing

$$\mu_i = \beta_0 + \beta_1 x_i \tag{9.1}$$

we can write

$$\mu_i = \exp\left(\beta_0 + \beta_1 x_i\right) \tag{9.2}$$

and this will force μ_i to be positive. In fact, you might recall that we have already used this kind of transformation in Section 8.3 to force variances to be positive (without calling it an inverse link function at that time).

Another equivalent way of writing (9.2) is

$$\log\left(\mu_i\right) = \beta_0 + \beta_1 x_i. \tag{9.3}$$

This happens because log and exp are inverse functions of each other, and the right-hand side can be simplified using $\log(\exp(x)) = x$. We can describe (9.3) in English as "the logarithm of the expected value has a linear functional form given by $\beta_0 + \beta_1 x_i$." So the link function allows us to use a linear functional form to predict a *transformation of the expected value*.

The conventional way of writing the link function is on the left-hand side as in (9.3). Thus log is the link function in (9.3). If we want to talk about a transforming function on the right-hand side, we call that the inverse link function as in (9.2). The inverse link function for (9.2) is therefore exp. We can describe (9.2) in English as "the expected value has a nonlinear functional form given by $\exp\left(\beta_0 + \beta_1 x_i\right)$."

In general, if we have a distribution for our outcome data and this distribution has certain restrictions on the parameter values, we create the link function to make sure that these restrictions are upheld. The inverse link function modifies our linear functional form so that we only obtain valid values for our outcome data distribution. Even so, we usually prefer to discuss our results in terms of the link function, i.e., we have a linear form predicting some function of the mean. The reason is that it makes our slope terms more interpretable. A brief example shows why this is the case.

Suppose that we have $\beta_1 = 0.2$. Then using (9.3) we see that for every one-unit change in x_i the logarithm of the expected value will increase by 0.2. This is reasonably easy to describe. However, if we use the interpretation implied by (9.2), the interpretation of $\beta_1 = 0.2$ becomes more difficult. There is a nonlinear relationship between β_1, x_i, and μ_i, so we cannot summarize the relationship succinctly. For example, suppose that $\beta_0 = 0$. Then suppose that x_i has a starting value of −0.4 and we make a one-unit increase to 0.6. The change in μ_i will be $\exp(0.2 * 0.6) - \exp(0.2 * -0.4) = 0.204$. However, if the starting value of x_i is 10 and we make a one-unit increase to 11, then the change in μ_i will be $\exp(0.2 * 11) - \exp(0.2 * 10) = 1.636$. This makes β_1 less easy to describe under the (9.2) approach. As a result, the link function is usually preferred over the inverse link function when we want to describe relationships in our data (to be fair to the log link function, it happens to have the semi-elasticity interpretation which does give an interpretation of $\exp(\beta_1)$ in terms of percent change—see Section 6.3).

One important special case of the family of link functions is the *identity function*. When we use the identity link function, we do not transform either side. For example, the identity function is used in (9.1). If you like, you can also think of the identity function as the result of multiplying both sides of the equation by 1, so that both sides stay the same. The identity function is the simplest link function and can sometimes be used in generalized linear models if we constrain our priors in certain ways or if we have data that support it. For example, we may have count data where all the counts are very large. If so, the chance of obtaining a negative count might be so remote that we can use the identity link function without any worries of crashing *WinBUGS* or other software.

9.2 COUNT DATA MODELS: POISSON REGRESSION

Our basic count data model of Section 3.7 allowed us to model only the mean of a set of counts. However, we can use the powerful linear functional form we have used in regression models to model the *conditional* mean of a set of counts. This allows us to see what predictor variables help us to explain a count outcome. The parameter λ is the rate parameter, that is, the expected count per unit of time. We extend our simple model of Section 3.7 to the regression context by writing the functional form as

$$\log\left(\lambda_i\right) = \beta_0 + \beta_1 x_{1,i} + \cdots + \beta_k x_{k,i}.$$

We previously used a gamma prior for λ to permit an easy conjugate analysis and to prevent λ from becoming negative. However, the log link will now keep λ from becoming negative. So we can continue to use our familiar broad normal priors for our β parameters. *WinBUGS* will sample from the posterior distributions using MCMC.

Putting our likelihood, functional form, and priors all together gives the following Poisson regression model for count data:

$$\begin{aligned} y_i &\sim \text{Poisson}(\lambda_i) \\ \log\left(\lambda_i\right) &= \beta_0 + \beta_1 x_{1,i} + \cdots + \beta_k x_{k,i} \\ \beta_0, \ldots, \beta_k &\sim \text{Normal}(0, 0.00000001). \end{aligned} \tag{9.4}$$

In Practice: The Poisson Regression Model We can check the model fit of generalized linear models using the residuals. Our residuals can be calculated as $y_i - \lambda_i$. It is natural to consider plotting the residuals against the observed values of the data, y_i. However, Cameron and Trivedi (1998, p. 144) caution against this because in Poisson regression the residuals and y_i will be correlated. They instead recommend plotting the residuals against the predicted values λ_i because these will not be correlated.

We should expect nonconstant variance (heteroscedasticity) in Poisson regression because the mean and the variance are equal under the Poisson distribution. So as the expected value (i.e., the conditional mean) increases, the conditional variance will also increase around the conditional mean. This motivates the use of the Pearson residual. The Pearson residual is the usual "raw" residual divided by the square root of the variance function. In Poisson regression, it is calculated as $(y_i - \lambda_i)/\sqrt{\lambda_i}$. There is a very close relationship between the Pearson residual in the Poisson model and the standardized residual in the normal model. The difference is that the denominator quantity varies for the Poisson Pearson residual (i.e., it involves λ_i), whereas in the homoscedastic normal model there is only one quantity, σ. The Pearson residual gives a consistent frame of reference across observations in a single Poisson model as well as across Poisson models.

Our count data regression example uses information appearing in Benaroch et al. (2006). Benaroch et al. (2006) investigated information technology (IT) project risk management in a large Irish financial services organization. They present data on 50 large-scale IT projects undertaken by the organization in their Appendix B. The length of each project in months, y, is a count which can be examined using Poisson regression. Benaroch et al. (2006) also provide information as to whether a project was discretionary or statutory. We use this information as a dummy-coded variable x_1, taking the variable to be 1 if the project was statutory and zero otherwise. The expected investment cost of the project in millions of euros is used as predictor x_2. A third variable, x_3, is the total risk score for each project. We also consider the interaction term between certain predictors. The following code (see also **WinBUGS Code 9.2 Poisson Regression.odc**) performs Poisson regression for Functional Form 1 (FF1) using this data. Straightforward modifications of the code can be used to estimate FF2 and FF3 (see Table 9.1 for the functional forms to be examined).

```
model
{
 for (i in 1:n) {
   y[i] ~ dpois(lambda[i])
   log(lambda[i]) <- beta[1] + beta[2]*x1[i]
       + beta[3]*(x2c[i]) + beta[4]*x1[i]*x2c[i]
   resid[i] <- y[i] - lambda[i]
   pearson.resid[i] <- (y[i] - lambda[i] )/sqrt(lambda[i])
   x2c[i] <- (x2[i] - x2.bar)/1000
   }
 #prior for beta
   for (j in 1:4) { beta[j] ~ dnorm(0, 0.000001) }
 #other calculated parameters
   x2.bar <- mean(x2[])
}
```

We can examine scatter plots of the two kinds of residuals versus λ using *WinBUGS*' `Compare` tool. Figure 9.1 displays these plots. The raw residuals appear in Figure 9.1(a). At higher values of λ, the model has a tendency to overpredict the length of a project. Thus the largest negative raw residual (a value of –8.5 corresponding to Project 36) appears at the right side of the left-hand plot in the figure.

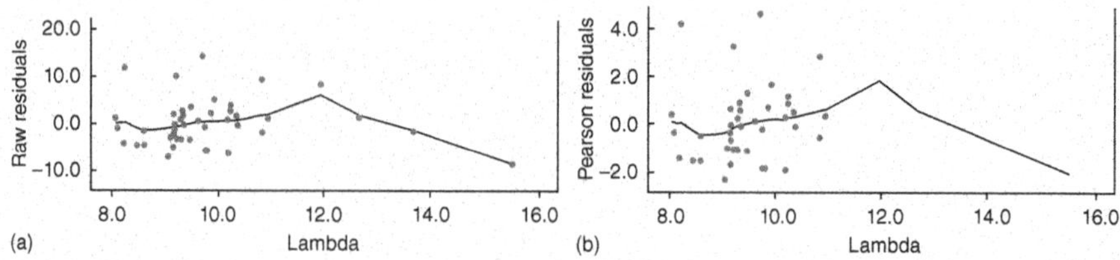

Figure 9.1 (a,b) Residual Scatter Plots versus λ for Poisson Regression Functional Form 1: IT Project Data

The Pearson residuals are plotted versus λ in Figure 9.1(b). Here, we see at far right that Project 36, which had the largest negative raw residual previously, does not have the largest negative Pearson residual. While this project still has a large negative residual, its magnitude is at least partly due to the fact that λ_i is itself large for this observation. Since the Pearson residuals are the raw residuals divided by the square root of the variance, we can approximately interpret the magnitude of the residuals in terms of the more familiar z-scores. This approximation can be rough because the Poisson distribution is skewed, but as a first-pass approximation it may be tolerable. Under this approximation, the Pearson residuals between −2 and +2 would give less cause for concern but the ones in the vicinity of +3 and +4 would be more cause for concern. Overall, the impression from Figure 9.1 is that an appreciable percentage of the 50 observations are not being predicted very well by the current simple model. The curving solid line also indicates that the model is overpredicting the data for high values of λ and underpredicting the data for moderate values of λ.

Results for the β coefficients used in FF1 through FF3 are listed in Table 9.1. Coefficients with 95% credible intervals that exclude the value of zero appear in boldface. In FF1, only x_2 (expected cost) would be considered influential. Using the semi-elasticity interpretation of coefficients under the log link described in the previous section, we could conclude that expected number of months increases about 6.3% for every additional one million euro increase in project investment cost. In FF2 we remove the nonsignificant interaction term and add the total risk score (x_3). This appears to have negligible impact, so the risk does not have a direct relationship on the project length.

We add an interaction term between cost and risk in FF3. We now find that project length increases by about 24% for each additional million euros in cost. However, this conclusion is tempered by the interaction term, because the 95% credible interval of the interaction coefficient does exclude zero and is negative. We can interpret this coefficient as follows. Since cost has been centered around its mean, projects with above-average cost will have positive values and projects with

TABLE 9.1 β Coefficient Posterior Means and 95% Credible Intervals: IT Project Data

Coefficient	Functional Form 1	Functional Form 2	Functional Form 3
Discretionary/statutory (x_1)	−0.071	−0.134	−0.082
	(−0.623, 0.441)	(−0.380, 0.105)	(−0.332, 0.160)
Expected cost (x_2)	**0.063**	**0.071**	**0.249**
	(0.014, 0.109)	(0.018, 0.122)	(0.117, 0.378)
$x_1 * x_2$	0.092		
	(−0.675, 0.826)		
Total risk score (x_3)		−0.031	0.0026
		(−0.122, 0.057)	(−0.087, 0.089)
$x_2 * x_3$			**−0.053**
			(−0.089, −0.016)

below-average cost will have negative values. The negative interaction coefficient indicates that projects with below-average cost and high risk scores will tend to be more lengthy than we might predict (if we were to predict strictly on the basis of cost). The interaction term also indicates that high cost, high risk projects will tend to have shorter lengths than we might otherwise predict strictly on cost. A good example of this is Project 36. This project has a high mean-centered cost of 7.038. Based on cost, we would expect this project to have a higher length (the change in $\log(\lambda_i)$ is 0.2491 * 7.038 = 1.753). However, it also has a total risk score of 4.681. As a result, the interaction term's influence (a change in $\log(\lambda_i)$ of −0.05259 * 4.681 * 7.038 = −1.733) almost completely counteracts the direct cost influence on the predicted length. Low-risk projects are less affected by the impact of the interaction term.

In addition to the credible intervals, we can examine the models' *DIC* values. The *DIC* values for FF1 (*DIC*=318.5) and FF2 (*DIC*=318.1) indicate no substantial preference for one versus the other. FF3 (*DIC*=312.2) is preferred according to *DIC*. FF3 also has an improved picture in terms of Pearson residuals (Figure 9.2). We see that Project 36 at the far right has a Pearson residual nearer to zero than before (Pearson residual = −1.029). The solid line is approximately horizontal in Figure 9.2, which indicates that the model is no longer overpredicting for some values of λ and underpredicting for other values of λ. Still the existence of fairly extreme Pearson residuals suggests that the model has room for improvement. One possibility is that changes to the functional form will improve the fit sufficiently to make the Poisson model adequate. The other possibility, which we explore in Section 9.6, is that the Poisson distribution is not adequate to describe the data and that a different distribution should be used.

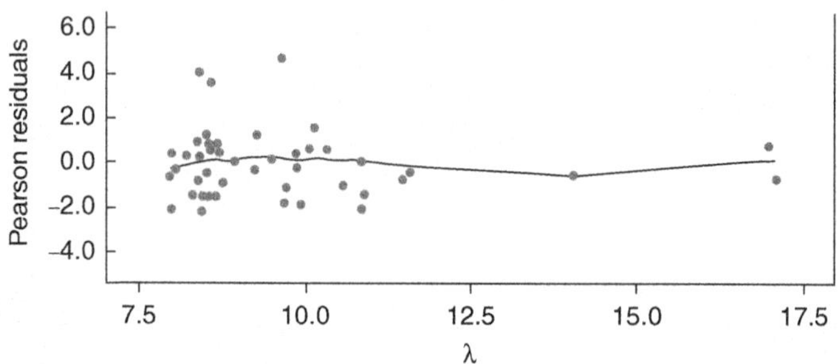

Figure 9.2 Pearson Residuals Versus λ for Functional Form 3: IT Projects Data

9.3 MODELS FOR BINARY DATA: LOGISTIC REGRESSION

We examined data with binary outcomes previously in Section 5.5. There we considered success/failure data, where y was the number of successes out of n, the total number of trials. We found that we could model this kind of data by estimating the proportion of successes, π, from y/n. However, our functional form in Section 5.5 was the very simple $\pi = \pi$. It would be useful to have a regression-like functional form so that many explanatory variables could be used to predict the proportion of successes.

One obstacle is that the proportion is limited to values between 0% and 100% whereas our regression functional form can take on any value in the set of real numbers. We can overcome this obstacle by using a link function that transforms values from 0–100% to the set of real numbers. One way to do this is to first find the odds ratio $\pi/(1-\pi)$, and then take the logarithm of the odds ratio. This operation is sometimes called the *logit* transformation. Odds ratios greater than 1 : 1 will have a logit (log odds ratio) that is positive, while those less than 1 : 1 will have a negative logit.

Now that we have addressed our obstacle, we can write out a functional form for π that can accommodate many explanatory variables. For priors, we can consider our broad normal distributions for coefficients as in other regression models. The model specification for logistic regression is then as follows:

$$\begin{aligned} y_i &\sim \text{Binomial}(\pi_i, n_i) \\ \log\left(\frac{\pi_i}{1-\pi_i}\right) &= \beta_0 + \beta_1 x_{1,i} + \cdots + \beta_k x_{k,i} \\ \beta_0, \ldots, \beta_k &\sim \text{Normal}(0, 0.00000001). \end{aligned} \tag{9.5}$$

We can calculate the Pearson residuals for logistic regression using the binomial mean $n\pi$ and the binomial variance $n\pi(1-\pi)$. It is also possible to specify this model with the Bernoulli likelihood in *WinBUGS* if all $n_i = 1$. In this case, *WinBUGS* will not need the values of n_i to be supplied as data.

If we have a value of βx, we can find the corresponding value of π by using the inverse link function. The inverse link function for the logit transformation is

$$\pi = \frac{\exp(\beta x)}{\exp(\beta x) + 1}. \tag{9.6}$$

This formula can be useful for understanding the outcome probabilities in different groups. If we have different groups with the corresponding dummy variables, we can calculate the probability of the outcome in each group. We can then compare the outcome probability of one group with that of another.

In Practice: Logistic Regression Analysis Mata and Freitas (2012) studied why firms choose to exit markets. Foreign firms may be more likely to exit a market because of the difficulties of adapting to the local market. Alternatively, foreign firms may exit because they are "footloose." This explanation suggests that foreign firms do not have deep ties to the local country and may feel freer to end their operations in one country when another more promising option arises. To investigate these possibilities, they examined data on firm exits that occurred in Portugal (Mata and Freitas, 2012, p. 623). We reexamine the data appearing in their work here.

The authors provide information on three groups of firms: domestic firms (nonmultinationals), foreign firms, and domestic multinationals. Domestic firms are the most numerous of the three groups in the Portuguese market. The authors list exit data by firm age, where age is given in brackets. For example, one firm age bracket is 0–4 years. One way to approach this data would be to treat it as a grouping variable. To get a clearer view of the effect of age, we have instead decided to convert the grouping variables to a continuous variable using an approximation. Firm age has been approximated here by setting all firms in a given age bracket to the bracket mean age. The mean ages for the six available brackets are 2 years (0–4 age bracket), 7.5 years (5–10 age bracket), 15.5 years (11–20 age bracket), 25.5 years (21–30 age bracket), 40.5 years (31–50 age bracket), and 75 years (51–99 age bracket). Mata and Freitas (2012) also provide the number of firms per bracket per group, n_i, as well as the exit rate in percentage form. Since we will need y_i, that is, the actual number of exits per category, it has been approximated here by multiplying n_i by the exit rate and rounding the result to the nearest integer.

As a first model, we examine the influence of firm age and firm type. Since firm group is a categorical variable while firm age is represented with a continuous variable, we will have a functional form just like the ANCOVA functional form in Section 5.8. We use domestic firms as the reference group. We create two dummy variables to allow each of the other two firm group types (`fgn` for foreign and `dm` for domestic multinational) to have its own particular exit rate. Interaction terms using the dummy variables and firm age allow us to estimate the impact of age on the exit rate for each firm type. The program listing appears below (see also **WinBUGS Code 9.3 Logistic Regression.odc**). The firm age (`x1c`) has been centered around its mean to improve MCMC performance.

```
model
 {
 for (i in 1:n) {
   y[i] ~ dbin(pi[i], n[i])
   logit(pi[i]) <- beta[1] + beta[2]*x1c[i]
      + beta[3]*fgn[i] + beta[4]*fgn[i]*x1c[i]
      + beta[5]*dm[i] + beta[6]*dm[i]*x1c[i]
   resid[i] <- y[i] - n[i]*pi[i]
   pearson.denom[i] <- sqrt(n[i]*pi[i]*(1-pi[i]))
   pearson.resid[i] <- resid[i]/pearson.denom[i]
   x1c[i] <- age[i] - mean.age
   }
 #prior for beta
 for (j in 1:6) { beta[j] ~ dnorm(0, 0.00000001) }
 #other calculated parameters
 mean.age <- mean(age[])
 }
```

Results from estimating Model 1 appear in Table 9.2. The intercept is –2.874. This is the logit transformation of the estimated exit percentage for the reference group (domestic firms) at the mean firm age. We can solve for the exit rate percentage using the inverse logit transformation in (9.6). The reference group has a constant value of $x = 1$, corresponding to the intercept. Substituting $\beta_1 \times 1 = -2.874$ into (9.6), we find the estimated exit percentage is 5.3% for domestic firms at the mean firm age of 27.66. As a note, we do not have the actual raw data on firm ages needed to precisely find the mean firm age. However, taking the mean of the bracket means gives 27.66 years and the intercept corresponds to the exit rate at this point.

Table 9.2 shows that the slope for firm age in domestic firms is –0.0167, and the 95% credible interval for this slope excludes zero. Interpreting the sign of the coefficient is straightforward. Negative coefficients indicate a declining trend in the outcome percentage (π) with increases in x, while positive coefficients indicate an increasing trend in π with increases in x. A more detailed interpretation of the slopes involves the odds ratio. Calculating $\exp(\beta)$ gives the odds ratio per unit change in x. Looking at the coefficient for age—domestic firms, we see $\exp(-0.0167) = 0.983$. Hence the odds of a domestic firm exiting Portugal declines with each passing

TABLE 9.2 β Coefficient Means and 95% Credible Intervals: Firm Exit Data, Binary Logit Model

Coefficient (WinBUGS Variable Name)	Functional Form 1	Functional Form 2
Intercept	**−2.874** (−2.898, −2.849)	**−2.763** (−2.786, −2.739)
Age—domestic firms (x1c)	**−0.0167** (−0.0179, −0.0154)	**−0.0397** (−0.0424, −0.0370)
Foreign (dummy) (fgn)	**0.632** (0.506, 0.753)	**0.521** (0.397, 0.642)
Age—foreign firms (fgn*x1c)	**0.0225** (0.0162, 0.0286)	**0.0455** (0.0388, 0.0521)
Multinational (dummy) (dm)	−0.132 (−0.650, 0.332)	−0.214 (−0.744, 0.260)
Age—multinational (dm*x1c)	**0.0342** (0.0118, 0.0549)	−0.00854 (−0.0914, 0.0723)
Age-squared—domestic firms (x1sqc*dom)		**0.00043** (0.00039, 0.00047)
Age-squared—multinationals (x1sqc*dm)		0.00082 (−0.00016, 0.00184)
DIC	586.5	242.1

year (odds ratio of exit is 0.983 : 1). Domestic firms in this sample appear to become better able to remain in the market as they grow older.

The foreign dummy variable is 0.632 with a 95% credible interval that excludes zero. As in our previous ANCOVA model, we can apply (9.6) to the sum of the reference-group dummy variable and the dummy variable of interest (the sum being found as $-2.874 \times 1 + 0.632 \times 1$) to get the foreign firm estimated exit percentage at the mean firm age. Substituting this sum into (9.6) gives an exit percentage of 9.6% for foreign firms at the mean age. We conclude that foreign firms have a higher exit percentage than domestic firms. The age—foreign firms slope is 0.0225. Since the credible interval excludes zero, we conclude that foreign firms are more likely to exit with increasing years than domestic firms. Adding the foreign firms' slope with that of the reference group gives a slope of $-0.0167 + 0.0225 = 0.0058$. The odds of exit for foreign firms are estimated to increase at a rate of exp(0.0058) (1.0058 : 1) per year. The dummy variable for multinationals is not credibly different from that of domestic firms. However, the slope for age in multinationals is credibly different from zero. Repeating the calculations (exponentiating $-0.0167 + 0.0342$) shows that the odds of a multinational exiting increases at a rate of 1.0177 : 1 per year of firm age.

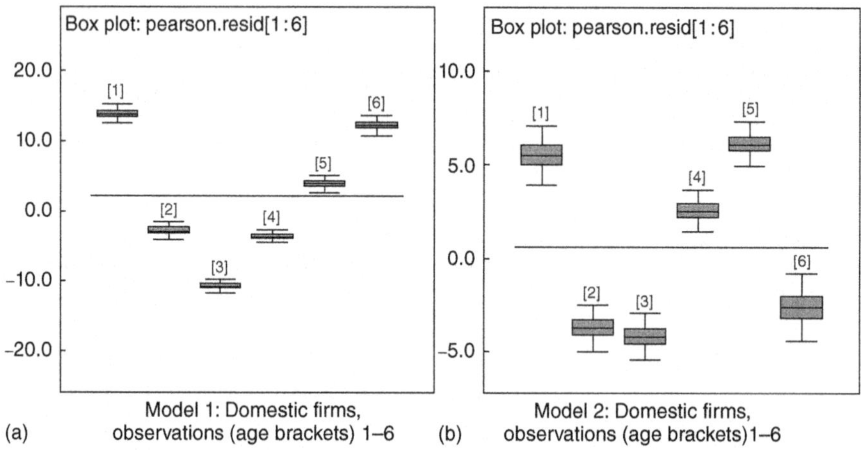

Figure 9.3 (a,b) Pearson Residual Distributions for Age Brackets 1–6: Firm Exit Data

Some concerns arise in judging the adequacy of this model. The Pearson residuals for domestic firms show that our current functional form is inadequate (Figure 9.3(a)). The values of most of the Pearson residuals are considerably outside the −2 to +2 range that we would expect. Part of the explanation for the extreme values of the Pearson residuals is the very large sample sizes in the domestic firms data, `n`. These values of `n` range from 6390 to 104,433. These large sample sizes make the `pearson.denom` terms become very small and so the Pearson residuals become large.

Because the data observations are sorted by firm age bracket from low to high, we can also use Figure 9.3 to understand the relationship between age and the Pearson residuals. The U-shaped pattern in the Pearson residuals suggests that there is a curvilinear (quadratic) relationship between the expected count $n_i\pi_i$ and age above and beyond the linear relationship captured by our *WinBUGS* variable `x1c`. Further examination of the Pearson residuals for domestic multinationals also shows a U-shaped pattern as seen in Figure 9.3, but in a less dramatic manner. We therefore examine fitting a quadratic term for both these groups in a follow-up Model 2. We create a variable where the value of `age` has been squared. The resulting variable after centering around the mean is called `x1sqc`. The Pearson residuals for foreign firms did not show a quadratic change (see **WinBUGS Code 9.3 Logistic Regression.odc**), so there would be little point in estimating a quadratic term for this group. The *WinBUGS* code for Model 2 appears in **WinBUGS Code 9.3 Logistic Regression.odc**.

The results for Model 2 appear in Table 9.2. The coefficient for the domestic firm quadratic variable (`x1sqc*dom`) has a 95% credible interval which excludes zero. The sign of this coefficient is positive.

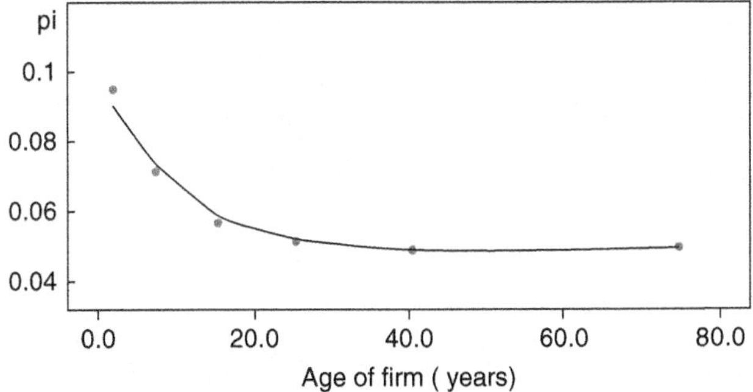

Figure 9.4 Scatter Plot of Age Versus π for Domestic Firms: Model 2 Firm Exit Data

Domestic firms also continue to have a negative linear coefficient for age (`x1c`), which has now become even more steeply negative. We can better visualize the relationships by instructing *WinBUGS* to create a scatter plot using the `Compare` tool. After running the Markov chain where `pi` has been monitored, we enter `pi[1:6]` into the `node` dialog box in the tool. We type `age[1:6]` in the `axis` dialog box and click `scatterplot` to produce the graphic. The output appears in Figure 9.4.

The dots indicate the posterior means of the estimated `pi` parameters. Although there are only six discrete age brackets, *WinBUGS* extrapolates the linear and quadratic trends by age (in years) to create the plot. We see that the probability of a firm's exit is estimated to decline steeply in earlier years. The exit probability reaches a minimum at the 40-year age bracket. It is then estimated to increase for firms that are in the highest age bracket.

9.4 THE PROBIT MODEL

Logistic regression is perhaps the most commonly encountered model for binary data. However, it is not uncommon to see a similar model called the *probit* model. The only difference between the two models is the link function (and, by necessity, the inverse link function). In the probit model, we sum up all the values of the functional form. This summed value can be thought of as a z-score from the normal distribution. The probability of a success, y/n, is estimated by the cumulative probability associated with the z-score. As an example, suppose our functional form is $\beta_0 + \beta_1 x_1$. Imagine that we find $\beta_0 = -1$, $\beta_1 = -0.1$, and $x_1 = 9.6$ for a certain data point. Then summing up the functional form gives $-1 + -0.1 \times 9.6 = -1.96$. You may recall that a z-score of -1.96 is the lower limit of the classical 95% confidence interval, and that there is 2.5% probability that a random observation from a standard normal distribution will be less than or equal to the value of $z = -1.96$. So, because our functional form sums

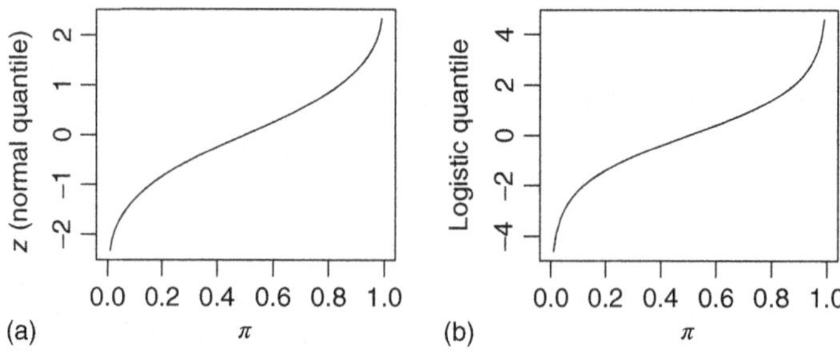

Figure 9.5 (a,b) Normal and Logistic Quantile Functions for $0.01 \leq \pi \leq 0.99$

up to −1.96, we would estimate a 2.5% probability of success for y/n data corresponding to this value of x.

To visualize matters, Figure 9.5(a) shows the quantile function for the normal distribution plotted as a function of the probability of success, π. As the probability gets closer to zero, the quantile function becomes more and more negative. We have limited our plotted values of π to the range 0.01–0.99 here because the quantile function goes to positive infinity when π is 1 and negative infinity when π is zero. The normal distribution quantile function is the link function in the probit model. We can compare the normal quantile function with the logistic quantile function in Figure 9.5(b). These look similar except that the logistic quantile function has a larger standard deviation. As a result, the range on the y-axis is greater for the logistic quantile function. The logistic distribution also has somewhat heavier tails than the normal distribution. Albert and Chib (1993) showed that the logistic distribution closely resembles the t-distribution having eight degrees of freedom.

The logistic quantile function is the link function from (9.5), namely $\log(\pi/(1-\pi))$. The normal quantile function, however, cannot be expressed by a simple formula. Symbolically, it is usually written as $\Phi(\pi)^{-1}$; so, for example, $\Phi(0.025)^{-1} = -1.96$. The coefficients of the probit model do not have an easy interpretation in terms of the logarithm of the odds ratio. However, a positive coefficient means that the probability of success is increasing with larger x for both the logit and probit models, while a negative coefficient means that the probability of success is decreasing with larger x for both the logit and probit models. Thus the signs of the coefficients should be the same for both the logit and the probit model. McCullagh and Nelder (1989) describe that the probit model was historically the first to be developed (Bliss, 1937), while the logit model was introduced later. Given the resemblance between the logit and the probit model visible in Figure 9.5, it is natural to wonder which one to use. For a single binary y outcome as we consider here, both models typically lead to similar conclusions unless sample sizes are large and the data has extreme values of π (Chambers and Cox,

1967). However, some evidence suggests that differences may be more noticeable when there are several outcome variables (Hahn and Soyer, 2005).

The model specification for the probit model is written as follows:

$$\begin{aligned} y_i &\sim \text{Binomial}(\pi_i, n_i) \\ \Phi(\pi_i)^{-1} &= \beta_0 + \beta_1 x_{1,i} + \cdots + \beta_k x_{k,i} \\ \beta_0, \ldots, \beta_k &\sim \text{Normal}(0, 0.00000001). \end{aligned} \tag{9.7}$$

The only change from the logit model is the new probit link function. We can again calculate the Pearson residuals using the binomial mean $n\pi$ and the binomial variance $n\pi(1-\pi)$.

We may sometimes want to find π for specific values of x. For example, we may wish to predict π for a new observation with its particular values of x. We can use the inverse link function for this task. The inverse link function is the cumulative distribution function for the normal distribution. The cumulative distribution function is usually written as $\Phi(\cdot)$. Inserting the summed-up value of the functional form into $\Phi(\cdot)$ gives the predicted probability of success π for that observation. The *R* function `pnorm()` can be used to obtain $\Phi(\cdot)$ as can the Microsoft Excel function `normsdist()`. This function can be difficult to calculate (as we will see in the next section), so using *R* is recommended over Microsoft Excel for serious work.

In Practice: Probit Analysis We reexamine the data of Mata and Freitas (2012) using probit analysis. We reestimate the simpler Model 1 from Table 9.2 for comparative purposes. In theory, we only need to change the link function in our previous *WinBUGS* code to estimate the probit model. However, *WinBUGS* can run into numerical problems fairly easily under the probit model. This is because calculating the normal quantile function or cumulative density function for extreme numbers can cause *WinBUGS* to crash. Consider the calculation of the cumulative distribution function for $z = -6$. The cumulative probability that $z \leq -6$ is a very small number, namely 0.000000001973175. While *WinBUGS* can handle this particular number, the Markov chain can propose values that are outside of *WinBUGS'* ability to handle. This seems to be even more common with vague priors.

Ntzoufras (2009, p. 265) found a computational way of avoiding this problem. He proposed truncating the values supplied to *WinBUGS'* probit functions so that they did not exceed some predefined constant. He provides an example using a constant of 5, so that the truncation is in the range $-5 \leq z \leq 5$. Values greater than this range are changed to +5, and values are smaller than this range are changed to −5. Additional testing shows that this constant can be raised to 7.5. We find that, at the value of 7.5, *WinBUGS* produces values of π that are slightly greater than zero (for $\Phi(-7.5)$) and slightly less than one (for $\Phi(7.5)$), as would be expected for cumulative probabilities. Also, *WinBUGS* produces results at this value which are equal to those computed by *R* to at least three significant digits. Raising the constant further to 8 leads to an agreement with *R* for $\Phi(-8)$ of only one significant digit. When the constant is raised to 9, *WinBUGS* will crash. Bearing this in mind, we add three extra lines of code plus a comment to implement Ntzoufras' truncation.

```
...
 for (i in 1:n) {
  y[i] ~ dbin(pi[i], n[i])
  mu[i]<- beta[1] + beta[2]*x1c[i]
   + beta[3]*fgn[i] + beta[4]*fgn[i]*x1c[i]
   + beta[5]*dm[i] + beta[6]*dm[i]*x1c[i]
  #Ntzoufras' truncation with wbc as WinBUGS constant
  ntz[i] <- mu[i]*(1-step(abs(mu[i])-wbc))
   - wbc*step(-wbc -mu[i]) + wbc*step(mu[i] -wbc)
  probit(pi[i]) <- ntz[i]

  resid[i] <- y[i] - n[i]*pi[i]
...
```

We discuss briefly how the Ntzoufras truncation works in *WinBUGS*. We first encountered *WinBUGS'* `step` function in Section 7.2.1. The `step` function will return a value of 1 if its argument is greater than or equal to zero, and will return zero otherwise. The first product term in `ntz[i]` allows the value of `mu[i]` to be unchanged as long as its absolute value is less than or equal to the *WinBUGS* probit constant, `wbc`. Otherwise, this product is set to zero because of the `step` function. The second product returns `-wbc` if the value of `mu[i]` is less than `-wbc`, and returns zero otherwise. The third product returns `wbc` if the value of `mu[i]` is greater than `wbc`, and returns zero otherwise. Together, these product terms have the desired truncating effect.

We supply a value of `wbc` equal to 7.5 in the data listing for the program. Full code appears in **WinBUGS Code 9.4 Probit Analysis.odc**. Results from the probit model appear in Table 9.3. We find that the same coefficients that were credibly different from zero in the logit model are credibly different here. The values of the probit coefficients are all closer to zero than their counterparts in the logit model. This is attributable to the larger standard deviation of the logistic distribution versus the normal distribution as was seen in Figure 9.5.

9.5 IN DETAIL: MULTINOMIAL LOGISTIC REGRESSION FOR CATEGORICAL OUTCOMES

We have been using the binomial distribution to model binary outcome data. For outcome data that have more than two categories, we can use an

TABLE 9.3 β Coefficient Means and 95% Credible Intervals: Firm Exit Data, Probit Model

Predictor Name (WinBUGS Variable Name)	Functional Form 1
Intercept	**−1.604** (S−1.615, −1.594)
Age—domestic firms (`x1c`)	**−0.0075** (−0.0080, −0.0069)
Foreign (dummy) (`fgn`)	**0.302** (0.239, 0.364)
Age—foreign firms (`fgn*x1c`)	**0.0105** (0.0073, 0.0138)
Multinational (dummy) (`dm`)	−0.052 (−0.279, 0.162)
Age—multinational (`dm*x1c`)	**0.0155** (0.0055, 0.0251)

extension of the binomial distribution called the *multinomial distribution*. The logistic regression framework can then be applied to our outcome data that has a multinomial distribution. The resulting model is called *multinomial logistic regression* or *multinomial logit*.

Multinomial logit proceeds by blending our existing binary logit model with the dummy coding concept introduced in our ANCOVA model of Section 5.8.2. We will choose one of the categorical outcomes as the reference group, just as in dummy coding. The reference group is often called the *baseline category* in multinomial logit modeling (Agresti, 2002; Powers and Xie, 2000). The comparison categories are the categories other than the baseline category. Our model coefficients are then interpreted as the log odds that an observation appears in a particular comparison category versus the baseline category.

More formally, suppose we have J categories and that the last category (category J) is the baseline category. We will estimate $J - 1$ logistic regressions to understand the differences between the $J - 1$ comparison categories and the baseline category. The functional form for comparing category j to category J is

$$\log\left(\frac{\pi_j}{\pi_J}\right) = \beta_{0,j} + \beta_{1,j}x_{1,j} + \cdots + \beta_{k,j}x_{k,j}. \tag{9.8}$$

Another possibility is that we estimate J logistic regressions. If we do this, we will have an additional regression for comparing the baseline category to itself. This can be accomplished if we fix all of the β coefficients for the "baseline category versus itself" regression to zero. While this may seem to be an unusual approach, it has a benefit. In particular, this approach gives us more flexibility in assigning the baseline category to any of our outcomes. We are no longer tied to the requirement that category J is the baseline category. Instead, we set coefficients to zero for a particular category that we would like to be the baseline depending on what is relevant for a particular analysis. This approach makes a particular piece of *WinBUGS* code more easily adaptable for different analyses of different datasets, and will be illustrated in Section 9.5.1.

Suppose we would like to see how two comparison categories are different from one another. For example, we wish to see how Category 1 differs from Category 2. We can show using the basic rules of logarithms that

$$\log\left(\frac{\pi_1}{\pi_2}\right) = \log\left(\frac{\pi_1}{\pi_J}\right) - \log\left(\frac{\pi_2}{\pi_J}\right)$$

for the left-hand side of (9.8). This means that the needed linear predictor and β coefficients can also be found by the same type of subtraction on the right-hand side of (9.8). With MCMC, we can estimate these coefficients within the run and obtain the desired posterior distributions without having to respecify and reestimate the model. We may also wish to find the probabilities directly. Assuming that we have set all of the coefficients to zero for the baseline category, we calculate

$$\pi_j = \frac{\exp\left(\beta_{0,j} + \beta_{1,j}x_{1,j} + \cdots + \beta_{k,j}x_{k,j}\right)}{\sum_{j=1}^{J} \exp\left(\beta_{0,j} + \beta_{1,j}x_{1,j} + \cdots + \beta_{k,j}x_{k,j}\right)}. \tag{9.9}$$

This formula generalizes the binary logit inverse link function of (9.6) that we previously encountered.

For priors on coefficients, broad normal distributions can be used as in other related models. We will typically have many rows of data indexed by i from 1 to I, where I is the total number of rows of data. The i subscript will therefore need to be included along with our j subscript which differentiates the categories. We will use $\boldsymbol{y}_i$ to indicate that we are referring to an entire row of outcome category data. This can be contrasted with a particular outcome category for a particular row of data, $y_{i,j}$. As an example, suppose we conduct a survey on a particular date and the replies to a question are 17 "Yes" replies, 23 "No" replies, and 8 "Don't Know" replies. Then $\boldsymbol{y}_i = (17, 23, 8)$ while $y_{i,2} = 23$. We can also write $\boldsymbol{\pi}_i$ to represent the entire row of probabilities. The total number of outcomes in row i is n_i, that is, $n_i = \sum_{j=1}^{J} y_{i,j}$.

With this notation, the multinomial logit model can be written as

$$\begin{aligned}
\boldsymbol{y}_i &\sim \text{Multinomial}(\boldsymbol{\pi}_i, n_i) \\
\log\left(\frac{\pi_{i,j}}{1-\pi_{i,j}}\right) &= \beta_{0,j} + \beta_{1,j}x_{1,i,j} + \cdots + \beta_{k,j}x_{k,i,j}, \quad j = 1, \ldots, J-1 \\
\beta_{0,j}, \ldots, \beta_{k,j} &\sim \text{Normal}(0, 0.00000001), \quad j = 1, \ldots, J-1 \\
\beta_{0,J}, \ldots, \beta_{k,J} &= 0.
\end{aligned} \tag{9.10}$$

Here, category J has been used as the baseline category to make the model specification more tidy and to help emphasize the family resemblance with the binary logit model. However, this is not required as described above.

9.5.1 In Practice: Multinomial Logit for Contingency Tables

We first examine the multinomial logit model in the context of quarterly earnings announcements. Companies disclose earnings quarterly but sometimes these announcements are late. A delay in earnings announcements may signal that usual operating procedures are not proceeding as planned, and thus might be considered as a bad sign for earnings and hence for the stock price. Kross and Schroeder (1984) examined whether the timing of announcements was related to whether the resulting news was bad or good. The authors collected 12 quarters worth of data on 296 firms listed on the New York Stock Exchange (NYSE) and the American Stock Exchange. This produced 3552 observations which were categorized as in Table 9.4 (Kross and Schroeder, 1984, p. 166). A dataset such as that of Table 9.4 where outcomes fall into mutually exclusive categories is called a *contingency table.*

We analyze this data using the multinomial logit model where the outcomes are early, on time, or late. We wish to see whether the type of news predicts the timing of the announcement. We select Category 1 (early) as the baseline category to illustrate the setting of coefficients to zero. Our code and data for the model appears below (see also **WinBUGS Code 9.5.1 Multinomial Logistic Regression.odc**).

```
model
{
 for (i in 1:news) { # type of news
   y[i,1:timing] ~ dmulti(pi[i,1:timing], n[i])
   n[i] <- sum(y[i,])
   for (j in 1:timing) { # timing of report
     pi[i,j] <- exp.lp[i,j]/sum(exp.lp[i,])
     log(exp.lp[i,j]) <- alpha[j] + beta[i,j]
     }
  }
 #priors for alpha
  alpha[1] <- 0 # compare vs. early
```

TABLE 9.4 Timing of 3552 Earnings Announcements Versus Announcement News

	Timing		
Type of news	Early	On time	Late
Bad	577	508	510
Good	728	611	618

```
  for (j in 2:timing) {alpha[j] ~ dnorm(0, 0.000001)}
 #priors for beta
  for (j in 1:timing) {beta[1,j] <- 0 } #compare vs. early
  for (i in 2:news) {
    beta[i,1]<- 0 # compare vs. bad news
    for (j in 2:timing) {beta[i,j] ~ dnorm(0, 0.000001)}
    }
 #other calculated parameters
  for(j in 1:timing) {diffpi[j] <- pi[1,j] - pi[2,j]}
}
#DATA
list(news = 2, timing=3,
y = structure(.Data =
    c(577, 508, 510,
      728, 611, 618), .Dim = c(2,3))
```

We see from the code listing that our outcome variable y has a multinomial distribution (dmulti). We define n[i] for dmulti to follow the formula given in our previous discussion. We next estimate the values of pi for each row as required using the formulas in (9.9). In the second half of the code, we set baseline category coefficients to zero and give broad priors to other coefficients. We also use the ability of MCMC to find the posterior distributions of functions of parameters. Here, we might be interested comparing the chances of getting a bad early announcement versus a good early announcement. So we simulate from the difference of pi[1,1] (bad early announcement) and pi[2,1] (good early announcement). We also calculate these differences for on-time and late announcements, and store them in a variable called diffpi[j].

Results appear in Table 9.5. The table begins with the estimated probabilities of each of the outcomes, π. We can draw conclusions about the research question from the differences in π, which appear next. The posterior distributions of the differences in π show that the chances of good versus bad news do not seem to be related to the timing of an announcement. All of the differences in probabilities have credible intervals, which include zero.

Although the differences in probabilities are probably the easiest to interpret, we can also interpret the α and β coefficients if we are interested in the conclusions they provide. For example, α_2 is the log odds ratio of the probability of a bad announcement being on time versus a bad announcement being early. *WinBUGS* estimates this as −0.1271 which can be compared to calculation from the raw data of log(508/577) = −0.1274. The value of α_3 from *WinBUGS* is −0.1240, which can be compared to calculation from the raw data of log(510/577) = −0.1234. The α coefficients are less relevant to our research question since they refer to how prevalent the different timings are when the news is bad.

The β coefficients are the *changes* in log odds when moving from bad news to good news. For example, we just observed that α_2 is the log odds

TABLE 9.5 Multinomial Logit Parameter Means and 95% Credible Intervals: Earnings Announcements

Coefficient	Interpretation	Mean	95% Credible Interval
$\pi_{1,1}$	$\pi_{\text{Bad Early}}$	0.3618	(0.3378, 0.3857)
$\pi_{1,2}$	$\pi_{\text{Bad On-Time}}$	0.3185	(0.2963, 0.3413)
$\pi_{1,3}$	$\pi_{\text{Bad Late}}$	0.3197	(0.2971, 0.3430)
$\pi_{1,1}$	$\pi_{\text{Good Early}}$	0.3720	(0.3506, 0.3933)
$\pi_{2,2}$	$\pi_{\text{Good On-Time}}$	0.3121	(0.2921, 0.3327)
$\pi_{2,3}$	$\pi_{\text{Good Late}}$	0.3158	(0.2954, 0.3364)
$\pi_{1,1} - \pi_{2,1}$	$\pi_{\text{Bad Early}} - \pi_{\text{Good Early}}$	−0.0102	(−0.0428, 0.0219)
$\pi_{1,2} - \pi_{2,2}$	$\pi_{\text{Bad On-Time}} - \pi_{\text{Good On-Time}}$	0.00636	(−0.0236, 0.0360)
$\pi_{1,3} - \pi_{2,3}$	$\pi_{\text{Bad Late}} - \pi_{\text{Good Late}}$	0.00382	(−0.0270, 0.0350)
α_2	$\log(\pi_{\text{Bad On-Time}}/\pi_{\text{Bad Early}})$	−0.1276	(−0.2442, −0.00586)
α_3	$\log(\pi_{\text{Bad Late}}/\pi_{\text{Bad Early}})$	−0.1240	(−0.2434, −0.00309)
$\beta_{2,2}$	(see text)	−0.0480	(−0.2058, 0.1090)
$\beta_{2,3}$	(see text)	−0.0398	(−0.2048, 0.1220)

ratio of the probability of a bad announcement being on time versus a bad announcement being early. In changing from bad announcements to good announcements while still focusing on the log odds of on-time versus early, the log odds dip by −0.0480 as given by $\beta_{2,2}$. Overall, the log odds for good news being on-time versus early is $-0.1276 + -0.0480 = -0.1756$. This can be compared with $\log(611/728) = -0.1752$. In classical statistics, we would have to make do with interpreting the α and β coefficients. However, in a Bayesian approach to this particular model, we can choose to use the more intuitive differences in π provided by MCMC.

We have considered a contingency table with two dimensions here, Type of News and Timing. Contingency tables with more than two dimensions can also be examined with the multinomial logit model. The linear predictor will need to be extended for such tables. For example, we estimated α for the first dimension and β for the second dimension. Higher dimensions will require subsequent parameters. A contingency table with three dimensions has been examined in *WinBUGS* by Spiegelhalter et al. (1996c, pp. 51–54).

9.5.2 In Practice: Multinomial Logit with Continuous Predictors

We can also perform multinomial logit with continuous predictors. We examine data from Suzuki (1980), who described the diversification of the business models of leading Japanese industrial companies. Companies were classified into one of three categories based on the diversification of the company's product line, the industry in which the company operated, and year (we have omitted a fourth category of diversification because of low sample sizes). The data has been resummarized into the quantities

TABLE 9.6 Product Diversification of Leading Japanese Companies by Year

	Diversification Level		
Year	Less diversified (D)	Moderately diversified (R)	Highly diversified (N)
1950	14	13	3
1960	8	20	5
1970	1	25	7

appearing in Table 9.6. Less diversified companies were denoted with a (D) in Suzuki (1980, p. 279, Table 5), while moderately and highly diversified companies were denoted with an (R) and (N), respectively. The data suggest that Japanese firms moved increasingly to diversified strategies during this time period. We examine the data with a multinomial logit model to see whether the model corroborates our intuition.

We center the continuous variable, Year, around the midpoint of 1960 for our analysis. We select the least diversified category as the baseline category. We will estimate an intercept and a slope for each diversification level (column in Table 9.6) except for the baseline category. The code and data for this model appears below (see also **WinBUGS Code 9.5.2 Multinomial Logistic Regression.odc**.)

```
model
{
 for (j in 1:years) {
   y[j,1:level] ~ dmulti(pi[j,1:level], n[j])
   n[j] <- sum(y[j,])
   for (k in 1:level) { #diversification level
     pi[j,k] <- exp.lp[j,k]/sum(exp.lp[j,])
     log(exp.lp[j,k]) <- alpha[k]+beta[k]*yearctr[j] }
   }
 #priors for alpha
  alpha[1] <- 0 #compare vs. least diversified
  for (k in 2:level) { alpha[k]~dnorm(0, 0.00000001) }
 #priors for beta
  beta[1] <- 0 #No year slope for baseline
  for (j in 2:level) {beta[j] ~ dnorm(0, 0.00000001)
          expbeta[j] <- exp(beta[j]) }
}
list(years=3, level=3, yearctr=c(-10,0,10),
 y=structure(.Data = c(14,13, 3,8,20,5,1,25,7),
 .Dim = c(3,3)) )
```

Since the slopes will be estimated based on `yearctr`, the slopes estimate the annualized change in log odds of being in some other category versus the less diversified baseline category. We may also want to examine

TABLE 9.7 Multinomial Logit Parameter Means and 95% Credible Intervals: Firm Diversification

Coefficient	Mean	95% Credible interval
α_2	1.264	(0.673, 1.948)
α_3	−0.170	(−0.959, 0.623)
β_2	0.146	(0.0713, 0.229)
β_3	0.159	(0.0632, 0.261)
$\exp(\beta_2)$	1.158	(1.074, 1.257)
$\exp(\beta_3)$	1.173	(1.065, 1.298)

the change in the odds instead of the change in the log odds. We instruct *WinBUGS* to simulate from `expbeta[j]` so as to produce the distribution of the annualized change in the odds. A 5000-iteration burn-in and 50,000 iterations of MCMC produced the estimates appearing in Table 9.7. The 95% credible intervals for the β coefficients are greater than zero, which indicates that companies are moving to higher diversification levels over time.

We can also refer to the $\exp(\beta)$ distributions. The annualized odds of companies being in the moderately diversified category versus the less diversified category are estimated to be 1.158 : 1, while the annualized odds of companies being in the highly diversified category versus the less diversified category are estimated to be 1.173 : 1. The credible intervals for these parameters exclude the value of 1 as we would expect given the findings for the non-exponentiated parameters. This allows us to conclude that the odds are greater than 1 : 1 for these parameters. It is probably easier to discuss the odds than the log odds in most contexts just as it was easier to discuss the differences in π in the previous example. So, again, we see that MCMC can help us communicate findings by allowing us to draw conclusions about quantities that are easier to discuss.

9.6 HIERARCHICAL MODELS FOR COUNT DATA

In Chapter 8, we saw that adding a hierarchical component to regression models based on the normal distribution can provide several benefits. Hierarchical models can be used to provide more appropriate models for clustered data, and the introduction of an additional set of random terms can make these regression models more flexible in fitting the data. We now take a look at extending these useful properties to generalized linear models. This class of models is often called *hierarchical generalized linear models* or *generalized linear mixed models*, where the word *mixed* refers to the inclusion of both random and fixed coefficients. We will consider hierarchical models for count data first, and then consider hierarchical models for binary data.

9.6.1 The Negative Binomial Regression Model

The Poisson regression model of Section 9.2 is useful for predicting a count based on our knowledge of other variables. Yet, a limitation of the Poisson distribution is that the mean and the variance are assumed to be equal. This assumption may or may not be realistic. A distribution called the *negative binomial distribution* expands on the Poisson distribution and relaxes this assumption. The negative binomial distribution allows the variance of count data to be greater than or equal to the mean.

Regression modeling using the negative binomial distribution has become increasingly common for business and management data given the limitations of the Poisson distribution. For example, it has been used for examining knowledge-building as measured by patent citations (Rosenkopf and Nerkar, 2001), (Almeida et al., 2002), global diffusion of ISO 9000 quality practices (Guler et al., 2002), Internet page views (Danaher, 2007), and expansion strategies of multinationals (Jiménez, 2010).

One succinct way to describe the negative binomial distribution is in terms of a hierarchical model (Cameron and Trivedi, 1998, pp. 100–101). Our data y that has a Poisson distribution with constant mean λ can receive a new source of random variation as follows: Suppose y now has a mean λv where v is a random multiplier term. We give v a gamma distribution where both gamma parameters are set to the same (positive) value a. This will cause our multiplier to be equal to 1 on average because the mean of our gamma distribution is $a/a = 1$. However, sometimes v will be multiplying λ by a number smaller than 1, and sometimes by a number larger than 1.

The resulting data y will no longer have a Poisson distribution because the random multiplier v will be spreading the data out to a greater extent than before. Instead, y will have a negative binomial distribution. It turns out that the variance of y will be $\lambda + \lambda^2/a$ when it has the negative binomial distribution. We examine the relationship between the Poisson and the negative binomial distribution from an MCMC perspective in the next section.

9.6.2 In Practice: Simulating from the Negative Binomial Distribution

We ask *WinBUGS* to simulate 1000 Poisson observations with constant mean $\lambda = 5$ using the code below (see also **WinBUGS Code 9.6 Negative Binomial Hierarchical.odc**).

```
model
 {
```

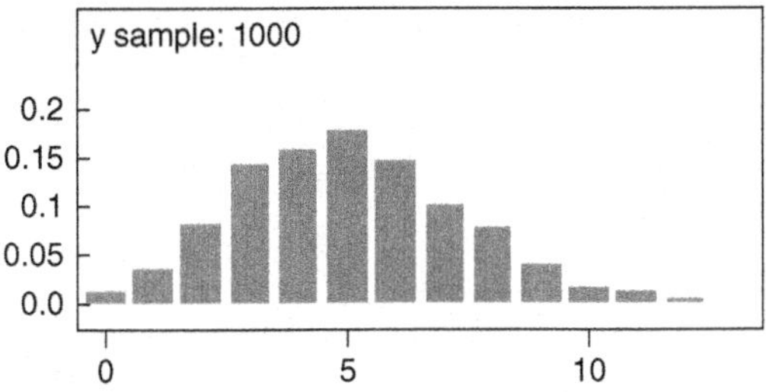

Figure 9.6 One-Thousand *WinBUGS* Simulations from Poisson Distribution: $\lambda = 5$

```
  y ~ dpois(lambda)
 }
list(lambda=5)
#inits
list(y=5)
```

Figure 9.6 displays a histogram of 1000 simulations from our *WinBUGS* code. We can estimate the mean and variance of our Poisson simulations in y in the usual manner. We monitor y in *WinBUGS* for 50,000 iterations (for accuracy), and we find the mean of y is 4.989, just a nudge under the true theoretical Poisson mean of 5 when $\lambda = 5$. The standard deviation is 2.246, so the variance is estimated as 5.044. The variance of the simulations is also very close to its theoretical value of 5.

Next we let the data have a varying mean `product`. We keep `lambda` at 5 but simulate from nu which has a gamma distribution where $a = 1$. We ask *WinBUGS* to simulate 1000 observations with the code below. We can see that the data y will be the result of a hierarchical structure where the first-level source of variation is from the Poisson distribution, and the second-level source of variation is from the gamma distribution.

```
model
 {
  y ~ dpois(product)
  product <- lambda*nu
  nu ~ dgamma(a,a)
 }
list(lambda=5,a=1)
#inits
list(y=5,nu=1)
```

Figure 9.7(a) shows a histogram of 1000 simulations of y. This histogram looks considerably different from that of Figure 9.6. It also clearly has a larger variance. We can see the distribution of nu on Figure 9.7(b). This distribution gives some intuition into the way in which the histogram for y has changed. We see that there are many simulated values of nu that are below 1. So multiplying our mean by a number below 1

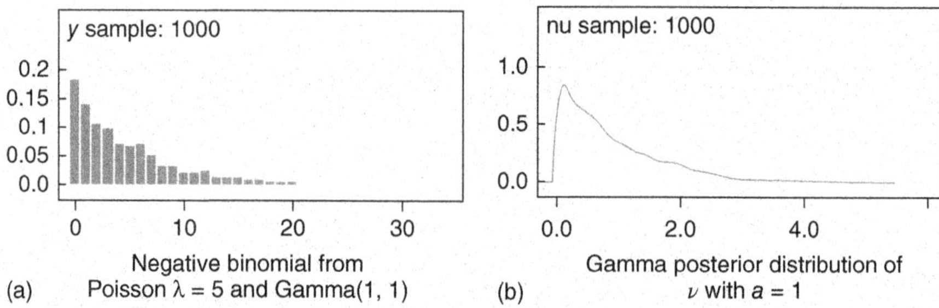

Figure 9.7 (a,b) One Thousand *WinBUGS* Simulations from the Poisson–Gamma Hierarchical Model (Negative Binomial Distribution with$\lambda = 5$)

will tend to produce y values closer to zero. Sometimes, we will also get values of nu that are greater than one. These will inflate the mean and cause y to tend to be larger than if the mean were 5. Both these influences cause the histogram of y to have a different shape by expanding the observations away from the mean value.

We can again check our simulation results against the values we would expect from our theoretical results for the negative binomial distribution. Monitoring y in *WinBUGS* for 50,000 iterations gives a mean for y of 5.007, in line with theory. The standard deviation of y is 5.51, so the variance is estimated as 30.36. The negative binomial variance formula indicates that our variance would be 30, so this is also reasonably close to what we would expect.

Since the variance of y when it has a negative binomial distribution is $\lambda + \lambda^2/a$, we can see that, as a gets larger and larger, the variance will get closer and closer to the mean λ. So, for interpretation if we have large values of a, we will have results similar to those under the Poisson distribution. If we have smaller values of a, then our results will be more overdispersed than the Poisson distribution.

From our discussion, we can now write our negative binomial regression model with a few modifications of our Poisson regression model. The only thing to consider is our new parameter a. Since we will not know the value of a in the data, we will have to estimate it. This means we will have to assign it a (hyper-)prior. We must ensure that a is positive, or else we will not have a valid gamma distribution. The prior for a requires some thought just as we have encountered for the hierarchical standard deviation σ_b. One possibility that we use here is a uniform prior in the range 0.001–1000. We could also put a log-normal prior on a as in Section 8.3, or else a log-uniform prior on a. The log-normal and log-uniform priors will place much weight on small values of a because half of their weight will be favoring values of $a < 1$. As always, sensitivity analysis for parameters is recommended (Section 9.9).

Putting our likelihood, functional form, and priors together leads to the following model for negative binomial regression:

$$\begin{aligned} y_i &\sim \text{Poisson}(\lambda_i v_i) \\ \log\left(\lambda_i\right) &= \beta_0 + \beta_1 x_{1,i} + \cdots + \beta_k x_{k,i} \\ \beta_0, \ldots, \beta_k &\sim \text{Normal}(0, 0.00000001) \\ v_i &\sim \text{Gamma}(a, a) \\ a &\sim \text{Uniform}(0.001, 1000). \end{aligned} \tag{9.11}$$

Now that the model has been specified, we can turn toward using it to analyze real-world data.

9.6.3 In Practice: Negative Binomial Regression

Our *WinBUGS* code for the negative binomial model makes a few modifications to the code for the Poisson model. We must add our v_i terms and give them a gamma prior. The a parameter of the gamma prior also needs its own (hyper-)prior. The program listing appears below (see also **WinBUGS Code 9.6 Negative Binomial Regression.odc**).

```
model
 {
for (i in 1:n) {
   y[i] ~ dpois(product[i])
   product[i] <- lambda[i]*nu[i]
   log(lambda[i]) <- beta[1] + beta[2]*x1[i]
          + beta[3]*(x2c[i]) + beta[4]*x3[i]
          + beta[5]*x2c[i]*x3[i]
   nu[i] ~ dgamma(a,a) # random multiplier
   var[i] <- lambda[i] + pow(lambda[i],2)/nu[i]
   resid[i] <- y[i] - lambda[i]
   pearson.resid[i] <- (y[i] - lambda[i] )/sqrt(var[i])
   x2c[i] <- (x2[i] - x2.bar)/1000 #divide by 1000 to
       improve numerical issues
   }
#prior for beta & a
   for (j in 1:5) { beta[j] ~ dnorm(0, 0.000001) }
   a ~ dunif(.001, 1000)
#other calculated parameters
   x2.bar <- mean(x2[])
}
```

We apply this code to the IT project risk management data of Benaroch et al. (2006). The Pearson residuals for the negative binomial regression model show improvement compared to the Poisson regression model (Figure 9.8). The Pearson residuals all fall in the range −2 to +2, which

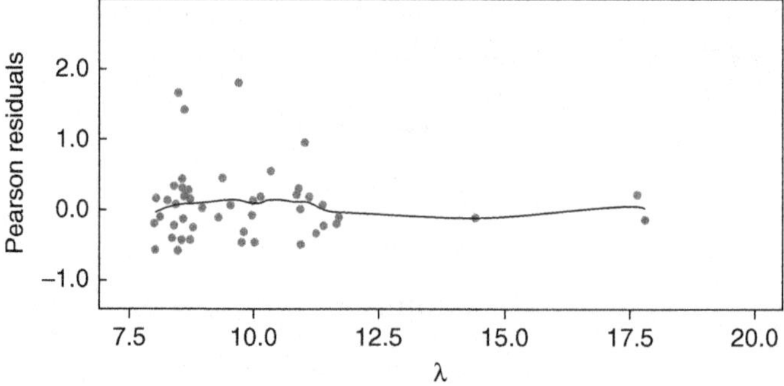

Figure 9.8 Pearson Residuals Versus λ: Negative Binomial Functional Form 3 with IT Projects Data

is the approximate range of what we would expect for a model that has appropriate fit. There are still several projects that could be explained better which cluster toward lower values of λ. Since these projects are underpredicted, a functional form that predicted these better would move these projects to higher values of λ where projects are currently sparse. However, overall the negative binomial version would be preferred over the Poisson version. We can see that the Pearson residuals appear to be skewed because the smallest ones are near −0.5 on the y-axis while the largest ones are near +2 on the y-axis. This is a common occurrence in count data models and is attributable in part to the skewness of non-normal distributions such as the Poisson (Figure 9.6).

Parameter estimates and 95% credible intervals appear in Table 9.8. These results can be compared with the Poisson results for FF3 in Table 9.1. The expected cost remains a relevant predictor but the $x_2 * x_3$ interaction term no longer has a 95% credible interval that excludes zero.

TABLE 9.8 β Coefficient Means and 95% Credible Intervals: Negative Binomial Model with IT Projects Data

Predictor Variable	Functional Form 3
Discretionary/statutory (x_1)	−0.082 (−0.427, 0.275)
Expected cost (x_2)	**0.255** (0.057, 0.478)
Total risk score (x_3)	0.0039 (−0.132, 0.145)
$x_2 * x_3$	−0.053 (−0.113, 0.002)

FF3 has appreciable autocorrelation in the Markov chain, so longer run times or reparameterizations are worth considering (Section 9.9).

9.7 HIERARCHICAL MODELS FOR BINARY DATA

We can also add random-effects terms to a logit regression model to produce a hierarchical logit model. A common approach for the logit model is to include one random additive term per observation. In doing so, we produce a random intercepts model as in Section 8.1. A random intercepts model with one random intercept per observation is often applied to data that has overdispersion, as we have seen in the previous section. The extra variability from the random intercepts allows the model to better account for the overdispersion.

A conventional choice is to assume the random intercepts b_i follow a normal distribution as in our hierarchical normal regression model (8.1). The mean of the distribution of the intercepts can be assumed to be zero, and the standard deviation can be written as σ_b. We can place a broad, uniform prior on σ_b to reflect our uncertainty about it. Merging these ideas with our logit model specification in (9.5), we arrive at the following:

$$\begin{aligned}
y_i &\sim \text{Binomial}(\pi_i, n_i) \\
\log\left(\frac{\pi_i}{1-\pi_i}\right) &= \beta_0 + \beta_1 x_{1,i} + \cdots + \beta_k x_{k,i} + b_i \\
\beta_0, \ldots, \beta_k &\sim \text{Normal}(0, 0.00000001) \\
b_i &\sim \text{Normal}(0, \tau_b) \\
\sigma_b &\sim \text{Uniform}(0, U).
\end{aligned} \tag{9.12}$$

While this model may run perfectly well, in some cases it will not. Convergence problems may arise in part because the binomial variance is more constrained than the normal variance. Normal variances can take on any positive value while the binomial variance is a strict function of π and n. As a result, a logit model with the linear predictor at hierarchical level 1 and random intercepts at hierarchical level 2 can experience difficulty.

Fortunately, convergence in hierarchical logit models may be able to be improved by using hierarchical centering (Roberts and Sahu, 2001). Doing so can prevent the different hierarchical levels from competing over the same variance. Previously, in (8.3), we saw that we can use hierarchical centering to place the mean of the intercepts at level 2. There we gave our random intercepts α_i their own mean μ_α and took away the first-level intercept for the model. This corresponded to pushing the first-level intercept up to the second level. For the logit model, we can go farther and "eliminate" the first level of the hierarchy by placing the entire linear predictor at the second level. In other words, we push up the entire linear predictor to the second level. We do this by defining a new set of random intercepts, a_i. The mean for a_i is then the linear predictor

$$\mu_{a_i} = \beta_0 + \beta_1 x_{1,i} + \cdots + \beta_k x_{k,i}.$$

The a_i terms are given a normal distribution with mean μ_{a_i} and precision τ_a. We have introduced a *conditional* hierarchical mean for the a_i such that the mean of a_i is conditional on the values of β and x. This strategy can also be applied to probit models as well as normal hierarchical regression models if desired.

The changes described lead to the following model:

$$\begin{aligned} y_i &\sim \text{Binomial}(\pi_i, n_i) \\ \log\left(\frac{\pi_i}{1-\pi_i}\right) &= a_i \\ a_i &\sim \text{Normal}(\mu_{a_i}, \tau_a) \\ \mu_{a_i} &= \beta_0 + \beta_1 x_{1,i} + \cdots + \beta_k x_{k,i} \\ \beta_0, \ldots, \beta_k &\sim \text{Normal}(0, 0.00000001) \\ \sigma_a &\sim \text{Uniform}(0, U). \end{aligned} \tag{9.13}$$

As in (8.1), we should select the upper bound U of σ_a to be large enough to accommodate the posterior distribution of σ_a (this also applies to σ_b depending on which model we are using). However, since we have proportion data in the logit model, U may not need to be quite as large as it may need to be in the normal regression model. We also will need to calculate $\tau_a = 1/\sigma_a^2$ as we did in (8.1) for the distribution of the a_i terms (again similar considerations apply to τ_b).

9.7.1 In Practice: Logistic Regression with Random Intercepts

We return to the firm exit data of Mata and Freitas (2012) and apply a hierarchical logit model to it. The firm exit data showed evidence of overdispersion with respect to our previous models. The extra variability of the random intercepts should hopefully address this issue. Portions of the *WinBUGS* code appear below, with modifications to previous code appearing in bold. The entire program listing can be found in **WinBUGS Code 9.7 Hierarchical Logit.odc**. We use the model of (9.13) for our estimates.

```
model
  {
```

```
 for (i in 1:n) {
   y[i] ~ dbin(pi[i], n[i])
   logit(pi[i]) <- a[i]
   a[i] ~ dnorm(mu[i], tau.a)
   mu[i] <- beta[1] + beta[2]*x1c[i]
     + beta[3]*fgn[i] + beta[4]*fgn[i]*x1c[i]
     + beta[5]*dm[i] + beta[6]*dm[i]*x1c[i]
     + beta[7]*x1sqc[i]*dom[i] + beta[8]*x1sqc[i]*dm[i]
   x1c[i] <- age[i] - mean.age
   ...
   }
#priors for beta and sigma.a
  for (j in 1:8) { beta[j] ~ dnorm(0, 0.00000001) }
  sigma.a ~ dunif(0,10)
#other calculated parameters
  tau.a <- 1/pow(sigma.a,2)
   ....
 }
```

The results in Table 9.9 can be compared with the results for Functional Form 2 in Table 9.2. Both models have the same functional form, and the difference is in whether or not random intercepts have been included in the model. The two sets of results lead to largely the same conclusions.

TABLE 9.9 β Coefficient Means and 95% Credible Intervals: Firm Exit Data, Hierarchical Binary Logit Model

Predictor Name (`WinBUGS Variable Name`)	Parameter Estimates
Intercept	**−2.751**
	(−2.877, −2.625)
Age—domestic firms	**−0.0315**
(`x1c`)	(−0.0506, −0.0119)
Foreign (dummy)	**0.514**
(`fgn`)	(0.300, 0.729)
Age—foreign firms	**0.0373**
(`fgn*x1c`)	(0.0163, 0.0580)
Multinational (dummy)	−0.227
(`dm`)	(−0.821, 0.285)
Age—multinational	−0.0171
(`dm*x1c`)	(−0.1051, 0.0687)
Age-squared—domestic firms	**0.00031**
(`x1sqc*dom`)	(0.00006, 0.00054)
Age-squared—multinationals	0.00083
(`x1sqc*dm`)	(−0.00018, 0.00187)
σ_a	0.138
	(0.063, 0.297)
DIC	125.7

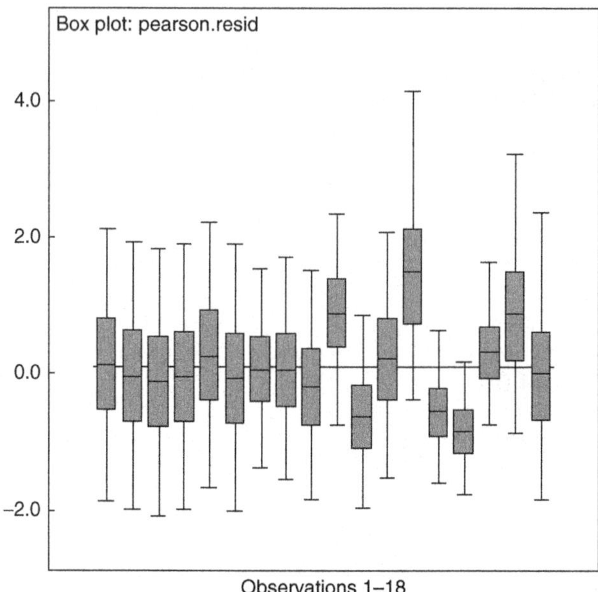

Figure 9.9 Pearson Residuals: Firm Exit Data, Hierarchical Logit Model

However, the hierarchical model coefficient values in Table 9.9 tend to be slightly smaller and have wider credible intervals. Since the Pearson residuals suggest that the current model has better fit (see the left side of Figure 9.9), it suggests that the results of Table 9.2 may have been overly sharp because we had not been modeling the extra dispersion at that time. The improvement in *DIC* suggests that this model should have the best short-term predictive performance of the models we have considered. One change from the previous results is that the coefficient for age—multinational, β_6, is no longer credibly different from zero. We also provide the estimated value of σ_a and its credible interval. The estimate for σ_a does not appear in bold since it cannot be negative.

The Pearson residuals appearing in Figure 9.9 represent a large improvement over those seen previously. All posterior means are within the ± 2 range, whereas in Figure 9.3 the residuals were well out of this range. The hierarchical model has been reasonably successful in modeling the overdispersion in the data.

9.8 SUMMARY

We have introduced generalized linear models in this chapter. Generalized linear models expand upon normal regression models by allowing

y to have some distribution other than the normal distribution. We have examined the use of the binomial, Poisson, multinomial, and negative binomial distributions for y. These distributions are the most commonly encountered in practice, but we need not limit ourselves to these distributions. *WinBUGS* supports other distributions that can be used for such models. We have also introduced the topic of hierarchical generalized linear models. Such models can add flexibility to a generalized linear model and are sophisticated tools for business insight.

We have seen that interpreting the classical parameters for generalized linear models is not always a straightforward affair. Fortunately, MCMC allows us to define parameters based on functions of other parameters that are tailored to our research interests, and obtain the posterior distributions of these parameters. Thus, MCMC helps us to communicate results that are more relevant to our business concerns.

9.9 EXERCISES

1. Reexamine the *WinBUGS* code for the probit model for the firm exit data (**WinBUGS Code 9.4 Probit Analysis.odc**). Set `wdc` equal to the value of 5 as proposed by Ntzoufras (2009). How, if at all, do the parameter estimates change in this particular dataset?

2. Find DIC for the model in **WinBUGS Code 9.5.1 Multinomial Logistic Regression.odc**. Next, estimate the model in **WinBUGS Code 9.5.1 Multinomial Logistic Regression.odc** without the β parameters and find the value of DIC for this model. Compare the two values of DIC. Which model do you prefer based on DIC? How, if at all, does your preference relate to the posterior distributions of β that were estimated initially?

3. Reexamine the *WinBUGS* code for the negative binomial model for the IT Projects data (**WinBUGS Code 9.6 Negative Binomial Regression.odc**) and create an informative prior for a. Give a a uniform prior between 500 and 1000. This should make the negative binomial model very similar to the Poisson model. Does it? Estimate this model and examine your results for β and a. How, if at all, have they changed? Instruct *WinBUGS* to produce the posterior density for a and see what it looks like. Comment on this distribution.

4. Reexamine the *WinBUGS* code for the negative binomial model for the IT Projects data (**WinBUGS Code 9.6 Negative Binomial Regression.odc**) and perform sensitivity analysis for a. Define the parameter `log.a` in the code and assign a normal prior to it. Next remove the

uniform prior for a and instead define it as `a <- exp(log.a)`. Estimate this model and examine your results for β and a. How, if at all, have they changed? Plot the Pearson residuals against λ and comment on your plot.

5. As discussed at the end of Section 9.6.3, attempt to reduce MCMC autocorrelation in **WinBUGS Code 9.6 Negative Binomial Regression.odc** by mean-centering `x3`. Does this improve the appearance of the autocorrelation function plots? Do the conclusions as measured by the 95% credible intervals change?

6. Reread Section 3.6.3 where we describe how the t-distribution can be formed from the normal distribution and a gamma distribution on the precision τ. In particular, examine the *R* code in the middle of Section 3.6.3 where `mg1.mu` was simulated. In your own words, compare what is happening in this code versus the negative binomial *WinBUGS* code in Section 9.6.2. What is similar and what is different?

7. Perform sensitivity analysis for σ_a in **WinBUGS Code 9.7 Hierarchical Logit.odc** by using the Root-Uniform(1.1, 10) distribution. How, if at all, are inferences affected?

8. Estimate the random effects model of (9.12) by making the appropriate changes to **WinBUGS Code 9.7 Hierarchical Logit.odc**. Examine the autocorrelation functions and comment on the convergence properties of the Markov chain. How, if at all, are inferences affected?

9. Re-examine the data of Section 5.8.1 using a negative binomial model with identity link function. Compare your results to those of Section 5.8.1.

9.10 NOTATION INTRODUCED IN THIS CHAPTER

Notation	Meaning	Example	Section Where Introduced
$\Phi(\cdot)^{-1}$	The normal quantile function (takes a cumulative probability and produces a z-score that corresponds to the cumulative probability)	$\Phi(0.025)^{-1} = -1.96$	9.4

(continued)

$\Phi(\cdot)$	The normal cumulative distribution function (takes a z-score and produces the cumulative probability of the normal distribution that is less than or equal to the z-score)	$\Phi(-1.96) = 0.025$	9.4
y_i	An entire row of outcome category data	$y_i = (17, 23, 8)$	9.5
v	Random multiplier term in the negative binomial model	λv	9.6

10

MODELS FOR DIFFICULT DATA

We have introduced three major families of models so far: the linear regression model for normal data, the generalized linear model for certain kinds of non-normal data, and the hierarchical model for data that has a hierarchical structure. These three families of models cover many situations that we encounter in business. Still we regularly come across data that do not conform to the assumptions that underlie the models we have discussed so far. This chapter investigates how to relax some of the assumptions of regression models. We examine extensions to regression models that handle common challenges such as the existence of outliers, the presence of heteroscedasticity, and the occurrence of certain kinds of missing data. We will focus primarily on the linear regression model but the extensions we discuss can often be more broadly applied.

10.1 LIVING WITH OUTLIERS—ROBUST REGRESSION MODELS

Outliers appear with some regularity in business data. Some firms or individuals may be truly exceptional while others may dramatically underperform. Statistical rules of thumb often recommend eliminating these

Bayesian Methods for Management and Business: Pragmatic Solutions for Real Problems, First Edition. Eugene D. Hahn.

cases according to some guideline, such as whether or not an observation departs from the mean by a certain number of standard deviations. Yet, these rules of thumb can be associated with their own problems. Venables and Ripley (2002, p. 120) list several reasons why routine deletion of outliers can be problematic. First, a binary yes/no decision to retain or delete an observation can be wasteful of the data. Barring a typographic error, an outlier contains at least some information about the business situation that would be lost if discarded. A more useful practice would be to down-weight these observations so that they contribute to our understanding, but to a lesser extent than more typical data. Second, an observation may be an outlier in a multivariate sense but not in a univariate sense. In this situation, an observation may appear to be non-outlying on any one variable; however, the observation may be outlying when multiple variables are considered. It can be difficult to detect such multivariate outliers especially as data complexity grows. As a result, only the easily detectable univariate outliers will be discarded, leading to any claims about the data now being outlier-free being incomplete. Finally, deleting outliers has an impact on the distribution theory for the outcome data. This has an impact on the likelihood function we select and may also artificially reduce the variance we observe.

Other complications can also arise. The choice of the cutoff for declaring an observation to be outlying is to an extent subjective and arbitrary. Different choices can lead an investigator to different conclusions arising for the same set of data, making it less clear about what should be reported. Alternatively, two teams of investigators using the same business data source may disagree about the treatment of outliers (McNamara et al., 2005; Hawawini et al., 2003). There is also the risk of assuming the data to be normally distributed when in fact it is not. For example, the inadequacy of the normal distribution for certain business data such as financial returns has been noted since Fama (1965) and Mandelbrot (1963). It is perhaps safe to say that normal models are more common than normal data in many business contexts. This may be due to either reduced awareness of alternatives or perceived difficulty in putting such alternatives to use. Yet, the situation for models is beginning to change, and fortunately we can use heavy-tailed models in *WinBUGS* with comparative ease after making a few small changes to our code. These so-called robust methods are less sensitive to outliers and so conclusions are less affected by their presence.

There are a number of ways of implementing robust methods. Here we investigate the use of the t-distribution as a replacement for the normal distribution. This approach directly builds on our currently existing concepts and *WinBUGS* code. As an alternative, one developing area for robust methods is Bayesian nonparametric statistics. This approach allows considerable freedom but with a higher technical cost (Dey et al., 1998; Chamberlain and Imbens, 2003).

10.1.1 Another Look at the t-Distribution

The t-distribution is almost exactly the same as the normal distribution when the degrees of freedom parameter ν is very large. However, as ν becomes smaller, the tails of the t-distribution become heavier. We can visualize the differences between the t-distribution and the normal distribution by plotting them together as in Figure 10.1. There we see the density of the normal distribution as well as that of the t-distribution with four degrees of freedom. The variance of the normal distribution has been matched to that of the t-distribution in the figure (the variances of both distributions are equal to 2).

The peak of the t density is higher than that of the normal distribution because more density is allocated to the peak and the tails. The normal distribution, by contrast, has more density in the "shoulders" of the distribution. The heaviness of the t-distribution's tails may seem underwhelming in Figure 10.1, but this is due to the magnitude of the small numbers involved. We can calculate the height of the density at different values to compare the heaviness of the tails. At the value $x = -4$ for example, $t(4)$ has a height that is 1.3 times that of the normal distribution. At the value $x = -6$, $t(4)$ has a height that is over 34 times that of the normal distribution.

The variance of the t-distribution is influenced by ν. In particular, it is equal to $\nu/(\nu - 2)$. If ν is allowed to go to 2 or below, then we will no longer be able to obtain a valid variance for the t-distribution. If ν is allowed to go to 4 or below, the kurtosis (a measure of the heaviness of the tails) will no longer be valid. Hence in practical applications we should ensure that ν is either greater than 4 if we want a valid kurtosis, or greater than 2 if we want a valid variance. For a regression model, we can examine replacing

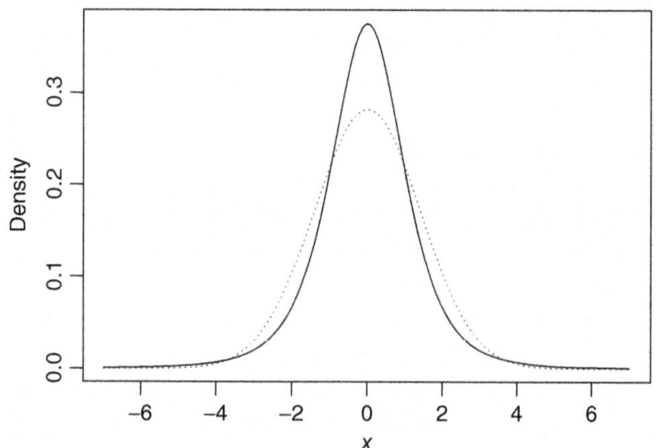

Solid: t with four degrees of freedom; dashed: normal density

Figure 10.1 Normal and $t(4)$ Densities

the normal likelihood function with the $t(4)$ likelihood. If a data source has outliers, we may find that the $t(4)$ likelihood gives a regression line that is less affected by the outliers. We explore this possibility in the next section.

10.1.2 In Practice: Robust Regression with the t-Distribution

The world economy is heavily dependent on the smooth flow of global supply chains for goods. China has become an integral part of many of these global supply chains. Large volumes of raw materials enter China and then later emerge as finished or semifinished goods. China could therefore be called an *outlier* among nations in terms of global supply chains because of its size in global supply chains. Yet, even though China may be an outlier on statistical grounds, omitting it from an inquiry into global supply chains could be seen as a curious decision. In the following, we examine the regression relationship between the population of a country and their involvement in global supply chains using 2010 data from the United Nations Conference on Trade and Development (UNCTAD, 2013).

The predictor variable is taken from UNCTAD's data table called *Total population, annual, 1950–2050*. The population data for 237 individual world economies was originally obtained for this variable. The raw data from UNCTAD gives each nation's population in thousands. The country populations for the year 2010 were retained.

The dependent variable was taken from the data table entitled "Container port throughput, annual, 2008–2010." This variable gives the number of twenty-foot equivalent unit (TEU) containers that arrived a given country's ports. The TEU container is a measure of cargo capacity. The TEU data for 2010 was retained and converted to indicate TEUs in thousands. The dependent variable had many missing values, particularly for smaller or less developed countries (landlocked but more advanced countries such as Switzerland were included). There were 124 observations remaining after missing values were removed and the data was combined.

Since the data contains countries with very large populations (China, India, and the United States were the three largest) as well as with very small ones (e.g., Aruba and the Cayman Islands), the log transformation was applied to the population predictor variable x_1. We would also usually apply the log transformation to the outcome variable to reduce outliers (Section 10.7) but will not do so in this example to illustrate the use of robust regression.

Our initial model is the conventional normal regression model that was presented in (5.4). The log-population predictor variable is centered around its mean, giving the following specification:

$$y_i \sim \text{Normal}(\mu_i, \tau)$$

$$\mu_i = \beta_1 + \beta_2 \left(\log(x_{i,1}) - \overline{\log(x_1)} \right)$$

$$\beta_1 \sim \text{Normal}(0, 0.00000001)$$

$$\beta_2 \sim \text{Normal}(0, 0.00000001)$$

$$\tau \sim \text{Gamma}(0.001, 0.001).$$

The *WinBUGS* code to estimate this normal regression model appears in **WinBUGS 10.1.1 Robust Regression.odc**. Graphical results from the estimation of this model appear in Figures 10.2 and 10.3. The *WinBUGS'* `Comparison` tool was used to create Figure 10.3's plot of the Pearson residuals as in previous chapters. Figure 10.2 shows that one observation has a very large Pearson residual. China's Pearson residual is greater than 9, confirming that it has a highly outlying number of TEUs of cargo passing through its ports after controlling for its log-population size. We also see that the Pearson residual of the United States (at far right) and Singapore (middle right) are somewhat large given their national populations—but nowhere near China's value.

We can also instruct *WinBUGS* to create a graphical display of the model fit. After monitoring the *WinBUGS* quantity for the predicted value (`mu`), the `Comparison` tool can be selected and used to create a

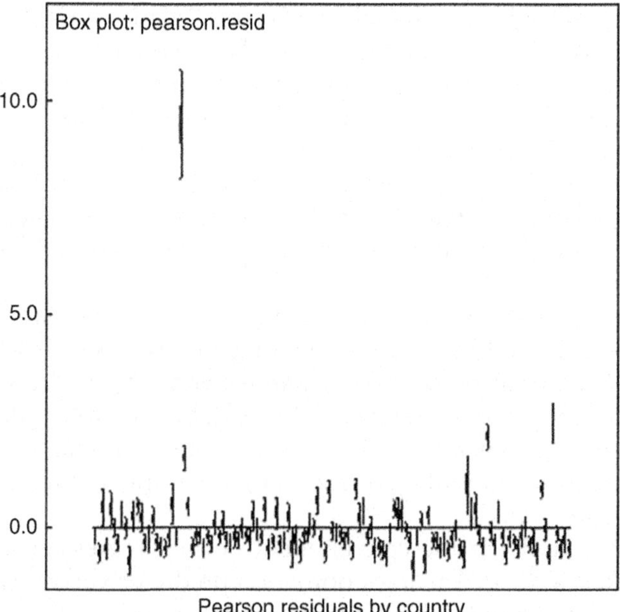

Figure 10.2 Pearson Residuals for Normal Regression: Global Supply Chain Data

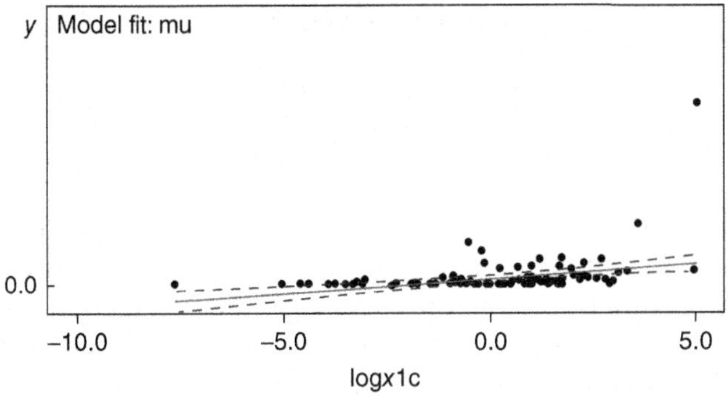

Figure 10.3 Model Fit Plot for Normal Regression: Global Supply Chain Data

plot such as that in Figure 10.3. In the tool, we enter `mu` in the `node` field, `y` in the `other` field, and `logx1c` (the predictor variable) in the `axis` field. Clicking `model fit` displays a model fit plot with the value of μ_i on the y-axis and the centered $\log x_i$ variable on the x-axis. The data appears as points on the plot as in Figure 10.3. The prediction line for μ_i appears as a solid line, and the 95% credible intervals for the prediction line appear as dashed lines. We see the prediction line is sloped and, as a result, it does not fit either the high or low values of `logx1c` particularly well. The higher values of `logx1c` exert a pull on the line, causing the low values of `logx1c` to be underpredicted. In fact, the low values of `logx1c` have negative values of `mu`, giving a nonsensical result for this dataset. Coefficient estimates for this model appear in Table 10.1 in the normal model likelihood portion.

We reconsider our model specification based on the results seen so far. We change the likelihood function to be the $t(4)$ distribution and write out our model as follows:

$$y_i \sim t(\mu_i, \tau, 4)$$

$$\mu_i = \beta_1 + \beta_2 \left(\log(x_{i,1}) - \overline{\log(x_1)} \right)$$

TABLE 10.1 β Coefficient Means and 95% Credible Intervals: Global Supply Chain Data

Model Likelihood	Parameter	Mean	Std. Dev.	95% Credible Interval
Normal	β_1	4309	1091	(2165, 6456)
	β_2	2201	515.4	(1186, 3209)
$t(4)$	β_1	1671	250.9	(1199, 2180)
	β_2	548.4	120.3	(324.7, 794.8)

$$\beta_1 \sim \text{Normal}(0, 0.00000001) \tag{10.1}$$

$$\beta_2 \sim \text{Normal}(0, 0.00000001)$$

$$\tau \sim \text{Gamma}(0.001, 0.001).$$

We next rewrite our code to implement the model of (10.1). The *WinBUGS* code for this model appears below (see also **WinBUGS 10.1.1 Robust Regression.odc**). The only change involves the likelihood function, which is changed to a *t*-distribution (see the bolded code below). The degrees of freedom parameter nu is given a value of 4 in the data portion of the *WinBUGS* code.

```
model
{
 for (i in 1:n) {
   y[i] ~ dt(mu[i], tau, nu)  # t likelihood
   mu[i] <- beta[1] + beta[2]*logx1c[i]
   logx1[i] <- log(x1[i])
   logx1c[i] <- logx1[i] - logx1bar
   }
 #priors
   for (j in 1:2) { beta[j] ~ dnorm(0, 0.00000001)  }
   tau ~ dgamma(0.001, 0.001)
 #other calculated parameters
   sigma <- 1/sqrt(tau)
   logx1bar <- mean(logx1[])
}
```

Table 10.1 shows that the posterior distributions of the parameters are considerably different in the robust regression with the $t(4)$ likelihood. The intercept and slope are much smaller, indicating that the outliers are having less of an effect on the regression coefficients. The posterior standard deviations and credible intervals are also much smaller, again because the outliers are being discounted. This allows the regression line to better reflect the bulk of the remaining observations. A model fit plot for the robust regression appears in Figure 10.4. The prediction line passes directly through the bulk of the observations and is less affected by the outliers. There is still a slight tendency to underpredict the lowest values of logx1c, but performance here is noticeably better. The 95% credible interval for the prediction line is much closer to the prediction line in this model and is almost indistinguishable from it in the figure.

We may also calculate the deviance information criterion (*DIC*) for the two models we have estimated. The *DIC* is 2688 for the normal regression while it is 2426 for the robust regression, supporting the robust regression model. The *DIC* has been constructed to be a short-term measure of predictive performance. We see that *DIC* selects the model that intuitively seems to fit the bulk of the data more closely. However, note that using

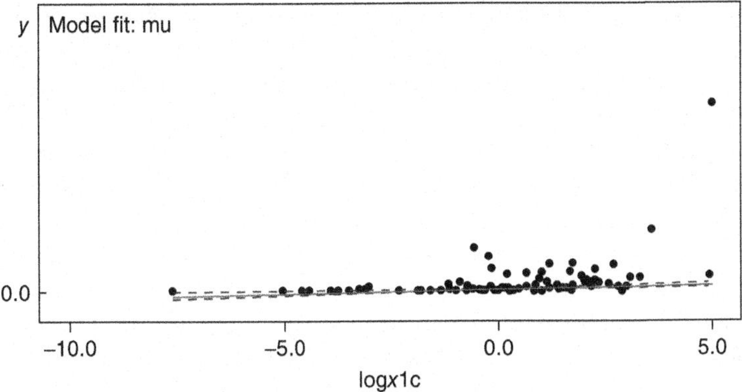

Figure 10.4 Model Fit Plot for Robust $t(4)$ Regression: Global Supply Chain Data

DIC to compare models with different likelihood functions is not always straightforward (Spiegelhalter et al., 2002) and should be done cautiously. A review of the Pearson residuals that have been transformed to the normal distribution from the t-distribution provides some corroborating evidence of better fit under the t-distribution (see Question 2, Section 10.7). We mentioned that it would be preferable if we had used the logarithm of a country's TEUs as the dependent variable in our regression. If we reestimate our models with y being the logarithm of a country's TEU, the conclusions based on *DIC* change (Section 10.7).

10.1.3 In Detail: Placing a Prior on ν

We have seen that changing the likelihood to the t-distribution can help reduce the impact of outliers on the regression parameters. However, we need to select a value for ν and this is not always an easy task. The Bayesian response to uncertainty about a parameter is to formulate a prior that represents our uncertainty about the parameter. We then add this prior to our model specification. Doing so has the effect of including our uncertainty in the overall model.

As pointed out by Spiegelhalter et al. (1996b, pp. 35–36), one way to create a prior for ν is to create a list of several ν values and use these as possible choices. For example, we may consider a *discrete prior* where ν is $4, 8, 20$, or 100. We have seen that when $\nu = 4$, the tails are quite heavy, but they will be less heavy when ν is 8 or 20. The tails of the t-distribution will be similar to the tails of the normal distribution when $\nu = 100$. We could give each of these candidate values a 25% prior probability. Updating in *WinBUGS* will produce the posterior of ν and we can see which value(s) were most supported by the data.

To use a discrete prior in *WinBUGS*, we can define a variable called `chooser`. Loosely speaking, the role of the `chooser` variable is to choose which value of ν to use at a given iteration of the Markov chain. This variable will be designed to take on the value 1 if the first value of ν is to be chosen, the value 2 if the second value of ν is to be chosen, and so on. We write a function to select the appropriate value from our list of possible ν values, and then assign this value to a *WinBUGS* variable called nu. For example, if `chooser` is 3, then the value 20 should be selected from the list of values mentioned above and assigned to nu. At the end of the Markov chain run, nu will provide our estimates of the posterior probabilities of the different values of ν.

We can use a categorical distribution in *WinBUGS* called `dcat` to implement our `chooser` variable. Variables with a `dcat` distribution can take on integer values from 1 to a predefined maximum. Each category has to be given a prior probability of being chosen at the current iteration of the Markov chain. It is required that all of these category probabilities sum up to 1.

We reexamine the global supply chain dataset using these concepts. The original data as examined in the previous section has extreme outliers, and it was suggested there that the logarithm of y be analyzed instead. Here we perform such an analysis, while also allowing for the possibility that a t-distribution may provide a better fit to the data than would the normal distribution. Selected code for this model appears below (see also **WinBUGS 10.1.2 Robust Regression Nu Prior.odc**).

```
model
{
 for (i in 1:n) {
   logy[i] <- log(y[i])
   logy[i] ~ dt(mu[i], tau, nu)  # t likelihood
   mu[i]   <- beta[1] + beta[2]*logx1c[i]
   logx1[i]  <- log(x1[i])
   logx1c[i] <- logx1[i] - logx1bar
   }
 #priors
 for (j in 1:2) { beta[j] ~ dnorm(0, 0.00000001)  }
 tau ~ dgamma(0.001, 0.001)
 chooser ~ dcat(nu.prior[])
 for (j in 1:n.nu) {  nu.prior[j] <- 1/n.nu }
 #other calculated parameters
 nu <- nu.val[chooser]
 ...
}
#data
list(nu.val=c(2,4,8,10,20,50,100,500,1000), n.nu=9, ...
```

Code has been added to calculate the logarithm of y and to use this as the outcome variable. We also see that `chooser` has been given a `dcat` (discrete) distribution. The prior probabilities for the `dcat` distribution are in a variable called `nu.prior`. Code has been set up to give equal prior probabilities for each category. Each prior probability is equal to `1/n.nu`, where `n.nu` is the total number of categories. The last line of code sets `nu` to a given value of `nu.val` based on the current value of `chooser`. The possible values of `nu` are listed in `nu.val` in the data statement. We consider a total of nine possible values, including the very heavy-tailed $t(2)$ distribution considered by Spiegelhalter et al. (1996b).

Figure 10.5 displays the trace of `nu` over 50,000 iterations after a 5000-iteration burn-in had been performed. We see the Markov chain primarily visited the `nu` values of 8, 10, and 20. We can also see that the trace appears to have a substantial amount of autocorrelation which can be confirmed by looking at the autocorrelation function (ACF) plots. We might expect to encounter higher autocorrelations in discrete variables, but a longer run time would be worth considering for this model.

The posterior distributions of `chooser` and `nu` appear in Figure 10.6. We see more clearly that the values of 8 and 10 have the most support, followed by 20, and then 4. We might consider several follow-up analyses based on these findings. We might consider removing the value of 8, which seems to be competing with 10, and then finding the posterior probability of $\nu = 10$ versus the posterior probabilities of the other values. Alternatively, we might be interested in placing more fine-grained values around the vicinity of 8 and 10 and omitting more distant values. Finally, we could examine a $t(10)$ regression without a prior on ν and compare it with the current model.

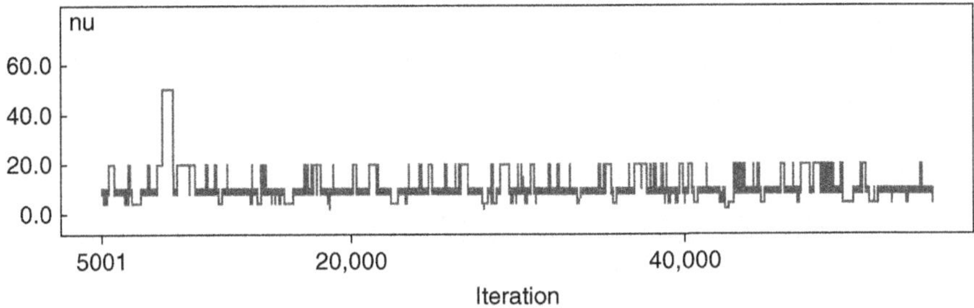

Figure 10.5 Trace Plot for `nu`: Global Supply Chain Data

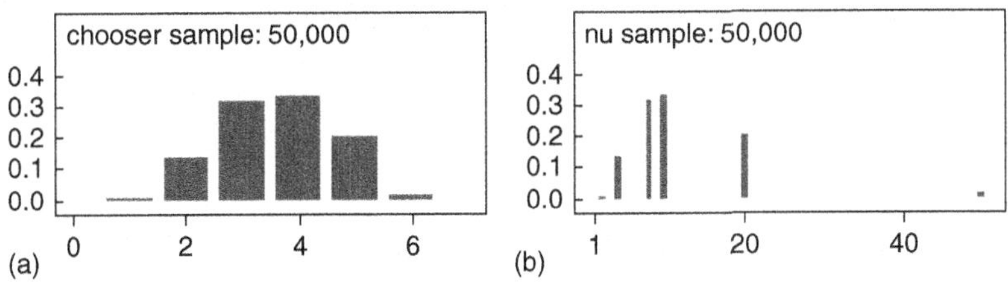

Figure 10.6 Posterior Distributions of (a) `chooser` and (b) `nu`: Global Supply Chain Data

10.2 HANDLING HETEROSCEDASTICITY BY MODELING VARIANCE PARAMETERS

Another situation we may encounter is that the regression error variance changes with respect to the value of x. This is known as *heteroscedasticity* or *nonconstant error variance*. We have assumed that our regression equation has constant variance σ^2, so the presence of heteroscedasticity indicates that our model may benefit from further refinement. We will see that it is straightforward to address heteroscedasticity in *WinBUGS*.

We may create a functional form that expresses how a parameter depends on other parameters and the data. This approach was used in Chapter 8. There we saw that intercepts and slopes could be predicted on the basis of the relationships that were present in the data. Here we have a very similar situation. Suppose we focus on the regression standard deviation σ. As in Chapter 8, we redefine this parameter to allow it to change on a per-observation basis. This gives us a new set of parameters σ_i. Our next step is to consider what functional form we believe to be appropriate. Figure 10.7 suggests that σ_i may vary along with x_i. We may therefore consider the following functional form:

$$\log(\sigma_i) = \alpha_\sigma + \beta_\sigma x_i. \tag{10.2}$$

We place $\log(\sigma_i)$ on the left-hand side to ensure that σ_i is positive. The data x_i then influences the size of $\log(\sigma_i)$ through the functional form.

Other functional forms are possible. Moreover, we can tailor our functional form to be relevant to other models besides regression. For example, we have already considered the unequal variances ANOVA (analysis of variance) model in Section 5.6.3. In that model, each group had its own separate variance σ_j to be estimated. We can accomplish the same outcome using a variation on (10.2). We could create a dummy

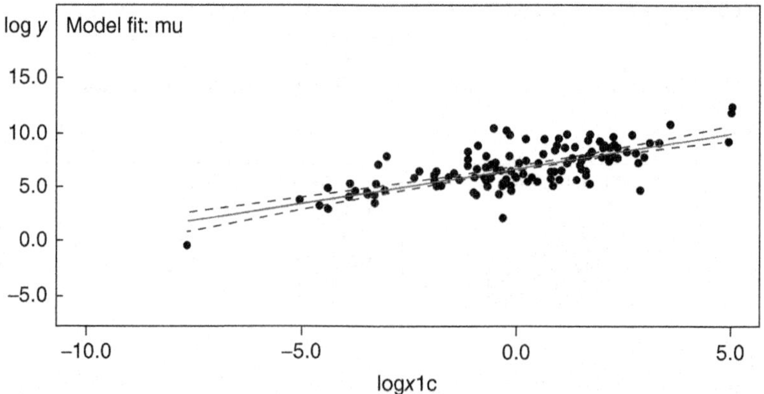

Figure 10.7 Fit Plot for Unknown ν Model: Global Supply Chain Data

variable $d_{i,j}$, taking on the value 1 if observation i was in group j and 0 otherwise. Then we would set σ_i to be equal to $\sigma_1 d_{i,1} + \cdots + \sigma_J d_{i,J}$. This will have the effect of equating σ_i to the appropriate σ_j based on the group information provided by $d_{i,j}$. We would not need to use the logarithm on the left-hand side in this case because all the quantities on the right-hand side are nonnegative.

10.2.1 In Practice: Modeling Heteroscedasticity

The Fortune 1000 list contains the top 1000 U.S. firms by gross revenue in a given time period. Here we examine companies on the 2005 edition of the list (Fortune, 2005). We use a normal regression model to predict 2005 profit (in millions of U.S. dollars). The predictor variable is the firms' total assets in millions of dollars. We would expect heteroscedasticity because firms with more assets are expected to have higher potential profits as well as higher potential losses. We cannot apply a log transformation to the dependent variable, profit, because negative values exist. However, we were able to apply the log-transformation the predictor variable and then center the result around the mean. These transformations were applied to generate a new predictor variable, `logx1c`, which is a firm's log-assets after centering around the mean. A small number of firms had missing values, so the total usable sample size was 996.

We first consider a regression model where the error variance is assumed to be constant (i.e., a homoscedastic regression). Two different views of the *WinBUGS* model fit plots for a homoscedastic normal regression model appear in Figure 10.8. The data is plotted as dots around the line of best fit in both views. The top view is the default view created by *WinBUGS* , while in the bottom view the y-axis has been zoomed into cover the range −1000 to +2000. As expected, the top view reveals that companies with greater amounts of assets can generate larger profits and tolerate larger losses. The bottom view reveals that the large profits of the bigger firms are also having an impact on the regression slope and intercept. There are more large profits than large losses, so the slope is pulled upward and the intercept shifted downward. Pearson residual plots (not shown here) confirm that the fit for smaller firms is poor in this model.

We can develop a heteroscedastic normal regression model where the values of `sigma` depend on the predictor `logx1c`. The *WinBUGS* code for this model appears below (see also **WinBUGS 10.2.1 Sigma Functional Form.odc**.) The functional form for `sigma` appears at the end of the main loop. Here, `log(sigma[i])` has its own intercept `alpha.sigma` and slope `beta.sigma`, both of which are estimated with reference to the predictor variable `logx1c[i]`.

```
model
{
 for (i in 1:n) {
```

```
    y[i] ~ dnorm(mu[i], tau[i])
    mu[i] <- beta[1] + beta[2]*logx1c[i]
    logx1[i] <- log(x1[i])
    logx1c[i] <- logx1[i] - logx1bar
    tau[i] <- 1/pow(sigma[i],2)
    pearson.resid[i] <- (y[i] - mu[i])/sigma[i]
    #Functional Form for Sigma
    log(sigma[i]) <- alpha.sigma + beta.sigma*logx1c[i]
    }
  #priors
    for (j in 1:2) { beta[j] ~ dnorm(0, 0.00000001)  }
    alpha.sigma ~ dnorm(0, 0.00000001)
    beta.sigma  ~ dnorm(0, 0.00000001)
  #other calculated parameters
    logx1bar <- mean(logx1[])
}
```

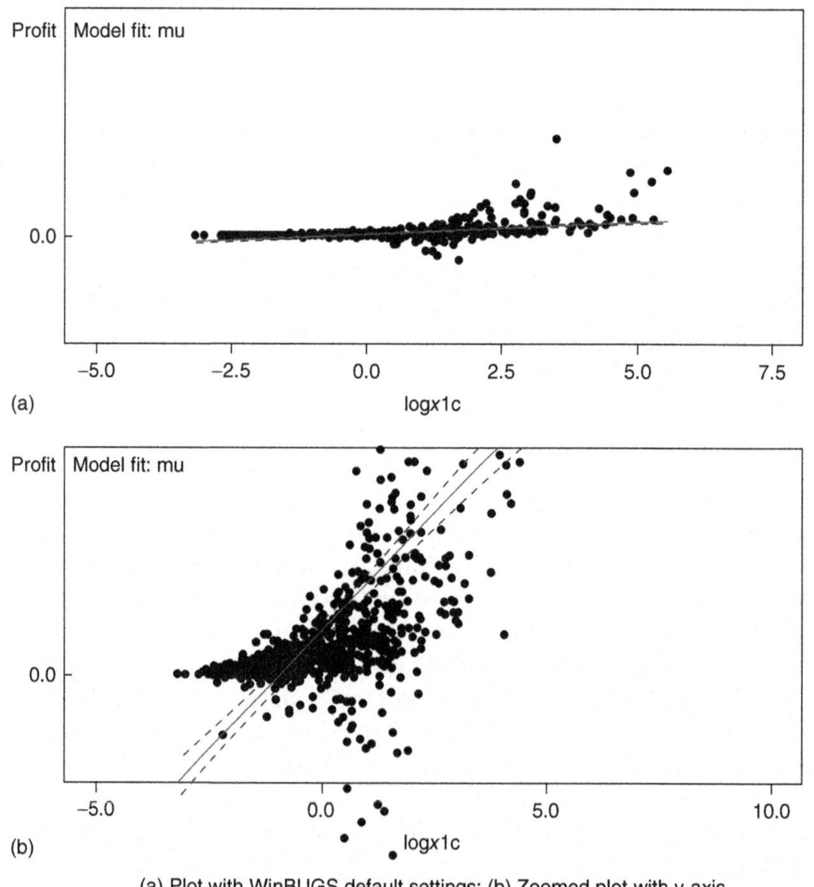

(a) Plot with WinBUGS default settings; (b) Zoomed plot with y-axis ranging from –1000 to +2000

Figure 10.8 (a,b) Model Fit Plots for Homoscedastic Regression: Fortune 1000 Data

Figure 10.9 shows the fit plot for this model. Improvements are clearly visible as compared to Figure 10.8. The line of fit now passes through the center portion of the bulk of the data, and the fit for the smaller and medium-sized members of the Fortune 1000 is noticeably better. The fit for the larger members of the Fortune 1000 is not as good because the values of `y` trend upward in a curvilinear manner as `logx1c` increases. We can address this issue by modifying our functional form for `y`. A quadratic term can be created by squaring the predictor and mean-centering the result. Adding this to the functional form should help us to capture the curvilinear aspect of the data. Details appear in **WinBUGS 10.2.1 Sigma Functional Form.odc**. The fit plot from the heteroscedastic regression model with the quadratic term appears in Figure 10.10. Again, improvements over previous model fit plots are noticeable.

A summary of the results for the different models appears in Table 10.2. We see that Model 2 has an intercept closer to zero compared to the

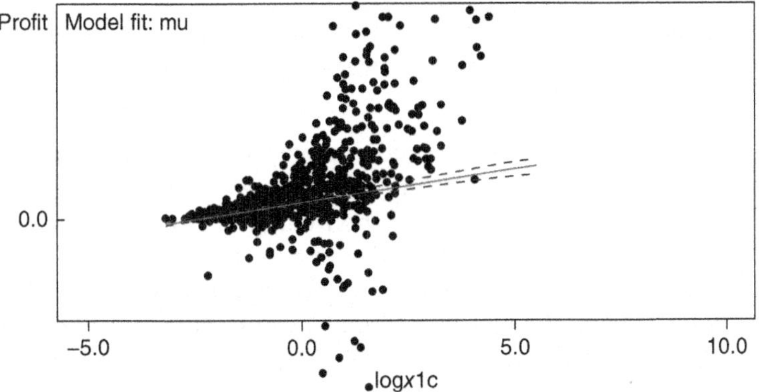

Figure 10.9 Model Fit Plot for Heteroscedastic Regression: Fortune 1000 Data

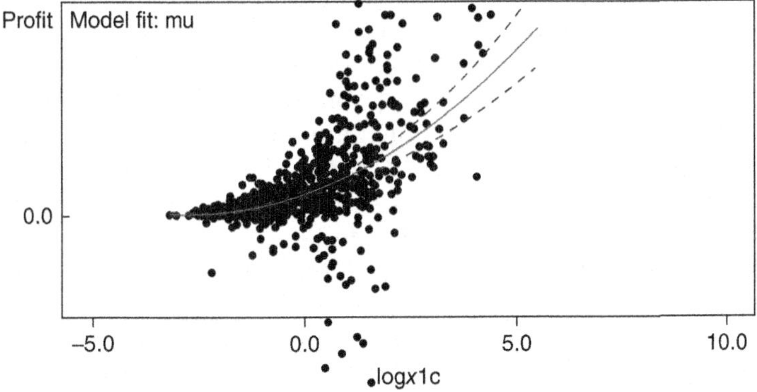

Figure 10.10 Model Fit Plot for Heteroscedastic Regression with Quadratic Term: Fortune 1000 Data

TABLE 10.2 Parameter Results: Fortune 1000 Data

Parameter	Model 1 (Homoscedastic)	Model 2 (Heteroscedastic)	Model 3 (Heteroscedastic)
Intercept (β_1)	**583.9** (481.2, 686.9)	**229.2** (202.6, 255.6)	**377.4** (326.7, 427.2)
Log assets (β_2)	**617.7** (546.7, 688.9)	**91.48** (77.36, 105.7)	**−460.4** (−622.7, −298.3)
Log assets-squared (β_3)			**38.69** (27.40, 49.99)
σ	**1661** (1590, 1737)		
α_σ		**5.980** (5.937, 6.024)	**5.959** (5.915, 6.003)
β_σ		**0.835** (0.808, 0.862)	**0.831** (0.804, 0.859)
DIC	17600.5	14743.8	14696.6

Parameter posterior means based on a run of 50,000 iterations after a 5000-iteration burn-in. 95% Posterior credible intervals appear in parentheses where appropriate.

intercept of Model 1. Since the predictor variable is mean-centered, the intercept estimates the profitability of the average firm in the sample. Model 1 gives a higher estimate of the average firm's profitability than does Model 2. We can reexamine the line of best fit at the bottom of Figure 10.8 and locate the approximate position of the line at the value of zero on the x-axis. We see that Model 1 appears to be overestimating the profitability of firms at `logx1c` = 0. Conversely, the line of best fit in Figure 10.9 seems to pass through the middle of the data at `logx1c` = 0.

The slope coefficient for log-assets is smaller in Model 2, as we saw in Figure 10.9. Results also appear regarding Model 2's two coefficients for modeling heteroscedasticity, α_σ and β_σ. The 95% credible interval for β_σ excludes the value of zero. This indicates that the error variance is related to the size of firm's log-assets. We would therefore conclude that the heteroscedastic model improves on the homoscedastic model. We can also compare the two models to see which model is preferred by *DIC*. At the bottom of Table 10.2, we see that *DIC* for Model 2 is substantially smaller than that of Model 1. This indicates extremely strong support for Model 2 over Model 1 according to *DIC*. In Model 3, we see that all coefficients have 95% credible intervals that exclude zero. This provides evidence that the quadratic coefficient for log assets-squared improves model fit. The *DIC* also corroborates this evidence because there is a sizable decline in *DIC* in Model 3.

While fit has improved considerably with the revisions to the models, additional changes could be considered. One possibility is that a robust regression using the t-distribution would be more appropriate given the data as appearing in Figure 10.10. Another possibility is that the functional form for log σ_i could be changed. Since the functional form for y in Model 3 has a quadratic term, it might also be plausible that the functional form for the error variance could benefit from a quadratic term. This possibility is explored in Section 10.7.

10.3 DEALING WITH MISSING DATA

Standard statistical analysis is designed to work with a rectangular dataset. In a rectangular dataset, each column represents a variable. Each row contains measurements on all the variables for a given respondent. Unfortunately, it is common to find that a dataset has missing values. This corresponds to various holes or irregularities appearing in the rectangular dataset. Missing values may occur because of incomplete recordkeeping or confidentiality requests. In surveys, respondents may refuse to answer for any number of reasons, or instead they may be unavailable at the time of the survey administration. Corporate filings may be incomplete because information is not known or has not been fully collated at the time of a report. A particular company or person may indicate that the question is "not applicable" and omit a response. Longitudinal data that follows entities over time may have a problem with dropouts—for example, a company may go bankrupt during the period of study. This will generate missing data for the bankrupt firm after the dropout.

There are a number of ways of addressing the problem of missing data, some of which are more satisfactory than others. One way of addressing missing data is to exclude a row of data if one or more of the data points on that row are missing. This strategy is called *listwise deletion* (or *casewise deletion*) and it was used in Section 10.2. In the Fortune 1000 data we analyzed there, we omitted 4 firms out of 1000 by excluding their rows from the dataset. We can therefore speak of the rate of missingness in that dataset as 0.4%. This is a low number which might not cause particular concern; however, the rate of missingness can be much higher. If we have a high rate of missingness, omitting cases has a number of negative consequences. First, our statistical power to detect relationships declines because each row of data omitted reduces the total sample size. Second, the greater the rate of missingness, the greater the chance that the missingness may produce bias in our estimates. An example illustrates the issue. Suppose we conduct a salary survey in a random sample from a population of interest. If all respondents report their salaries on the survey, then we can form an unbiased estimate of population salary quantities such as the mean. However, suppose that the higher the salary, the less likely a person will report the salary on the survey. Then the reported salaries will underestimate the actual salaries in the population, and consequently our estimates of quantities such as the mean will suffer from the same bias. A final consequence of listwise deletion is that it wastes the nonmissing data in the row. The nonmissing data in a deleted row may have required some effort to obtain but is now treated as statistically useless. To summarize, listwise deletion can have several drawbacks but it is commonly used because it is a simple way to handle missing data. Authors such as Schafer (1997) suggest that it is a worthwhile strategy when the rate of missingness is low (such as 5% or lower). In these situations, there is little wastage of data and bias is less likely.

A different way to address missing data is called *(unconditional) mean substitution*. In this method, we take the mean of a variable's nonmissing values. We then set all of the missing values of that variable equal to the mean of the nonmissing values. This method is simple but unattractive statistically. Thinking back on our salary example of the last paragraph, imagine that we take the missing values from the higher salary individuals and substitute in the biased mean of the lower salary individuals. We can see that this method does nothing to change the bias in the mean and only perpetuates the bias. Worse still, this method decreases the standard deviation of the variable by setting all the missing values to a constant. This approach would make us more falsely overconfident in our biased estimate than if we had performed listwise deletion. The covariances and correlations for this variable will also be falsely affected. For these reasons Little and Rubin (2002, p. 62) conclude that this method cannot be recommended.

Pairwise deletion is another strategy that is sometimes used. In this strategy, each variable's mean is calculated based on the available data, and the covariance between any two variables is calculated based on the available information from the two variables. Then, the parameters of the model of interest are estimated from these means and covariances. Pairwise deletion preserves a greater amount of information than listwise deletion because it does not require us to delete the nonmissing data in a row that has some missing values. We would therefore expect that the means and covariances under pairwise deletion would be less biased. Pairwise deletion thus can be attractive when it is available. However, covariance matrices have to satisfy a condition called *positive definiteness* to be valid (Section 8.5) and there is no guarantee that this will occur under pairwise deletion (Winglee et al., 2001). Also, the modeling software package would need to be programmed to accept the pairwise mean and variance summary statistics instead of the complete case data, and not all software does this.

The Bayesian Approach to Missing Data One observation we may make about the above approaches is that they are *ad hoc* approaches to deal with the problem of missing data. This means that various criteria based on partly statistical reasons (such as bias reduction) and partly nonstatistical reasons (such as convenience) can be applied with no clear guidance as to which should be used. The different approaches will tend to generate different sets of results which may be perplexing. Many other *ad hoc* approaches for addressing missing data exist (Little and Rubin, 2002, chs 3–4). These approaches are often plausible under certain conditions or for certain situations, but the diversity of approaches indicates that one would need to use them on a case-by-case basis after a subjective assessment of the particulars of the data and the properties of the approaches.

By contrast, there is only one way to estimate any unknown quantity in Bayesian inference—the use of Bayes' theorem. Bayes' theorem is applied in the same way for all unknown quantities, whether they are parameters or missing data. In this sense, approaching missing data from a Bayesian perspective is conceptually simple and no new Bayesian ideas are required.

There are also very few new ideas required to use *WinBUGS* with missing data if the missing data is confined to the dependent variable. Missing data is given the value NA in the *WinBUGS* data list. *WinBUGS* will simulate from the posterior distributions of the missing data points. This idea was originally termed *data augmentation* by Tanner and Wong (1987). Once we have the augmented (or simulated) data, we use it to perform the analysis of interest. Unlike real data, the augmented data will change from iteration to iteration as the Markov chain runs. However, we will see that this is actually a desirable property in the context of missing data.

Since we will use the Monte Carlo Markov chain (MCMC) to estimate our missing data, we will also need to specify initial values for the missing data just like we have been doing for the parameters. This means adding our variable with missing data to the initial values portion of the code. Here, the initial value of missing data is given a numeric value, while the initial value of *nonmissing* data is given the value NA because it is already known.

10.4 TYPES OF MISSING DATA

While we do not require new Bayesian ideas or much in the way of new *WinBUGS* ideas, we will need to think about the type of missing data we have. The type of missing data we have will potentially impact our model and hence our code. As such, there are some new modeling issues and the model we choose depends on the type of missing data we encounter.

10.4.1 Missing Completely at Random Data

One type of missing data we may encounter is known as *missing completely at random* (*MCAR*) data. If the data are MCAR, the missingness of the data do not depend on either the true values of the missing data or on any other observed data we might have. In other words, the missing values are independent of the observed data and the true values of the missing data. Under MCAR, we do not need to worry about the missing values creating bias. Instead, it is as if we were planning to collect a random sample from our population but we only collected a random subsample of the originally proposed random sample. Analyzing MCAR data with a more advanced approach to missing data will nonetheless tend to produce results similar to that of listwise deletion. Therefore, if we somehow knew in advance that our data was MCAR, listwise deletion would be an especially attractive option given its convenience and simplicity.

10.4.2 In Practice: Analyzing MCAR Data

The code below can be used to analyze univariate MCAR data with *WinBUGS* (see also **WinBUGS 10.4.1 MCAR Missing Data.odc**). The data for this code were artificially generated by simulating 20 values from the standard normal distribution. This dataset appears below as `y.true`. The first value was then arbitrarily set to missing, and the resulting data appears as `y.miss`. We can analyze both sets of data side by side for comparative purposes.

```
model
{
 for (i in 1:n) {
   y.true[i] ~ dnorm(mu[1], tau[1])
   y.miss[i] ~ dnorm(mu[2], tau[2])  }
 #priors & other calculated parameters
   for (j in 1:2) {
       mu[j]  ~ dnorm(0, 0.00000001)
       tau[j] ~ dgamma(0.001, 0.001)
       sigma[j] <- 1/sqrt(tau[j])    }
}
#data
list(n = 20, y.true=c(0.779,-1.177,-1.187,1.160,-0.481,
  -0.252,-1.551,0.178,1.028,-0.529,-0.768,-1.722,0.655,
  0.826,0.323,-0.679,-0.547,0.160,1.149,-1.699),
y.miss=c(NA,-1.177,-1.187,1.160,-0.481,-0.252,-1.551,
  0.178,1.028,-0.529,-0.768,-1.722,0.655,0.826,0.323,
  -0.679,-0.547,0.160,1.149,-1.699) )
#inits
list(mu=c(0,0), tau=c(1,1), y.miss=c(0,NA,NA,NA,NA,NA,
  NA,NA,NA,NA,NA,NA,NA,NA,NA,NA,NA,NA,NA,NA) )
```

The first value of `y.miss` has been changed to `NA` to represent a missing value to *WinBUGS*. We have added initial values for `y.miss` such that the initial value for our missing observation is zero. The remaining observations of `y.miss` are nonmissing. Therefore, we set these initial values to `NA`, which instructs *WinBUGS* not to try to estimate these as unknown parameters.

We have made changes to the `#data` and `#inits` sections, but notice that the model code looks exactly like our usual *WinBUGS* model for a normal mean. While we did not write anything differently in the model code: the model is now implicitly different. The model for the second set of data, `y.miss`, is

$$\begin{aligned} y_1 &\sim \text{Normal}(\mu_2, \tau_2) \\ y_2, \ldots, y_{20} &\sim \text{Normal}(\mu_2, \tau_2) \\ \mu_2 &\sim \text{Normal}(0, 0.00000001) \\ \tau_2 &\sim \text{Gamma}(0.001, 0.001), \end{aligned} \tag{10.3}$$

where y_1 is now an unknown. Since y_1 is an unknown, *WinBUGS* will estimate it using the posterior predictive distribution (2.7). In words, at each iteration of the Markov chain, *WinBUGS* will simulate a value from a normal distribution with mean μ_2 and precision τ_2 and then place this value in y_1. When we estimate a value for a data point, this is called *imputation*. Hence, *WinBUGS* is imputing values for y_1, which we can monitor with the standard MCMC tools.

When it is time for *WinBUGS* to estimate μ_2, *WinBUGS* will incorporate information from y_1 through y_{20} to estimate μ_2. However, since y_1 is simulated from its posterior predictive distribution, which itself has mean μ_2, *WinBUGS'* estimate of μ_2 will not be shifted up or down but will instead be unaffected by y_1. Importantly, the posterior standard deviation of μ_2 as well as the distribution of τ will also be appropriately wide. Our results for the posterior standard deviation of μ_2 and the distribution of τ can be compared with the results obtained under mean substitution. We will find that mean substitution causes these quantities to be inappropriately overprecise.

Table 10.3 displays the results from running the code above. The parameters estimated from the complete data in the code are μ_1 and σ_1, while the parameters estimated when y_1 is missing are μ_2, σ_2, and y_1. For completeness, we also present results estimated using two other methods in **WinBUGS 10.4.1 MCAR Missing Data.odc**. These are listwise deletion and mean substitution (see file for details). We see that the results using imputation are almost identical to those from listwise deletion. We also obtain the posterior distribution of y_1 under imputation. The posterior mean of y_1 is reasonably close to the mean of the nonmissing values of y_1 which is −0.269. We still have considerable uncertainty about y_1, and this is reflected in its standard deviation. Imputation and listwise deletion produce slightly different estimates of μ from those produced by the complete data, and the standard deviations of μ are slightly larger. However, the estimates of σ under all three methods are close to each other, particularly the posterior means of σ. Mean substitution produces estimates

TABLE 10.3 Comparing Methods for Handling MCAR Missing Data

Missing data Method	Posterior Estimates	Parameters		
		μ_1	σ_1	
(Complete data)	Mean	−0.216	1.004	
	Std. dev.	0.228	0.173	
		μ_2	σ_2	y_1
MCAR model	Mean	−0.271	1.004	−0.279
	Std. dev.	0.234	0.180	1.048
		μ	σ	
Listwise deletion	Mean	−0.270	1.003	
	Std. dev.	0.233	0.180	
		μ	σ	
Mean substitution	Mean	−0.270	0.974	
	Std. dev.	0.220	0.169	

Results based on a 5000-iteration burn-in and 20,000 posterior samples.

with the problems identified earlier. Under mean substitution, we set the value of y_1 equal to the mean of the nonmissing data, which is −0.2691. While the posterior mean of μ under mean substitution is consistent with imputation and listwise deletion, the posterior standard deviation of μ is too small. Results for σ under mean substitution are also too small.

10.4.3 Missing at Random Data

The second type of missing data is known as *missing at random* (*MAR*) data. This name can be slightly confusing, so it is worthwhile to remember the following informal definition. When a variable y is MAR, then its missingness is not independent of y but it *does depend* on other observed data we have. For example, suppose that we work for a corporation. Our customers with higher incomes are less likely to report their incomes, but we have customer addresses on file and we can obtain the square footage of their domiciles. Suppose further that people with larger domiciles in terms of square footage tend to have higher incomes. Then the missing data is MAR, and we could perform statistical inference on our missing data using a MAR model. Importantly, we can also impute customer incomes. While our imputations will be subject to error, under MAR our imputations will be more statistically plausible as compared to, say, values we might obtain by mean substitution.

In practice, what this means is that if we can create a functional form where the missing data is a function of observed predictors (such as in regression), we can model the missingness more appropriately. Suppose we split our dependent variable y into observed values y^o and missing values y^m. Then, if the data is MAR, we can use a model such as the following and obtain valid parameter inferences:

$$\begin{aligned} y_i^o &\sim \text{Normal}(\mu_i, \tau) \\ y_i^m &\sim \text{Normal}(\mu_i, \tau) \\ \mu_i &= \beta_0 + \beta_1 x_{i,1} + \ldots + \beta_j x_{i,j} \\ \beta_1, \ldots, \beta_j &\sim \text{Normal}(0, 0.00000001) \\ \tau &\sim \text{Gamma}(0.001, 0.001). \end{aligned} \tag{10.4}$$

Here we see in a more familiar notation that the missing data depend on the observed data x. The current model can be compared with (10.3), where the missing data were believed not to depend on an observed x or any other variable.

Little and Rubin (2002, p. 120) indicate that for valid Bayesian inference we need another assumption for MAR and MCAR imputation. Suppose there is a parameter ϕ which governs whether a data value is missing or not. We must assume that the *prior* for ϕ is independent of the other parameters in the model. We have made similar assumptions throughout most of this book (e.g., by assigning independent priors to β_1 and β_2 in a regression model). Schafer (1997, p. 11) indicates that this assumption will be reasonable in many cases.

10.4.4 In Practice: Analyzing MAR Data

We use a social media dataset to examine the modeling of MAR data. The data appears in Twitter, Inc. (2013, Oct. 3, p. 61). The worldwide number of Twitter users (in millions) over 14 consecutive quarters beginning March 2010 is $y = (30, 40, 49, 54, 68, 85, 101, 117, 138, 151, 157, 185, 204, 218)$. To create MAR data, we set the last value of y as missing. Here, the missingness is not independent of y because it intentionally only occurs at the highest value of y. However, this pattern of missingness is (positively) correlated with x. We examine this in *WinBUGS* with the following code (see also **WinBUGS Code 10.4.2 MAR missing data.odc**):

```
model
 {
 for (i in 1:n) {
   y.miss[i] ~ dnorm(mu[i], tau)
   mu[i]  <- b0 + b1*x1c[i]
   x1c[i] <- x[i] - mean.x
   }
 #priors & other calculated parameters
 b0 ~ dnorm(0, 0.00000001)
```

```
 b1 ~ dnorm(0, 0.00000001)
 tau ~ dgamma(0.001, 0.001)
 sigma <- 1/sqrt(tau)
 mean.x <- mean(x[])
}
#data
list(n=14, x=c(1,2,3,4,5,6,7,8,9,10,11,12,13,14),
y.miss=c(30,40,49,54,68,85,101,117,138,151,157,185,204,NA))
```

Examining our small dataset with four different missing data models produces the results in Table 10.4. The overall missingness rate is 7% so the missingness problem is not particularly severe. The MAR model replicates the β_0 and β_1 estimates produced by an analysis of the full data reasonably well. Listwise deletion replicates β_1 reasonably well; however, β_0 is biased downward versus the full data model even with this low missingness rate. In the current functional form, β_0 is also our estimate of the grand mean of y. Since the largest value of y was set to NA, it is not surprising that the mean of y is estimated as being too low under listwise deletion. In this simple example, reformulating the model so that x was not mean-centered would make listwise deletion look more competitive. However, this may lead to MCMC convergence issues in more complex models. Under mean substitution, the values for β_0 and β_1 are both biased away from their true values. Finally, we can compare the estimates of σ. Listwise deletion and the MAR model both seem to capture σ relatively well. However, mean substitution performs very poorly.

Table 10.4 also shows that the imputed (posterior predictive) distribution of y_{14} appears to be close to its original value. Under MAR, this is to be expected. However, we typically will not know whether our data

TABLE 10.4 Comparing Methods for Handling MAR Missing Data: Worldwide Twitter Users

Missing Data Method	Posterior Estimates	Parameters			
		β_0	β_1	σ	
(Complete data)	Mean	114.1	14.91	8.073	
	Std. dev.	2.205	0.550	1.827	
		β_0	β_1	σ	y_{14}
MAR model	Mean	113.4	14.63	8.052	208.5
	Std. dev.	2.237	0.615	1.914	9.618
		β_0	β_1	σ	
Listwise deletion	Mean	106.1	14.64	8.061	
	Std. dev.	2.293	0.615	1.921	
Mean substitution	Mean	106.1	11.72	28.17	
	Std. dev.	7.695	1.919	6.375	

Results based on a 5000-iteration burn-in and 50,000 posterior samples.

is truly MAR so this should not be expected to always occur. The critical issue is how well we can predict y. If we have a very good predictor of y (as is the case here), our distribution for missing values should tend to be reasonably accurate. Playing devil's advocate however, suppose the true value of y_{14} was 1000. Then the MAR model would perform poorly but so would have listwise deletion. If we do not know why missing values occurred, it is important that our functional form have as much explanatory power as possible through the inclusion of all relevant predictors of the missing data. Functional forms with low explanatory power are less likely to allow us to satisfy the MAR condition. Conversely, simulation studies by Collins et al. (2001, pp. 342–343) show that even modest predictive power of missingness (correlation = 0.4) can result in improved estimates of certain parameters under certain conditions as compared to no predictive power at all.

Since predictive power is an important issue in modeling missing data, we can reanalyze the data in the *R* software package to estimate the predictive power of the linear regression. A common measure of the predictive power of a regression functional form is through the estimation of the squared correlation explained by the functional form r^2. The r^2 measure is also called the *squared model correlation* by other authors. Estimating the linear regression in the *R* software package after omitting the missing value gives a "listwise deletion" r^2 value of 0.9844. Interestingly, we can estimate the MAR r^2 in *WinBUGS* by adding the formula to our model code. We may instruct *WinBUGS* to compute what is known as the *error sums of squares*, which involves `y.miss` as well as `mu` (Section 10.7). At each iteration, *WinBUGS* produces the error sums of squares based on the observed data and the current simulated value of the missing data point. We may also instruct *WinBUGS* to compute the total sums of squares. This also involves `y.miss` as well as its mean. Monitoring the final quantity allows us to obtain the posterior distribution of r^2 under MAR. The posterior mean of r^2 here is 0.9833 (slightly below that of *R*), and its 95% credible interval is 0.972 to 0.987. The posterior distribution of r^2 is skewed because an r^2 value of 1 is the upper limit indicating perfect prediction. The median value of the posterior distribution is 0.9847 and, given the skewness, we might prefer this as a summary measure.

10.4.5 Missing Not at Random Data

The final type of missing data is known as *not missing at random* (*NMAR*) data. If the missingness is not MCAR nor MAR, then it is NMAR. Under NMAR, the missingness of y is not independent of y but we lack the ability to model it with a predictor. This corresponds to having MAR data *without* having a predictor variable that is correlated with the missingness.

We can examine what happens to estimates when data is NMAR by reexamining our data on Twitter users. Failing to incorporate x in this dataset will make y NMAR. We first estimate μ and σ for y using the

TABLE 10.5 NMAR Missing Data: Worldwide Twitter Users

Missing Data Method	Posterior Estimates	Parameters	
		μ	σ
(Complete data)	Mean	114.1	66.76
	Std. dev.	18.23	14.43
		μ	σ
Missingness unmodeled (NMAR)	Mean	106.1	61.40
	Std. dev.	17.37	13.84

Results based on a 5000-iteration burn-in and 50,000 posterior samples.

full dataset. Next, we again set y_{14} as missing and estimate μ and σ for y without a predictive covariate. The results appear in Table 10.5 (see also **WinBUGS Code 10.4.2 MAR missing data.odc**).

We find that failure to model the missingness leads to biased estimates for both μ and σ. Our estimate for μ is too low when we allow y_{14} to be missing but do not predictively model the missingness. Here, the estimate of μ is identical to our listwise deletion and mean substitution estimates in Table 10.4. The estimate of σ is also too low, which means we will be overconfident in our biased estimate. If our data is NMAR and we do not believe we can satisfy the MAR assumption by adding covariates, there are more advanced models for this scenario (Little and Rubin, 2002, ch. 16). However, these models are outside the scope of this book.

10.5 MISSING COVARIATE DATA AND NON-NORMAL MISSING DATA

We have focused on normally distributed missing outcome data in the preceding discussion. We also may find that covariate data has missing values. It is possible to proceed in these cases if we can develop a covariate imputation model. These models are more complex than the ones we have discussed in this chapter. *WinBUGS* models for this scenario have been examined by Carrigan et al. (2007), Molitor et al. (2009), and Lunn et al. (2013, ch. 9.1.2). If we have missing covariates but complete outcome data, one possibility is to use the outcome data (as well as other covariates) to help impute the covariate with missing data. Thus we consider the covariate as an "outcome" variable, create the desired functional form to model the covariate as an outcome, and then impute it (Seaman et al., 2012).

Since the MAR assumption is more likely to hold when the functional form has a high r^2, we often want to consider adding many predictors to a model when there is missing data. Unfortunately, a large number of predictors can become a nuisance when we wish to examine a specific functional form implied by theory or to compare such theoretically implied functional forms. Fortunately we can develop a more extensive functional form solely for the purpose of producing imputations. We can then estimate a different specific functional form implied by theory where the imputed data is added to the observed data. The first more extensive functional form is often called an *imputation model,* while the second can be called an *estimation model* (Schafer, 1997, p. 31). *WinBUGS* can accommodate this kind of scenario using its `cut` function. This function is described by Spiegelhalter et al. (2003).

In addition to the normal distribution, *WinBUGS* can also handle missing Poisson data, binomial data, *t*-distributed data, and data from other likelihood functions. *WinBUGS* will simulate from the posterior predictive distribution for outcome data arising from these likelihood functions. For a more general discussion of this topic, see Schafer (1997, ch. 10). There is also the possibility of multilevel or hierarchical model-based imputation (Schafer, 1997, ch. 10). We have discussed in Chapter 8 how hierarchical models can help to better explain patterns in our data, particularly when our data is longitudinal. The additional explanatory power of hierarchical models should make satisfying the MAR condition more attainable and should also lead to better imputations of missing values.

How to best deal with missing data is a problem that has received attention from some of the leading minds in statistics, and we are only able to provide a very brief introduction to some of these ideas here. Required reading for those interested in this topic would almost certainly include Little and Rubin (2002). Further discussion of the many issues involved in handling missing data can be found in Schafer (1997), Gelman et al. (2004, ch. 21), Daniels and Hogan (2008), van Buuren (2012), and Carpenter and Kenward (2013).

10.6 SUMMARY

Real-world business data is often messier than we might like them to be. This chapter has reviewed different scenarios that may arise and models that can be used to address the kinds of data we might encounter. Outliers may appear as a result of many reasons, but one reason might be that the normal distribution is not the best likelihood function for the business situation we are examining. In this scenario, we can consider alternative likelihood functions which better fit our data such as the *t*-distribution. Next, it is convenient for us when variances do not change as a function of the data. However, if variances do change, we can create a functional form for them and this may lead us to additional insights. Finally, the most challenging situation we discussed is the occurrence of missing data. Missing data can lead to bias in estimates which may become potentially severe. If we can make the MAR condition plausible, however, then we may be able to reduce bias to negligible levels as in Table 10.4. A perennial caveat for missing data models is that we typically cannot know for sure whether our data satisfies the MAR condition. We can completely verify the MAR condition only by having the missing data. However, alternative simple methods for dealing with NMAR data such as listwise deletion will also tend to produce biased results, so they cannot be recommended either. NMAR models are considerably more complicated to use than MAR models, but outside reading on their use is recommended for a thorough understanding of their relevance and applicability given the advantages they offer. A careful and thorough examination of missing data using MAR or NMAR models represents the current best practice strategy for dealing with this statistical challenge.

10.7 EXERCISES

1. If outliers in the data are not a result of typos or errors, what are some of the limitations of automatically removing the outliers?

2. Name at least one data transformation that can be used to make outliers less problematic.

3. Removing outliers: Examine the effect of removing China from the global supply chain regression model in **WinBUGS 10.1.1 Robust Regression.odc**. China is Observation 23 in the dataset, with an x value of 1318193.985 and a y value of 130290.443. Reestimate the normal regression model with this observation removed, and examine your results. What has changed? Compare your findings with the normal regression and $t(4)$ results for the data where China was included.

4. Pearson residuals in t-regression: Estimate Pearson residuals in the global supply chain t(4) regression model of **WinBUGS 10.1.1 Robust Regression.odc** by adding the following line of code to the main loop: `pearson.resid[i] <- (y[i]-mu[i])/(sqrt(nu)*sigma).` The Pearson residual for China (Observation 23) seems quite large compared to China's Pearson residual in the normal regression model, but this is due in part to differences between the t-distribution and normal distribution. Find out how large this t Pearson residual would be under the normal distribution by using R. Enter the command `cPR <- *` in R after replacing the asterisk with the posterior mean of China's Pearson residual. Next, enter the command `qnorm(pt(cPR,4))`. This will convert the $t(4)$ Pearson residual to a normal Pearson residual. Compare the value you have just obtained

with China's Pearson residual in the normal regression model. What do you conclude about China's status as an outlier in the two models?

5. Reestimate the global supply chain regression model in **WinBUGS 10.1.1 Robust Regression.odc** with the dependent variable as the logarithm of y. Calculate *DIC* using a normal likelihood and a $t(4)$ likelihood. Which model do you now prefer—the normal or the robust regression model?

6. Reexamine the heteroscedastic regression model for the Fortune 1000 data in **WinBUGS 10.2.1 Sigma Functional Form.odc**. Modify the code for Model 3 of Table 10.2 so that the functional form for `log(sigma[i])` includes the quadratic predictor `logx1c2[i]` with its own coefficient `beta.sigma.2`. Do you prefer this model over Model 3?

7. Calculate r^2 for the MAR Twitter data in **WinBUGS 10.4.2 MAR missing data.odc** by adding the appropriate quantities. Hint: in the main loop, calculate the `sserror[i]` terms as the squared differences between each data point and its expected value, `mu[i]`. Also, in the main loop calculate the `sstotal[i]` terms as the squared differences between each data point and the overall mean of `y[i]`. Then, outside of the main loop calculate r^2 as `1 - sum(sserror[])/sum(sstotal[])`. How do your results compare with those of the *R* software package?

10.8 NOTATION INTRODUCED IN THIS CHAPTER

Notation	Meaning	Example	Section Where Introduced
r^2	The squared correlation explained by the functional form	$r^2 = 0.5$	10.4
y^o	Observed values of y		10.4
y^m	Missing values of y		10.4

11

INTRODUCTION TO LATENT VARIABLE MODELS

11.1 NOT SEEN BUT FELT

Many interesting business concepts are not directly observable. We cannot directly touch, hear, or see the loyalty of a customer or the creditworthiness of a firm. Fortunately, we can observe a customer's purchase history and use that to understand loyalty. For creditworthiness we can examine assets, liabilities, and past repayments. The observed information helps us to reduce our uncertainty about these latent concepts. We may not know for sure whether a firm will repay a given debt or not, but intuitively we should have a better idea by considering the data, and the more data we consider, the more sure we may feel. This intuitive approach corresponds well with Bayesian inference where we use information to go from a broad prior distribution to a more concentrated and precise posterior distribution. We might feel that there are some clear distinctions between estimating model parameters (Chapters 2–9), imputing missing data (Section 10.3), or inferring latent data (this chapter). However, from a Bayesian perspective these different tasks are all handled similarly.

In the previous chapter, we examined data that was partially missing. This chapter covers the analysis of data which is, in a sense, "entirely" missing. Latent data models are models for data that are not observed directly. While we do not observe the latent data directly, we do observe

Bayesian Methods for Management and Business: Pragmatic Solutions for Real Problems, First Edition. Eugene D. Hahn.

other data. We can then use the values of the observed data to make inferences about the latent data. This chapter examines how we can use the idea of latent data to shed light on situations we encounter in business.

11.2 LATENT VARIABLE MODELS FOR BINARY DATA

We considered the probit model (9.7) for binary data in Section 9.4. In this model, the linear functional form produces a quantile (i.e., a z-score) from the normal distribution. Next, the cumulative probability of the quantile is used to map the linear functional form onto the probability of a binary response. Albert and Chib (1993) recognized that this model could be written using the idea of latent data which can be produced by data augmentation. They observed that the functional form should give a higher probability (and a higher quantile) if the binary dependent variable y took on the value 1. Conversely, the functional form should give a lower probability (and quantile) if the binary dependent variable y took on the value 0.

For concreteness, suppose our functional form is currently equal to a value of $\mu = -0.5$ as in Figure 11.1. Let y^* denote the latent data. Albert and Chib (1993) proposed that, if $y = 0$, we simulate a value of y^* from the shaded portion of the figure. If $y = 1$, then we simulate a value of y^* from the unshaded portion of the figure. Once we have simulated all the required values of y^*, we now have a complete set of simulated latent data. At this point, we can now estimate the needed values of β using the basic regression approach with Gibbs sampling, which we discussed in Section 4.4.

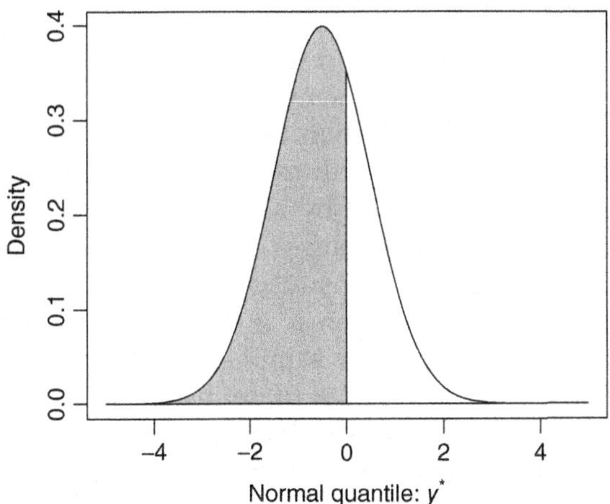

Figure 11.1 Simulating Latent Variables from a $\mu = -0.5$ Normal Distribution Given Binary Outcome

Albert and Chib's version of the probit model has an interesting advantage over the probit model in (9.7). We no longer need to evaluate the cumulative density function explicitly. In *WinBUGS* , this is attractive because the normal cumulative distribution function in *WinBUGS* is prone to generating errors (although there is a workaround as we discussed in Section 9.4). A limitation of the Albert and Chib approach is that we need to have individual-level 0/1 outcome data to proceed. This type of outcome data is also known as *Bernoulli-distributed data*. It can be contrasted with the more aggregated binomial type of data in which we have observed y successes among n events. We examined binomial data with the probit model in **WinBUGS Code 9.4 Probit Analysis.odc**. Fortunately, we can always expand binomial data to Bernoulli data by making n copies of a given row. We would then take these copies and code y of them to be successes and $n - y$ of them to be failures.

Assuming we have Bernoulli data, the Albert and Chib approach has two other smaller advantages. First, we would expect that the Gibbs sampling under the Albert and Chib approach will tend to run faster than the Metropolis sampling required for (9.7). This is because all we need to do is to simulate from the necessary conditional distributions. Simulating from the conditional distributions usually takes less numerical power than evaluating the Metropolis acceptance ratios. Second, a certain percentage of Metropolis draws does not result in the chain moving. Hence, the Metropolis draws may have somewhat higher autocorrelations and be slightly less efficient than Gibbs sampling.

We will need the ability to draw from a *truncated* normal distribution to implement the Albert–Chib probit model. We must draw from the appropriate region of Figure 11.1, either the shaded region or the unshaded one depending on the value of y. We can express the idea of truncated normal sampling using the indicator function I. The indicator function I takes on the value 1 when a condition is true and the value of zero otherwise. To express the idea of sampling in the shaded portion of the figure, we can write the distribution as $\text{Normal}(\mu, 1)I(y^* < 0)$. In words this is 'the normal distribution with mean μ and precision 1, multiplied by 1 when $y^* < 0$ and multiplied by zero otherwise'. When $y^* \geq 0$, the indicator function makes the resulting density of the distribution equal to zero. Since the probability density is zero, we will not sample any values from the distribution when $y^* \geq 0$. To express sampling in the unshaded portion of the figure, we sample from $\text{Normal}(\mu, 1)I(y^* \geq 0)$.

A simple extension of this idea is to define a lower truncation bound l and an upper truncation bound u. Then we change l and u as needed based on y. This can be made to be equivalent to our expression above. Additionally, it allows us to sample from the desired portion of the normal distribution depending on the value of y without having to change the inequalities. For sampling in the shaded portion of Figure 11.1, we can write $\text{Normal}(\mu, 1)I(l \leq y^* < u)$ with $l = -\infty$ and $u = 0$. Sampling in the unshaded portion of the figure can now be done by redefining $l = 0$ and $u = \infty$.

With these foundations, the latent data version of the probit model can be written as

$$
\begin{aligned}
y_i^* &\sim \text{Normal}(\mu_i, 1) I(l_i < y_i^* \leq u_i) \\
\mu_i &= \beta_0 + \beta_1 x_{1,i} + \cdots + \beta_k x_{k,i} \qquad (11.1) \\
\beta_0, \ldots, \beta_k &\sim \text{Normal}(0, 0.00000001).
\end{aligned}
$$

Here, $l_i = 0$ if $y_i = 1$, and $-\infty$ otherwise. Also $u_i = \infty$ if $y_i = 1$; otherwise it is 0.

11.2.1 In Practice: The Probit Model Using Latent Variables

Ingram (1984) studied differences in U.S. state government accounting practices. He noted that legislators and government officials are in a position to influence accounting practices used to monitor state government financial activities. Some states may have economic incentives to be more rigorous in their accounting while others may have incentives to be less so. Ingram (1984, Table 3) lists which generally accepted accounting principles (GAAP) practices were adopted by each state. For the purposes of our analysis here, we examine Practice 2 (Statements of Revenue and Expenses–Enterprise Funds) as our outcome variable y. States could either follow this practice or not, giving a Bernoulli outcome variable for our probit analysis. The original source of this information appears to be a 1980 publication by the Council on State Governments. Ingram was interested in whether various economic factors were related to state accounting practices. Here we use two economic predictor variables for our probit model. The first is the 1980 gross state product (GSP), which can be used to see whether states with larger overall economic output tend to adopt or not adopt GAAP Practice 2. The GSP data is taken from U.S. Bureau of the Census (1996, Table 689). Another possibility is that states with wealthier citizens may be able to generate more revenue per capita through taxation. The second predictor variable is therefore the 1980 personal income per capita (PIPC) (U.S. Bureau of the Census, 1996, Table 699). These variables are conceptually related, but the Pearson correlation between them is a more moderate $r = 0.33$.

Our model of (11.1) has an infinite upper bound for y^* when $y = 1$. It also has a lower bound of negative infinity for y^* when $y = 0$. However, we can set our lower and upper bounds to only finite values in *WinBUGS*. In practice, what we can do is to set the lower and upper bounds to reasonably extreme values. For example, we can set the upper bound to +10, +25, or +50. These would correspond to z-scores of 10, 25, or 50, all of which are considerably larger than what we might expect to see in practice. These values are also much larger than the internal limit of Φ that *WinBUGS* has (see Section 9.4). Below we set a constant named `bound.const` to 50. This produces a lower bound of −50 for y^* when $y = 0$ and an upper bound of +50 when $y = 1$. Implementation of the latent variable probit model appears in **WinBUGS Code 11.1 Probit Analysis.odc** as well as below.

```
model
 {
 for (i in 1:n) {
   y.star[i] ~ dnorm(mu[i], 1)I(L[i], U[i])
   mu[i] <- beta[1] + beta[2]*x1c[i] + beta[3]*x2c[i]
   x1c[i] <- gsp[i] - mean.gsp
   x2c[i] <- pipc[i] - mean.pipc
   L[i] <- -bound.const*(1-y[i])
   U[i] <- bound.const*(y[i])
   }
#prior for beta
  for (j in 1:3) { beta[j] ~ dnorm(0, 0.00000001) }
#other parameters
  mean.gsp <- mean(gsp[])
  mean.pipc <- mean(pipc[])
  bound.const <- 50
 }
```

Our latent variables `y.star[i]` are simulated from a normal distribution with mean `mu[i]`. Lower and upper bounds are given by `L[i]` and `U[i]`, and these in turn are influenced by our choice of `bound.const`. We do not have a link function in this model, as this is handled by `y.star`. The predictor variables `gsp` and `pipc` have been centered. Results of the model estimation also appear in **WinBUGS Code 11.1 Probit Analysis.odc**. The posterior distribution for `beta[2]` has a mean of 0.0072 and a 95% credible interval of (−0.001, 0.018). This suggests that GSP does not appear to be strongly predictive of the choice of accounting Practice 2. A similar nonsignificant conclusion is reached for `beta[3]`. The posterior mean of this distribution is −0.00011 and the 95% credible interval is (−0.00037, 0.00017). Thus neither of these two economic factors seems to predict the adoption of this accounting practice. We can compare the run times of the latent variable probit model of (11.1) to our previous probit model (9.7) in Section 9.4. The file **WinBUGS Code 11.1 Probit Analysis.odc** contains code listings for both of these approaches to the probit model. It took 59 s for *WinBUGS* to complete 100,000 iterations using the model (11.1), whereas it took 69 s for *WinBUGS* to complete 100,000 iterations using the model (9.7). This suggests an approximate 15% faster run time for the model (11.1) using the current data.

We can also use the latent variable approach to estimate other models related to the probit model. Albert and Chib (1993) showed that a logit model can be approximated by the latent variable method. They observed that the logistic distribution is closely related to a t-distribution with eight degrees of freedom. Therefore, it is possible to approximate a logit model by drawing the latent variables from a $t(8)$ distribution in (11.1). The standard deviations of the two distributions are different. This will cause the coefficients based on the latent $t(8)$ distribution to have magnitudes different from the logistic coefficients. However, Albert and Chib (1993) indicate that dividing the $t(8)$ coefficients by 0.634 should approximately produce the logistic coefficients (Section 11.6). It would also be possible to simulate from the logistic distribution directly in (11.1). However, Gibbs sampling is no longer possible. This means that the attractive computational properties of Gibbs sampling for the latent variable probit model will not be shared by the latent variable logit model. Even worse, experience indicates that *WinBUGS* may encounter difficulties with its Metropolis sampler for this model and generate error messages.

11.3 STRUCTURAL BREAK MODELS

Suppose there is a business environment that has stable properties for a certain length of time. However, at some point a change occurs, leading to some kind of transformation of the environment. Business activities may increase or decrease as a consequence. In practice, this means that certain parameter values will do a good job of predicting business activity at certain times, but at some point there is a break where the parameter values themselves change. The structural break model is a model that can be used for this kind of scenario. Typically, the exact timing of the breakpoint is not known perfectly. This means we have uncertainty in the model. The structural break model will try to locate the breakpoint and account for its uncertainty based on the available data.

The idea behind a structural break model can be applied to many business contexts. For example, product life cycles indicate that demand for a product may go through a number of distinct phases. There may be a growth phase with accelerating interest and sales. Next may be a maturity phase with a plateau, followed by a decline phase. A structural break model allows the different parameters related to the different life cycle phases to be estimated along with the unknown breakpoints (Hahn and Bunyaratavej, 2011).

We can draw parallels between the structural break model and a model with a latent dummy variable (Section 5.8). Suppose we have business events y_t unfold over time where t indicates the time period. Figure 11.2 graphically shows an example of this kind of data. Suppose we have a dummy variable X_t that takes on a value of zero prior to a known breakpoint. We use the capitalized variable X_t here to indicate that the breakpoint is known. Also, suppose X_t takes on a value of 1 at

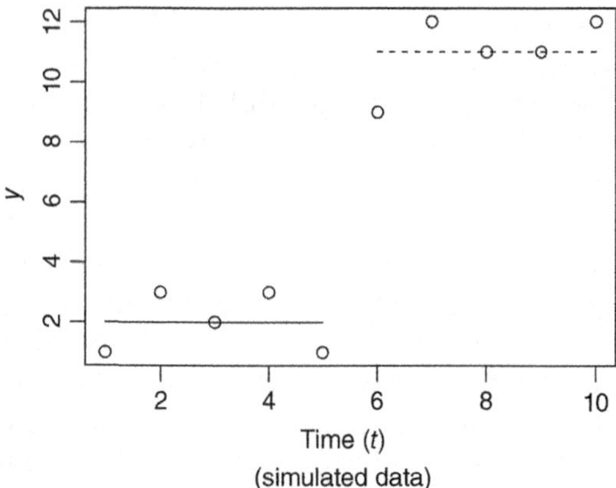

Figure 11.2 Business Outcomes Over Time

the known breakpoint and afterward. We can write a functional form such as the following for this kind of data:

$$y_t = \beta_0 + \beta_1 X_t.$$

Estimating this model for the data in Figure 11.2 would give a value of β_0 equal to 2 and a value of β_1 equal to 9 (which is the difference between 2 and the group mean). We can see that our model so far is equivalent to our ANOVA (analysis of variance) model with two groups.

This model is sensible if there is a known explanatory event or factor that occurs at the breakpoint which gives us the ability to assign 0's and 1's to X_t. For example, the data in Figure 11.2 might represent the number of sales transactions completed per day before and after a sales training course, or it might be the number of new products released per month before and after a new CEO was named at a firm. However, if we do not have a known before-and-after factor, assigning values to this variable becomes more of a guessing game. It would also overstate our certainty about the business data at hand. As a result our estimates could become overly precise. We encountered a similar problem in Section 10.3 when we discussed mean substitution. There we saw that, if we replaced an unknown value with a known number such as the mean, our results could become misleading.

A Bayesian approach would be to treat the breakpoint as an unknown quantity to be estimated. Suppose we use κ to symbolize the location of the breakpoint. We next introduce a latent dummy variable x_t, where the lowercase variable indicates that the breakpoint is no longer known. We can now use the indicator function to generate x_t based on κ. For values t less than or equal to κ, we set $x_t = 0$. Remaining values of x_t are set to 1. We can use the indicator function I to write this more compactly. Setting $x_t = I(t > \kappa)$ gives us the needed values of x_t.

We now need a prior for κ in order to perform a Bayesian analysis. Suppose t goes from 1 to its maximum value T. Also, suppose we believe there is one breakpoint in the data. We can give κ a uniform distribution on the integer values of t ranging from 1 to $T-1$. Since we are using the integer values of t from 1 to $T-1$, we call this kind of distribution the *discrete uniform distribution*.

Notice that if we set the upper limit of κ as T, then we will lose the structural break interpretation whenever κ is equal to T. We need to set the upper limit of κ to be $T-1$ to allow at least one observation after the breakpoint.

Our last consideration is what kind of data we might expect to observe over time. As always, the choice of the likelihood function will depend on the particulars of the data we are interested in. For some metrics, we might use the normal distribution as our likelihood function. It is very common, however, to observe counts of outcomes over time in business data. For this kind of data, the Poisson likelihood (or the negative binomial) would be more appropriate. So we can borrow elements from the Poisson regression model (9.4) and use them for our structural break model. The Poisson regression model makes use of a link function. The most common link function for this model is the log link as in (9.4). Putting it all together, we can now write a structural break model for count data that occurs over time, as in Figure 11.2, as follows:

$$\begin{aligned} y_t &\sim \text{Poisson}(\lambda_i) \\ \log(\lambda_t) &= \beta_1 + \beta_2 x_t \\ \beta_1, \beta_2 &\sim \text{Normal}(0, 0.00000001) \\ \kappa &\sim \text{Uniform}(1, T-1) \\ x_t &= I(t > \kappa). \end{aligned} \tag{11.2}$$

11.3.1 In Practice: Estimating Structural Break Models

We use targeted keyword advertising impression data from the Twitter social media network for our structural break model. A targeted advertisement may be shown to a Twitter user if two conditions are met. First, a user must have an interest in the keyword. Second, advertisers may place a bid amount for their advertisement to be shown. Higher bid amounts will result in a greater number of advertisements shown. When an ad is shown to a user, it is called an *impression*. The data consists of impressions by hour over a 24-hour period. For this campaign, ads were unlikely to be shown between the hours of midnight and 8 am. Ads were more likely to be shown in the afternoon and evening. We can conclude that there may be two qualitatively different time periods for ad impressions. A structural break in the counts seems to occur in the late morning. We analyze this data using code in **WinBUGS Code 11.2 Structural Break.odc**. The model code appears below.

```
model
 {
 for (t in 1:time) {
   y[t] ~ dpois(lambda[t])
   log(lambda[t]) <- beta[1] + beta[2]*x[t]
   x[t] <- step(t-1-kappa)
   kappa.post[t] <- equals(kappa,t)
   }
 #prior for beta and kappa
   for (j in 1:2) { beta[j] ~ dnorm(0, 0.00000001) }
   kappa ~ dcat(prior.p[])
 }
```

Here, the dependent variable `y[t]` is assumed to follow the Poisson distribution with expected value `lambda[t]`. The right-hand side of the functional form for `lambda[t]` involves the `beta` coefficients and the latent variable `x[t]`. The latent `x[t]` is generated based on `kappa` on a line toward the middle of the `for` loop. We can use the `step` function in *WinBUGS* to calculate the inequality. The `step` function is a "greater than or equal to" function, so it is necessary to make an adjustment to its argument. This adjustment appears in the code. We also create the variable `kappa.post`. This variable will provide the posterior probability that the breakpoint occurs at a given time for all possible breakpoints. We can use this to examine where the most likely location of the breakpoint is.

A 5000-iteration burn-in was used for this model, and estimation was based on 40,000 additional iterations of the Markov chain. Figure 11.3 shows trace plots for selected parameters in the model. We see that the posterior distributions occasionally make large shifts away from their typical locations. This is happening because of the uncertainty in κ. The best values of β depend strongly on whether κ is larger or smaller. We can see this on Figure 11.4(a). This portion of the figure shows a scatter plot of samples from the posterior distributions of `beta[1]` and `kappa`. When `kappa` is 11 or greater, `beta[1]` is almost always positive. However, when `kappa` is 9 or lower, `beta[1]` has a good chance of being negative.

Figure 11.4(b) shows the posterior distribution of κ. The most likely location for the breakpoint is 10 in the figure, which corresponds to 11 am. The estimated value of `kappa.post[10]` is 0.771. This means that the posterior probability of the breakpoint occurring at 11 am is 77.1%. We can see from Figure 11.4(b) that the next most likely breakpoint is 10 am. This breakpoint has a posterior probability of 13.8%. The posterior mean for β_1 is 0.371, and its 95% credible interval is (−0.959, 1.018). The posterior mean for β_2 is 2.771, and its 95% credible interval is (2.136, 4.021). The posterior mean for κ is 9.768, and its 95%

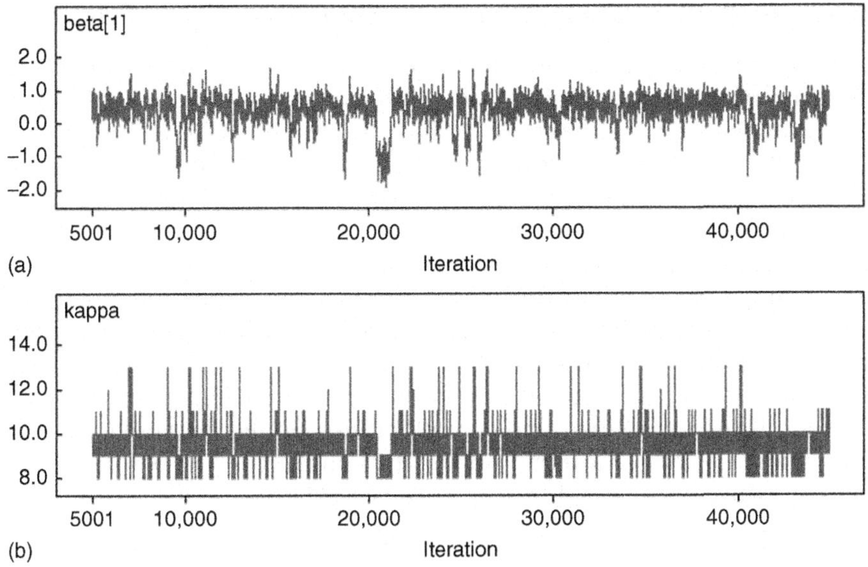

Figure 11.3 (a,b) Trace Plots for Twitter Structural Break Model 1

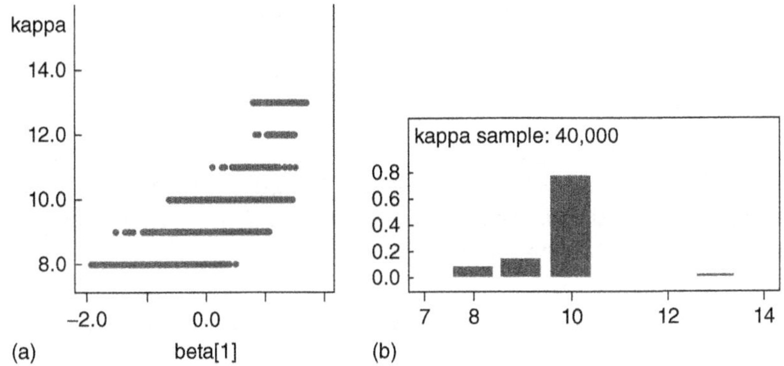

Figure 11.4 (a,b) Scatter Plot of β_1 Versus κ and Posterior Distribution of κ

credible interval is (8, 10). We can also simulate from the values of `lambda[t]`. This will give us the expected number of impressions each hour.

Finally, it is worth mentioning that latent variable models such as this one may have a greater chance of convergence problems. So it is especially worthwhile to use the full suite of convergence diagnostics for this class of models. Running parallel chains and using the Brooks–Gelman–Rubin diagnostic from Section 6.4 can help show whether there are any alternative modes to which chains might converge. Investigating this model with three parallel chains appears in Section 11.6.

11.3.2 In Practice: Adding Covariates to Structural Break Models

Our current model could do a better job of explaining our Twitter data. Our current model is designed to fit data such as appearing in Figure 11.2. However, the data is more complex than this (Figure 11.5). There appears to be a trend in the data, with increasing number of impressions as the afternoon progresses. We will now model this trend by adding a covariate to our functional form.

Since our current model has a latent variable, we want to spend a little time thinking about what would be the best way to extend the functional form. Our latent variable x_t takes on the value 1 when t is greater than κ, and zero otherwise. The trend in the data appears to largely increase with time t. This means that the product term $x_t * t$ will take on the value t when t is greater than κ, and zero otherwise. So the product term will produce a trend after the break, and no trend at the break or before it. This kind of covariate would address our needs. Updating our model with this new term gives the following specification:

$$\begin{aligned} y_t &\sim \text{Poisson}(\lambda_i) \\ \log(\lambda_t) &= \beta_1 + \beta_2 x_t + \beta_3 x_t t \\ \beta_1, \ldots, \beta_3 &\sim \text{Normal}(0, 0.00000001) \\ \kappa &\sim \text{Uniform}(1, T-1). \\ x_t &= I(t > \kappa). \end{aligned} \tag{11.3}$$

Our *WinBUGS* code needs only minor changes to estimate this model. We change the functional form so that the covariate is included along with a new coefficient. The change is as follows:

```
log(lambda[t]) <- beta[1] + beta[2]*x[t] + beta[3]*x[t]*t.
```

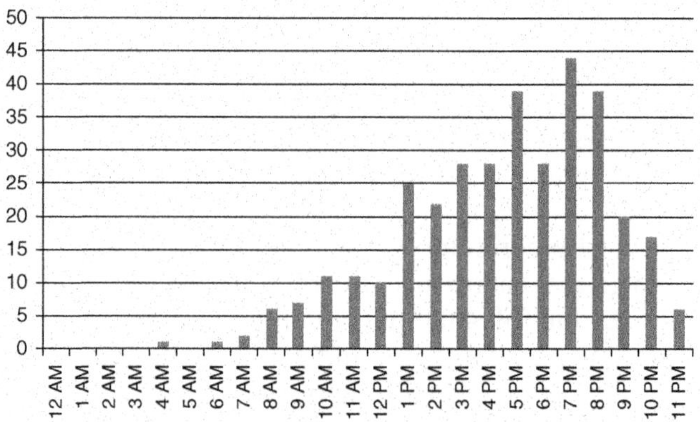

Figure 11.5 Counts of Ad Impressions by Hour

We also need to revise our prior code loop and the initial values to estimate the new coefficient. A run of 50,000 iterations after a 5000-iteration burn-in was used for parameter estimates. The posterior distribution of `kappa` is now strongly concentrated at 7 am. The posterior probability that the break occurs at 7 am was 97.2%. This seems more consistent with the data of Figure 11.5 than the results from our previous model.

The trace plots for our `beta` parameters appear to be stable. However, some evidence of slow mixing is visible. Clicking on the autocorrelation button in *WinBUGS* shows that the parameters have a high amount of autocorrelation. We can use the `Correlations` tool in the `Inference` menu to examine this further. Entering `beta` in the first field and clicking on `scatter` produces a plot such as that in Figure 11.6. We see that the `beta[1]` and `beta[2]` are highly correlated. This is preventing the chain from mixing as well as it could. As a result, our Monte Carlo error is higher than it would be if we had a chain that was mixing well. We can use the `print` button on the `Correlations` tool to estimate the correlation between the two parameters. The correlation is high at –0.929.

11.3.3 In Detail: Improving Parameter Mixing in Structural Break Models

We examined models that suffered from slow mixing in Section 6.3. Many times we are able to avoid slow mixing by centering our covariates around the mean. However, in this model one of our covariates is latent. We therefore may wish to avoid centering a latent variable around its latent mean. Our other covariate is t. However, centering this variable may not directly address the problem (Section 11.6). Our problem is the correlation between β_1 and β_2, while t is associated with a different coefficient, β_3.

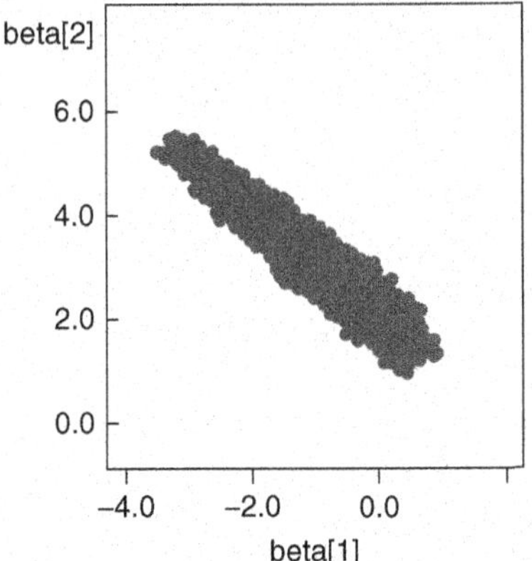

Figure 11.6 Scatter Plot of `beta[1]` Versus `beta[2]`: Structural Break Model with Covariate

In Section 6.3, we discussed another possible solution to the slow mixing problem. We can address the problem by transforming our priors to produce dependence. The goal is to have our original priors be independent of each other. Then we apply a functional form to them to produce the dependence. In Section 6.3 and in **WinBUGS Code 6.3 re-parameterizing.odc**, we saw that using a regression-like functional form could solve this problem for us. However, we are not limited to this particular functional form. We can experiment with others that we think might be relevant. One simple "trick" is to use β_1 and β_2 to make α_1 and α_2 as follows:

$$\alpha_1 = \beta_1 + \beta_2,$$
$$\alpha_2 = \beta_1 - \beta_2.$$

This transformation is capable of producing a strong correlation between α_1 and α_2 from reasonably independent β_1 and β_2. Then β_1 and β_2 can have fast mixing, which will improve our results. Further improvement can also be achieved when we center t around its mean. Since the linear trend occurs mainly in the afternoon, we can take the "afternoon mean" to be 3 pm ($t = 16$). The model can then be revised as follows:

$$\begin{aligned}
y_t &\sim \text{Poisson}(\lambda_i)\\
\log(\lambda_t) &= \alpha_1 + \alpha_2 x_t + \beta_3 x_t t\\
\beta_1, \ldots, \beta_3 &\sim \text{Normal}(0, 0.00000001)\\
\kappa &\sim \text{Uniform}(1, T-1)\\
x_t &= I(t > \kappa)\\
\alpha_1 &= \beta_1 + \beta_2\\
\alpha_2 &= \beta_1 - \beta_2.
\end{aligned} \tag{11.4}$$

For our *WinBUGS* code, we make the required changes to the functional form for `lambda[t]`. We also need to add the functional forms for `alpha[1]` and `alpha[2]`. Table 11.1 compares selected results from Model (11.3) and Model (11.4). Each line of the table represents the same quantity, but may have different names depending on the model. For example, the β_1 coefficient in Model (11.3) corresponds to the α_1 coefficient in Model (11.4), and so on. We see that the Monte Carlo error is considerably lower in Model (11.4). The revised Model (11.4) decreases the Monte Carlo error by over 90% for the first two parameters in Table 11.1. For the last parameter, β_3, Model (11.4) decreases the Monte Carlo error by 75%.

The term *structural break* can be traced back to at least Kuh (1956). These models are also called *changepoint models*. Carlin et al. (1992) performed an early Bayesian analysis of this kind of model using MCMC. The

TABLE 11.1 Coefficient Means and Monte Carlo Errors: Structural Break Model for Twitter Ad Impressions

Model (11.3)			Model (11.4)		
Parameter	Posterior Mean	MC Error	Parameter	Posterior Mean	MC Error
β_1	−0.854	0.0287	α_1	−0.809	0.0028
β_2	2.971	0.0305	α_2	3.806	0.0028
β_3	0.055	0.000299	β_3	0.055	0.000066

existence of structural breaks can pose a challenge for understanding and predicting business activities. However, modeling can provide insights about these issues. Our discussion in this section hopefully emphasizes the modular nature of Bayesian modeling. We see that we can take concepts from different models and snap them together to form a new model. Here we see that the structural break model can be "built" from a Poisson regression model and a missing data model. We are not limited to this particular model, however. We can build other kinds of structural break models by putting together other models. This modularity is not limited to structural break models, of course. The excitement and potential of Bayesian methods is that you can come up with your own model and "build" it to your own specifications.

11.4 IN DETAIL: THE ORDINAL PROBIT MODEL

Sometimes we will find that our data consists of ordered or sortable categories. This type of data is known as *ordinal data*. Ordinal data has information about which items are greater (or lesser) than others on a certain scale. For example, our company may make sweaters that are sized as small, medium, large, or extra large. We know the relative sizes of the sweaters and we can rank them from the smallest to the largest (or vice versa). Customers may respond to a marketing survey indicating that they are active participants, somewhat active participants, or inactive participants in a hobby. Stocks may be rated as buy, hold, or sell by financial analysts. Financial analysts may also say they are overweight, underweight, or market-weight to describe their positions in different sectors of a stock market. We see that ordinal data gives us relative information instead of absolute information. For example, a particular size of sweater is considered Small relative to other sweater sizes. However, the category Small does not provide us with information on the absolute size of the sweater.

The normal regression model is not designed for ordinal data. Instead, we can model ordinal outcome data with the ordinal probit model. This model is perhaps the most complex model we will discuss in this book.

However, it may be simpler to understand when we recognize that it blends ideas from our latent variable probit model of Section 11.2 with ideas from the structural break model of the previous section. For our latent variable probit model, recall that we simulated a latent y^* from a portion of the normal distribution. Referring back to Figure 11.1, we would simulate a value of y^* from the shaded portion of the distribution if our binary outcome y was zero. If our binary outcome y was 1, we would simulate y^* from the unshaded portion of the distribution. We then performed a regression analysis using the simulated y^* as our outcome variable. We can look at Figure 11.1 again and see that a sort of "breakpoint" occurs in the normal distribution. The shaded portion of the distribution is "structurally" different from the unshaded portion because the two locations correspond to qualitatively different outcomes in y.

11.4.1 Posterior Simulation in the Ordinal Probit Model

Simulating y^* in the Ordinal Probit Model Albert and Chib (1993) pointed out that the idea of simulating latent y^* outcomes can be extended to the case of ordinal data. Figure 11.7 shows graphically how we would simulate latent variables based on three ordinal categories. Suppose we have three sortable categories such as buy, hold, and sell for a stock. Also, suppose our functional form μ is currently equal to a value of $\mu = 0.5$ as in Figure 11.7. Our first breakpoint is at $\kappa_1 = 0$. If the stock has a buy rating, then we would simulate y^* from the shaded left part of the normal distribution in the figure. As in Section 11.2, we will simulate y^* on the range from a lower boundary constant (such as -15) up to zero.

Our second breakpoint is at $\kappa_2 = 2.5$. If the stock has a hold rating, we would simulate y^* from the central unshaded portion of Figure 11.7. This means that we will simulate y^* in the range κ_1 to κ_2 if the stock has a hold rating. If the stock has a sell rating, we would simulate y^* from the right portion of the Figure 11.7 that is shaded with lines. This means we will simulate y^* in the range κ_2 to an upper boundary constant such as $+15$. Once we have simulated values of y^* for all observations, we can then proceed to the next step.

Simulating κ in the Ordinal Probit Model Just as in the structural break model, the location of the breakpoints is not known in advance. We therefore have to estimate some of the κ parameters from the data. Fortunately, we can continue to assume that the first breakpoint, κ_1, is always equal to zero. The location of additional breakpoints (κ_2 and higher) must be estimated.

Figure 11.8 shows a close-up diagram of the previous normal distribution in Figure 11.7. Vertical tick marks along the x-axis depict the value of y^* that we have just simulated. We would now like to update κ_2. In this diagram, κ_2 can only move a small distance around its current location at

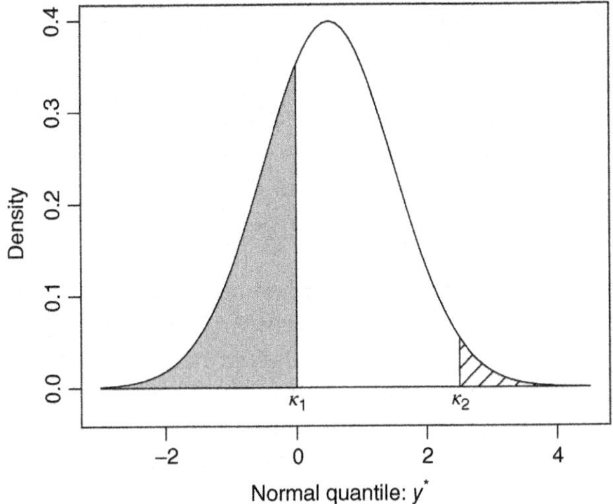

Figure 11.7 Simulating Latent Variables from the $\mu = 0.5$ Normal Distribution Given the Ordinal Outcome

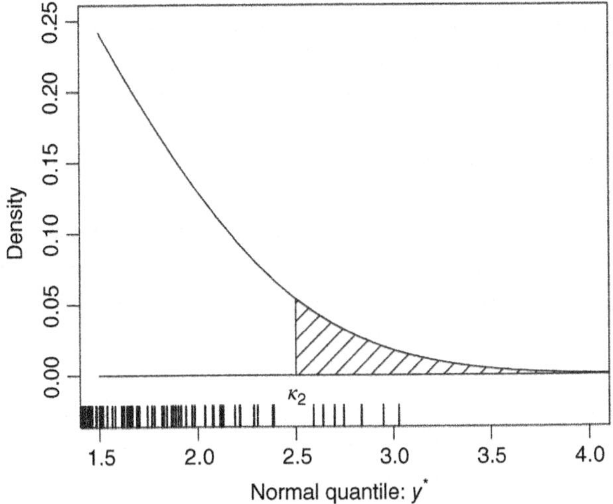

Figure 11.8 Updating κ_2 Given y^*

2.5 on the x-axis. This is because we do not want κ_2 to be larger than the smallest y^* under the line-shaded portion of the normal distribution. Let us look closer at this value of y^* which appears at about 2.6 on the x-axis. If κ_2 were to become larger than this y^*, this would be as if we changed the rating on that particular stock from sell to hold. We do not want our model to somehow change the actual stock ratings, so κ_2 must stay below this y^*.

Similarly, we do not want κ_2 to be smaller than the largest y^* under the unshaded portion of the normal distribution. This value of y^* appears at about 2.4 on the x-axis. If κ_2 were to become smaller than this y^*, it would be as if we changed the rating on that particular stock from hold to sell. Again, we do not want to change the actual ratings, so κ_2 must stay above this y^*. To summarize, we must ensure that κ stays within the appropriate boundaries when we simulate from its posterior distribution. The upper boundary of κ is the smallest value of y^* from the category to the right of κ. The lower boundary of κ is the largest value of y^* from the category to the left of κ.

Since κ_1 is a constant, it does not need a prior. All other κ parameters need priors. We need to make the prior mean larger for each subsequent κ parameter to ensure that the ordering is retained. We might want to have the prior mean of κ_2 to be equal to 1. If we had a κ_3, we could set its prior mean equal to 2, and so on. The prior precisions for the κ parameters can be a value such as 0.1. In practice, the values of y^* will often tend to limit the movement of κ more than the prior precisions.

Simulating β in the Ordinal Probit Model Once we have the appropriate values of y^* and κ, the posterior distribution of β is found just like in ordinary regression. The Gibbs sampler can be used as if we had normally distributed y outcome data. *WinBUGS* is able to detect that Gibbs sampling can be performed and will use this method to draw samples from the posterior distributions of the β parameters.

We can now consider how to write out the model specification for the ordinal probit model. Suppose we have J ordinal categories where $J \geq 3$. We let i be the index for the observations within each category. Our covariate is $x_{i,j}$ (extensions to more covariates are straightforward). Our latent variables are $y^*_{i,j}$. From our discussion above, we know $y^*_{i,j}$ will be simulated in a range depending on which category j we are in. The ranges involve the lower boundary constant, one of the κ parameters, or the upper boundary constant depending on j. We write the lower end of a range as l_j and the upper end of a range as u_j. To simulate κ, we need the largest and the smallest latent variables in adjacent categories. Let $y_j^{*\max}$ be the largest latent variable in category j and $y_j^{*\min}$ the smallest. κ_1 will be a constant equal to zero, and we will estimate $j - 1$ additional κ parameters. When we put it all together, we get the following model:

$$
\begin{aligned}
y^*_{i,j} &\sim \text{Normal}(\mu_{i,j}, 1) I(l_j < y^*_{i,j} < u_j) \\
\mu_{i,j} &= \beta_1 + \beta_2 x_{i,j} \\
\beta_1, \beta_2 &\sim \text{Normal}(0, 0.00001) \\
\kappa_j &\sim \text{Normal}(j - 1, 0.1) I(y_j^{*\max} < \kappa_j < y_{j+1}^{*\min})
\end{aligned}
\tag{11.5}
$$

$$\kappa_1 = 0, l_1 = -15, u_J = 15$$

$$l_2 = u_1 = \kappa_1$$

$$\dots$$

$$l_J = u_{j-1} = \kappa_J.$$

11.4.2 In Practice: Modeling Credit Ratings with Ordinal Probit

Credit ratings agencies use various factors to assess the risk of lending. Sovereign credit ratings indicate the relative level of risk associated with lending to different nations. Standard and Poor's (S&P) is one such agency that issues sovereign credit ratings. The highest tier of S&P ratings can range from "AAA," the highest level, to "A" which is a moderately high rating. The ratings continue with "BBB" to "B," "CCC" to "C," and "D" which indicates default. Finer gradings of the above ratings are created by adding a plus sign or a minus sign to the end of the letters. Ratings of "BB" or lower indicate higher risk countries and bonds from these countries are speculative grade (sometimes called *junk* bonds).

We examine whether GDP (gross domestic product) per capita is related to S&P credit rating in 117 countries. We use the 2011 GDP per capita data from World Bank (2013). S&P credit ratings were obtained from Standard and Poor's (2013). To simplify matters, we recode the data into four ordinal rankings. Category 1 contains 'AAA' and 'AA', the highest tier. Category 2 contains "A" and "BBB," so this tier is still investment grade. Category 3 contains speculative grade countries with "BB" or "B" ratings. Category 4 contains countries with "CCC" ratings and below.

The data is arranged into columns by ordinal category so that y^* can be simulated more effectively. For example, all the countries with a Category 1 credit score appear in column `y[,1]` (see **WinBUGS Code 11.3 Ordinal Probit Analysis.odc** for a full data listing). We supply the number of countries in a particular category with the data `n[1]` to `n[4]`. Countries do not evenly fall into categories, so `NA` values are added to complete the columns in the dataset. The program code uses the values of `n[1]` to `n[4]` to ensure that these `NA` values are not analyzed. Portions of the code listing appear below.

```
model
{
 for (j in 1:4) { # number of ordinal categories
  for (i in 1:n[j]) {
```

```
    lower[i,j] <- L[y[i,j]]; upper[i,j] <- U[y[i,j]]
    y.star[i,j]~dnorm(mu[i,j],1)I(lower[i,j],upper[i,j])
    mu[i,j] <- beta[1] + beta[2]*x.ctr[i,j]  }
    }
#prior for beta
 for (j in 1:2) { beta[j] ~ dnorm(0, 0.0001) }
#updating kappa
 for (i in 1:n[2]) {ystar2[i] <- y.star[i,2]}
 for (i in 1:n[3]) {ystar3[i] <- y.star[i,3]}
 for (i in 1:n[4]) {ystar4[i] <- y.star[i,4]}
 ystar.max[2] <- ranked(ystar2[],n[2])
 ystar.max[3] <- ranked(ystar3[],n[3])
 ystar.min[3] <- ranked(ystar3[],1)
 ystar.min[4] <- ranked(ystar4[],1)
 kappa[1] <- 0
 kappa[2] ~ dnorm(1,.1)I(ystar.max[2],ystar.min[3])
 kappa[3] ~ dnorm(2,.1)I(ystar.max[3],ystar.min[4])
#other quantities
 L[1] <- -15;    U[1] <- kappa[1]
 L[2] <- kappa[1]; U[2] <- kappa[2]
 L[3] <- kappa[2]; U[3] <- kappa[3]
 L[4] <- kappa[3]; U[4] <- 15
}
```

We take the log of GDP per capita and center it around the mean to produce `x.ctr`, the covariate. We use the *WinBUGS* function `ranked` to find the maximum and minimum values of `y.star`. This function sorts the `y.star` values from the smallest to the largest. The `kappa` parameters typically cannot move very far at each iteration because they are bounded by the `ystar.max` and `ystar.min` values. This means that the autocorrelation for these parameters is relatively high. We allow the chain to run for 100,000 iterations (after 5000 burn-in iterations) to produce the estimates. These estimates appear in Table 11.2.

The most important parameter in Table 11.2 is β_2. This measures the relationship between log GDP per capita and credit risk ranking. The parameter is negative, and the 95% credible interval is well away from zero. Thus, as log GDP per capita goes up, the credit score risk ranking goes down (lower risk). β_1 is the intercept, and we can interpret this as the expected y^* for a country with an average log GDP per capita. Since β_1 is

TABLE 11.2 Coefficient Estimates: Sovereign Credit Ratings Ordinal Probit Model

Parameter	Mean	Std. Dev.	95% Credible Interval
β_1	1.163	0.175	(0.830, 1.514)
β_2	−0.829	0.107	(−1.042, −0.624)
κ_1	1.310	0.197	(0.952, 1.721)
κ_2	3.787	0.350	(3.135, 4.494)

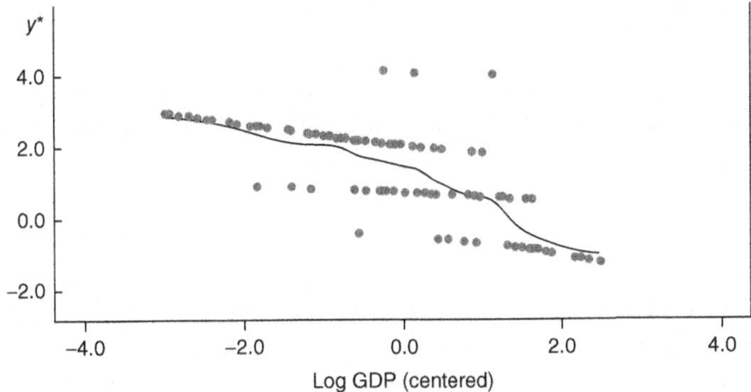

Figure 11.9 Model Fit Plot: Sovereign Credit Ratings Ordinal Probit Model

1.163 (>0), we know that an average country would not tend to be in the safest risk category. Instead, we see that β_1 is close to κ_1. This means that a country with an average log GDP per capita would tend to be near the border between Category 2 and Category 3 risk rankings.

Model fit can be examined in Figure 11.9 (Section 11.6). We see that there is room for improvement in the model. Log GDP per capita appears to be a good starting point for predictive capability. However, there are some highly risky countries that are not predicted well. We also see that countries may be in any of the four risk categories around the mean (centered log GDP = 0). Therefore, on its own, log GDP per capita best differentiates the countries that are at more extreme values.

11.5 SUMMARY

Latent variable models greatly expand the scope of our modeling capabilities. Bayesian MCMC gives us the interesting option of simulating a latent variable based on the parameters and the data. Once we have the simulated variable, we can then use a standard model to analyze it as if we had the actual data. This greatly simplifies some models which would otherwise be more complex to estimate.

We also saw (especially in Section 11.3) that Bayesian methods can be very modular. We can easily fit together different features of different models. This allows us to develop new models that are more appropriate for our data and for our modeling goals. Although the book ends here, the journey continues because there are even more modeling possibilities using Bayesian methods. There is no shortage of important business and management problems that can be analyzed, and no shortage of insights to be found. Enjoy the journey!

11.6 EXERCISES

1. Obtain an approximate logistic regression model by modifying the code in **WinBUGS 11.1 Probit Analysis.odc**. Change the distribution for `y.star` to be a $t(8)$ distribution with a precision of 1 and a mean `mu[i]` using *WinBUGS'* `dt()` function. Next create variables called `beta.rescale` that are equal to the `beta` variables divided by 0.638. Finally, modify the link function of the probit model at the bottom of **WinBUGS 11.1 Probit Analysis.odc** to obtain a "true" logit model. Compare your `beta.rescale` coefficients with those obtained from the "true" logit model and comment on their resemblance.

2. Consider the models in Question 1. Calculate the value of the precision needed for `dt()` so that the rescaling of `beta` is no longer necessary. Insert this value into the `dt()` function and estimate the model. Comment on the results you obtain and whether they match with the "true" logit coefficients.

3. Reestimate Model (11.2) using the code from **WinBUGS Code 11.2 Structural Break.odc** and the Twitter advertising data, but use three parallel chains. Consider initial values of 0, 5, and −5 for both `beta` parameters. Review the Brooks–Gelman–Rubin diagnostic and the trace plots. Comment on performance in a shorter run of 10,000 iterations. Compare with performance in a longer run of 50,000 iterations.

4. Reestimate Model (11.3) using the code from **WinBUGS Code 11.2 Structural Break.odc** and the Twitter advertising data, but now center the time covariate `t` around its mean. How does this affect the correlation between `beta[1]` and `beta[2]` that was discussed in Section 11.3.2? In Section 11.3.2, the correlation was −0.929. Is the correlation now smaller than −0.929? Has chain mixing improved?

5. Many times, ordinal data can be assumed to arise from an underlying continuous variable that gives rise to discrete outcomes. For example, the discrete ordered outcomes "good," "better," "best" could easily arise from a continuous underlying dimension of "quality." Consider the credit outcome data for the ordinal probit model in Section 11.4.2. What underlying continuous variable might we assume underlies the discrete credit scores?

6. Generate the model fit plot in Figure 11.9 by adding some extra code to **WinBUGS Code 11.3 Ordinal Probit Analysis.odc**. Add some loops to write out the 117 values of `y.start[i,j]` to a new variable `ystar.fit[i]`. Also, write out the values of `x.ctr[i,j]` to a new

variable `x.fit[i]`. Then produce the model fit plot as usual with the `Compare` tool.

7. Modify **WinBUGS Code 11.3 Ordinal Probit Analysis.odc** to perform the following analysis. Simulate the posterior predictive probability that a country has the safest credit rating if its log GDP per capita is 10.

APPENDIX A

COMMON STATISTICAL DISTRIBUTIONS

For ease of reference, the main statistical distributions used in this book appear in Tables A.1 and A.2.

TABLE A.1 Continuous Univariate Distributions

Name	Parameters	Density Function
Beta	$\alpha > 0, \beta > 0$ mean $= \alpha/(\alpha+\beta)$, var. $= \dfrac{\alpha\beta}{(\alpha+\beta)^2(\alpha+\beta+1)}$	$\dfrac{\Gamma(\alpha+\beta)}{\Gamma(\alpha)\Gamma(\beta)}\pi^{\alpha-1}(1-\pi)^{\beta-1}$
Normal (using precision)	mean μ, precision $\tau > 0$ var. $= 1/\tau$	$\sqrt{\dfrac{\tau}{2\pi}}\exp\left(-\dfrac{\tau}{2}(y-\mu)^2\right)$
Normal (using variance)	mean μ, variance $\sigma^2 > 0$	$\dfrac{1}{\sqrt{2\pi\sigma^2}}\exp\left(-\dfrac{1}{2\sigma^2}(y-\mu)^2\right)$
Gamma	$\alpha > 0, \beta > 0, y > 0$ mean $= \alpha/\beta$ var. $= \alpha/\beta^2$	$\dfrac{\beta^\alpha}{\Gamma(\alpha)}y^{\alpha-1}\exp(-y\beta)$

(continued)

Bayesian Methods for Management and Business: Pragmatic Solutions for Real Problems, First Edition. Eugene D. Hahn.

TABLE A.1 (*Continued*)

Name	Parameters	Density Function
Uniform	$a < b$ mean $= (a+b)/2$ var. $= \dfrac{(b-a)^2}{12}$	$\dfrac{1}{b-a}$

TABLE A.2 Discrete Univariate Distributions

Name	Parameters	Density Function
Binomial	$0 \geq \pi \geq 1, y \geq 0, n \geq y$ mean $= n\pi$, var. $= n\pi(1-\pi)$	$\binom{n}{y} \pi^y (1-\pi)^{n-y}$
Multinomial	$0 \geq \pi \geq 1, y_i \geq 0$ $n = \sum_{i=1}^{I} y_i, \sum_{i=1}^{I} \pi_i = 1$ mean $= n\pi_i$, var. $= n\pi_i(1-\pi_i)$	$\dfrac{n!}{y_1! \cdots y_I!} \pi_1^{y_1} \cdots \pi_I^{y_I}$
Negative binomial	$\lambda, a > 0$ mean $= \lambda$, var. $= \lambda + \lambda/a^2$	$\dfrac{\Gamma(y+a)}{\Gamma(a)\Gamma(y+1)} \left(\dfrac{a}{\lambda+a}\right)^a \left(\dfrac{\lambda}{\lambda+a}\right)^y$
Poisson	$\lambda > 0$ mean $= \lambda$, var. $= \lambda$	$\dfrac{\lambda^y}{y!} \exp(-y)$

REFERENCES

Agresti, A. (2002). *Categorical Data Analysis*. Hoboken, NJ: John Wiley & Sons, Inc., 2nd edition.

Akaike, H. (1973). Information theory and an extension of the maximum likelihood principle. In *Proceedings of the Second International Symposium on Information Theory*, B. N. Petrov and F. Csàki, eds., pp. 267–281. Budapest: Akadémiai Kiadó. Reprinted in (1992) *Breakthroughs in Statistics*, vol. 1, S. Kotz and N. L. Johnson, eds., pp. 610-624, New York: Springer-Verlag.

Akaike, H. (1983). Information measures and model selection. *Bulletin of the International Statistical Institute*, **50**, 277–290.

Albert, J. H. and Chib, S. (1993). Bayesian analysis of binary and polychotomous response data. *Journal of the American Statistical Association*, **88**(422), 669–679.

Almeida, P., Song, J., and Grant, R. M. (2002). Are firms superior to alliances and markets? An empirical test of cross-border knowledge building. *Organization Science*, **13**(2), 147–161.

Altman, M., Gill, J., and McDonald, M. P. (2004). *Numerical Issues in Statistical Consulting for the Social Scientist*. Hoboken, NJ: John Wiley & Sons, Inc.

Ando, T. (2010). *Bayesian Model Selection and Statistical Modeling*. Boca Raton, FL: Chapman and Hall/CRC Press.

Ansari, A., Jedidi, K., and Jagpal, S. (2000). A hierarchical Bayesian methodology for treating heterogeneity in structural equation models. *Marketing Science*, **19**(4), 328–347.

Arora, N., Allenby, G. M., and Ginter, J. L. (1998). A hierarchical Bayesian model of primary and secondary demand. *Marketing Science*, **17**(1), 29–44.

Bayesian Methods for Management and Business: Pragmatic Solutions for Real Problems, First Edition. Eugene D. Hahn.

Arregle, J.-L., Beamish, P. W., and Hébert, L. (2009). The regional dimension of MNEs' foreign subsidiary localization. *Journal of International Business Studies*, **40**(1), 86–107.

Baber, W. R., Daniel, P. L., and Roberts, A. A. (2002). Compensation to managers of charitable organizations: an empirical study of the role of accounting measures of program activities. *Accounting Review*, **77**(3), 679–693.

Bayarri, M. J. and Berger, J. O. (1998). Quantifying surprise in the data and model verification. In *Bayesian Statistics 6*, J. M. Bernardo, J. O. Berger, A. P. Dawid, and A. F. M. Smith, eds., pp. 53–82. Oxford: Clarendon Press.

Bayarri, M. J. and Berger, J. O. (2000). *p* values for composite null values. *Journal of the American Statistical Association*, **95**(452), 1127–1142.

Bayarri, M. J. and Berger, J. O. (2004). The interplay of Bayesian and frequentist analysis. *Statistical Science*, **19**(1), 58–80.

Bayes, T. (1763). An essay towards solving a problem in the doctrine of chances. *Philosophical Transactions*, **53**, 370–418.

Benaroch, M., Lichtenstein, Y., and Robinson, K. (2006). Real options in information technology risk management: an empirical validation of risk-option relationships. *MIS Quarterly*, **30**(4), 827–864.

Berg, A., Meyer, R., and Yu, J. (2004). Deviance Information Criterion for comparing stochastic volatility models. *Journal of Business and Economics Statistics*, **22**(1), 107–120.

Bernardo, J. M. (1979). Reference posterior distributions for Bayesian inference. *Journal of the Royal Statistical Society, Series B*, **41**(2), 113–147.

Bernardo, J. and Smith, A. (1994). *Bayesian Theory*. Chichester: John Wiley & Sons, Ltd.

Birley, S. (1986). The role of new firms: births, deaths, and job generation. *Strategic Management Journal*, **7**(4), 361–376.

Bliss, C. (1937). The calculation of the dosage-mortality curve. *Annals of Applied Biology*, **22**(1), 134–167.

Board of Governors of the Federal Reserve System (2012). Consumer credit–G.19. Current release Oct. 2012. Release date: Dec. 7, 2012. http://www.federalreserve.gov/releases/g19/current/default.htm, accessed Dec. 30, 2012.

Box, G. E. P. and Tiao, G. C. (1973). *Bayesian Inference in Statistical Analysis*. New York: John Wiley & Sons, Inc.

Bromiley, P. and Marcus, A. (1989). The deterrent to dubious corporate behavior: profitability, probability and safety recalls. *Strategic Management Journal*, **10**(3), 233–250.

Brooks, S. P. and Gelman, A. (1998). General methods for monitoring convergence of iterative simulations. *Journal of Computational and Graphical Statistics*, **7**(4), 434–455.

Broschak, J. P. (2004). Managers' mobility and market interface: the effect of managers' career mobility on the dissolution of market ties. *Administrative Science Quarterly*, **49**(4), 608–640.

Bureau of Labor Statistics (2013). Arizona, unemployment rate, male, 16+- gpu00400000r0062 and Arizona, unemployment rate, female, 16+- gpu00400000r0084. http://data.bls.gov/cgi-bin/surveymost?gp, accessed Apr. 09, 2014.

Burnham, K. P. and Anderson, D. R. (1998). *Model Selection and Inference: A Practical Information-Theoretic Approach*. New York: Springer-Verlag.

van Buuren, S. (2012). *Flexible Imputation of Missing Data*. Boca Raton, FL: CRC Press.

Cameron, A. C. and Trivedi, P. K. (1998). *Regression Analysis of Count Data*. Cambridge: Cambridge University Press.

Capelleras, J., Mole, K., Greene, F., and Storey, D. (2008). Do more heavily regulated economies have poorer performing new ventures? Evidence from Britain and Spain. *Journal of International Business Studies*, **39**(4), 688–704.

Carlin, B. P., Gelfand, A. E., and Smith, A. F. M. (1992). Hierarchical Bayesian analysis of changepoint problems. *Journal of the Royal Statistical Society. Series C*, **41**(2), 389–405.

Carlin, B. and Louis, T. A. (2000). *Bayes and Empirical Bayes Methods for Data Analysis*. Boca Raton, FL: Chapman & Hall/CRC, 2nd edition.

Carpenter, J. R. and Kenward, M. G. (2013). *Multiple Imputation and Its Application*. Hoboken, NJ: John Wiley & Sons, Inc.

Carrigan, G., Barnett, A. G., Dobson, A. J., and Mishra, G. (2007). Compensating for missing data from longitudinal studies using WinBUGS. *Journal of Statistical Software*, **19**(7), http://www.jstatsoft.org/v19/i07/paper, accessed Apr. 09, 2014.

Celeux, G., Forbes, F., Robert, C. P., and Titterington, D. M. (2006). Deviance information criteria for missing data models. *Bayesian Analysis*, **1**(4), 651–673.

Cenfetelli, R. T. and Schwarz, A. (2011). Identifying and testing the inhibitors of technology usage intentions. *Information Systems Research*, **22**(4), 808–823.

Chaloner, K. (1996). Elicitation of prior distributions. In *Bayesian Biostatistics*, D. A. Berry and D. K. Stangl, eds., pp. 141–156. New York: Marcel Dekker.

Chamberlain, G. and Imbens, G. W. (2003). Nonparametric applications of Bayesian inference. *Journal of Business & Economic Statistics*, **21**(1), 12–18.

Chambers, E. and Cox, D. (1967). Discrimination between alternative binary response models. *Biometrika*, **54**(3/4), 573–578.

Chan, K. S. and Geyer, C. J. (1994). Discussion: Markov chains for exploring posterior distributions. *Annals of Statistics*, **22**(4), 1747–1758.

Chen, M. H. (2006). Comments on article by Celeux et al. *Bayesian Analysis*, **1**(4), 677–680.

Chen, M.-H., Huang, L., Ibrahim, J. G., and Kim, S. (2008). Bayesian variable selection and computation for generalized linear models with conjugate priors. *Bayesian Analysis*, **3**(3), 585–614.

Collins, L., Schafer, J., and Kam, C.-M. (2001). A comparison of inclusive and restrictive strategies in modern missing data procedures. *Psychological Methods*, **6**(4), 330–351.

Cowles, M. and Davis, C. (1982). On the origins of the .05 level of statistical significance. *American Psychologist*, **37**(5), 553–558.

Cramér, H. (1946). *Mathematical Methods of Statistics*. Princeton, NJ: Princeton University Press.

Danaher, P. J. (2007). Modeling page views across multiple websites with an application to internet reach and frequency prediction. *Marketing Science*, **26**(3), 422–437.

Daniels, M. J. and Hogan, J. W. (2008). *Missing Data in Longitudinal Studies: Strategies for Bayesian Modeling and Sensitivity Analysis*. Boca Raton, FL: Chapman & Hall/CRC.

Dawid, A. P. (2002). Discussion on the paper by Spiegelhalter, Best, Carlin and van der Linde. *Journal of the Royal Statistical Society, Series B (Statistical Methodology)*, **64**(4), 624.

Dellaportas, P., Forster, J. J., and Ntzoufras, I. (2002). On Bayesian model and variable selection using MCMC. *Statistics and Computing*, **12**(1), 27–36.

Denrell, J., Fang, C., and Zhao, Z. (2013). Inferring superior capabilities from sustained superior performance: a Bayesian analysis. *Strategic Management Journal*, **34**(2), 182–196.

Department of Statistics Singapore (2013). Singapore's international trade in services 2011. Table 1, Feb. http: //www.singstat.gov.sg/publications/publications_and_papers/international_accounts/int-trade2011.pdf.

Dey, D., Müller, P., and Sinha, D. (1998). *Practical Nonparametric and Semiparametric Bayesian Statistics*. New York: Springer-Verlag.

Dong, M. and Stettler, A. (2011). Estimating firm-level and country-level effects in cross-sectional analyses: an application of hierarchical modeling in corporate disclosure studies. *International Journal of Accounting*, **46**(3), 271 – 303.

Edwards, A. W. F. (1972). *Likelihood*. Cambridge: Cambridge University Press.

Fama, E. F. (1965). The behavior of stock-market prices. *Journal of Business*, **38**(1), 34–105.

Field, L. C. and Karpoff, J. M. (2002). Takeover defenses of IPO firms. *Journal of Finance*, **57**(5), 1857–1889.

Fildes, N. (2010). Swimming fast in piranha-filled waters. The Times (London), p. 47, Sep. 20.

Fisher, R. A. (1922). On the mathematical foundations of theoretical statistics. *Philosophical Transactions of the Royal Society of London, Series A*, **222**, 309–368.

Fortune (2005). Fortune 500: a database of 50 years of fortune's list of America's largest corporations. http://money.cnn.com/magazines/fortune/fortune500_archive/full/2005/, accessed Sep. 19, 2013.

Francis, A. (2001). New method helps operators assess line integrity. *Oil and Gas Journal*, 73, Nov. 26.

Garthwaite, P. A., Kadane, J. B., and O'Hagan, A. (2005). Statistical methods for eliciting probability distributions. *Journal of the American Statistical Association*, **100**(470), 680–701.

Gelfand, A. E., Hills, S. E., Racine-Poon, A., and Smith, A. F. M. (1990). Illustration of Bayesian inference in normal data models using Gibbs sampling. *Journal of the American Statistical Association*, **85**(412), 972–985.

Gelfand, A. E., Sahu, S., and Carlin, B. P. (1995). Efficient parametrisations for normal linear mixed model. *Biometrika*, **82**(3), 479–488.

Gelfand, A. E., Smith, A. F. M., and Lee, T.-M. (1992). Bayesian analysis of constrained parameter and truncated data problems using Gibbs sampling. *Journal of the American Statistical Association*, **87**(418), 523–532.

Gelman, A. (2006). Prior distributions for variance parameters in hierarchical models. *Bayesian Analysis*, **1**(3), 515–533.

Gelman, A., Carlin, J., Stern, H., and Rubin, D. (2004). *Bayesian Data Analysis*. Boca Raton, FL: Chapman & Hall/CRC, 2nd edition.

Gelman, A., King, G., and Boscardin, W. J. (1998). Estimating the probability of events that have never occurred: When is your vote decisive? *Journal of the American Statistical Association*, **93**(441), 1–9.

Gelman, A. and Robert, C. P. (2013). "Not only defended but also applied": the perceived absurdity of Bayesian inference (with discussion). *American Statistician*, **67**(1), 1–5.

Gelman, A., Roberts, G., and Gilks, W. (1995).Efficient Metropolis jumping rules. In *Bayesian Statistics 5*, J. M. Bernardo, J. O. Berger, A. P. Dawid, and A. F. M. Smith, eds., pp. 599–607. Alicante: Oxford University Press.

Gelman, A. and Rubin, D. (1992). Inference from iterative simulation using multiple sequences. *Statistical Science*, **7**(4), 457–472.

Geweke, J. (1992). Evaluating the accuracy of sampling-based approaches to calculating posterior moments. In *Bayesian Statistics 4*, J. M. Bernardo, J. O. Berger, A. P. Dawid, and A. F. M. Smith, eds., pp. 169–193. Oxford: Clarendon Press.

Gilks, W. R., Richardson, S., and Spiegelhalter, D. J. (1996). Introducing Markov chain Monte Carlo. In *Markov Chain Monte Carlo in Practice*, W. R. Gilks, S. Richardson, and D. J. Spiegelhalter, eds., pp. 1–19. Boca Raton, FL: Chapman & Hall/CRC.

Gilks, W. R. and Roberts, G. O. (1996). Strategies for improving MCMC. In *Markov Chain Monte Carlo in Practice*, W. R. Gilks, S. Richardson, and D. J. Spiegelhalter, eds., pp. 89–114. Boca Raton, FL: Chapman & Hall/CRC.

Glinsky, M. E. and Gunning, J. (2011). Understanding uncertainty in CSEM. *World Oil*, **232**(1), 57–62, Jan. http://www.worldoil.com/Understanding-uncertainty-in-CSEM-January-2011.html, accessed Jan. 18, 2012.

Green, P. J. (1995). Reversible jump Markov chain Monte Carlo computation and Bayesian model determination. *Biometrika*, **82**(4), 711–732.

Guler, I., Guillén, M. F., and Macpherson, J. M. (2002). Global competition, institutions, and the diffusion of organizational practices: the international spread of ISO 9000 quality certificates. *Administrative Science Quarterly*, **47**(2), 207–232.

Gustafson, P., Hossain, S., and MacNab, Y. C. (2006). Conservative prior distributions for variance parameters in hierarchical models. *Canadian Journal of Statistics/La Revue Canadienne de Statistique*, **34**(3), 377–390.

Hahn, E. D. (2006). Re-examining informative prior elicitation through the lens of Markov chain Monte Carlo. *Journal of the Royal Statistical Society, Series A*, **169**(1), 37–48.

Hahn, E. D. and Bunyaratavej, K. (2011). Offshoring of information-based services: structural breaks in industry life cycles. *Service Science*, **3**(3), 239–255.

Hahn, E. D. and Soyer, R. (2005). Probit and logit models: differences in the multivariate realm. Unpublished manuscript, available at http://home.gwu.edu/soyer/mv1h.pdf, accessed Apr. 09, 2014.

Hansen, M. H., Perry, L. T., and Reese, C. S. (2004). A Bayesian operationalization of the resource-based view. *Strategic Management Journal*, **25**(13), 1279–1295.

Hardy, Q. (2013). Testing a new class of speedy computer. New York Times, p. B1, Mar. 22. http://www.nytimes.com/2013/03/22/technology/testing-a-new-class-of-speedy-computer.html?_r=0, accessed Mar. 28, 2013.

Hastie, T. (1987). A closer look at the deviance. *American Statistician*, **41**(1), 16–20.

Hastings, W. K. (1970). Monte Carlo sampling methods using Markov chains and their applications. *Biometrika*, **57**(1), 97–109.

Hawawini, G., Subramanian, V., and Verdin, P. (2003). Is performance driven by industry- or firm-specific factors? A new look at the evidence. *Strategic Management Journal*, **24**(1), 1–16.

Heidelberger, P. and Welch, P. (1983). Simulation run length control in the presence of an initial transient. *Operations Research*, **31**(6), 1109–1144.

Hogg, R. V. and Tanks, E. A. (1997). *Probability and Statistical Inference*. Upper Saddle River, NJ: Prentice Hall, 5th edition.

Hubbard, D. (2007). *How to Measure Anything: Finding the Value of "Intangibles" for Business*. Hoboken, NJ: John Wiley & Sons, Inc.

Ingram, R. W. (1984). Economic incentives and the choice of state government accounting practices. *Journal of Accounting Research*, **22**(1), 126–144.

Jeffreys, H. (1946). An invariant form for the prior probability in estimation problems. *Proceedings of the Royal Society of London, Series A*, **186**(1007), 453–461.

Jeffreys, H. (1961). *Theory of Probability*. Oxford: Oxford University Press, 3rd edition.

Jiménez, A. (2010). Does political risk affect the scope of the expansion abroad? Evidence from Spanish MNEs. *International Business Review*, **19**(6), 619–633.

Kass, R. E., Carlin, B. P., Gelman, A., and Neal, R. M. (1998). Markov chain Monte Carlo in practice: a roundtable discussion. *American Statistician*, **52**(2), 93–100.

Kass, R. E. and Raftery, A. E. (1995). Bayes factors. *Journal of the American Statistical Association*, **90**(430), 773–795.

Kass, R. E. and Wasserman, L. (1995). A reference Bayesian test for nested hypotheses and its relationship to the Schwarz criterion. *Journal of the American Statistical Association*, **90**(431), 928–934.

Kolmogorov, A. (1956). *Foundations of the Theory of Probability*. New York: Chelsea Publishing Co.

Kording, K. P. and Wolpert, D. M. (2004). Bayesian integration in sensorimotor learning. *Nature*, **427**(6971), 244–247.

Kross, W. and Schroeder, D. A. (1984). An empirical investigation of the effect of quarterly earnings announcement timing on stock returns. *Journal of Accounting Research*, **22**(1), 153–176.

Kuh, E. (1956). Five decades of United States saving. *Journal of Economic History*, **16**(2), 211–218.

Kull, T. J. and Wacker, J. G. (2010). Quality management effectiveness in Asia: the influence of culture. *Journal of Operations Management*, **28**(3), 223 – 239.

Kuo, L. and Mallick, B. (1998). Variable selection for regression models. *Sankhyā B*, **60**(1), 65–81.

Laird, N. M. and Ware, J. H. (1982). Random-effects models for longitudinal data. *Biometrics*, **38**(4), 963–974.

Lenz, R. T. and Engledow, J. L. (1986). Environmental analysis units and strategic decision-making: a field study of selected 'leading-edge' corporations. *Strategic Management Journal*, **7**(1), 69–89.

Lindley, D. (1965). *Introduction to Probability and Statistics from a Bayesian Viewpoint, Part 2, Inference.* Cambridge: Cambridge University Press.

Lipscomb, B., Ma, G., and Berry, D. A. (2005). Bayesian predictions of final outcomes: regulatory approval of a spinal implant. *Clinical Trials*, **2**(4), 325–33; discussion 334–9, 364–78.

Little, R. J. A. and Rubin, D. B. (2002). *Statistical Analysis with Missing Data*. Hoboken, NJ: John Wiley & Sons, Inc., 2nd edition.

Liu, J. and Hodges, J. S. (2003). Posterior bimodality in the balanced one-way random-effects model. *Journal of the Royal Statistical Society, Series B*, **65**(1), 247–255.

Loeb, S. E. (1971). A survey of ethical behavior in the accounting profession. *Journal of Accounting Research*, **9**(2), 287–306.

Lunn, D., Best, N., and Whittaker, J. C. (2009). Generic reversible jump MCMC using graphical models. *Statistics and Computing*, **19**(4), 395–408.

Lunn, D., Jackson, C., Best, N., Thomas, A., and Spiegelhalter, D. (2013). *The BUGS Book: A Practical Introduction to Bayesian Analysis*. Boca Raton, FL: CRC Press.

Lunn, D., Thomas, A., Best, N., and Spiegelhalter, D. (2000). WinBUGS – A Bayesian modelling framework: concepts, structure, and extensibility. *Statistics and Computing*, **10**(4), 325–337.

Lunn, D., Whittaker, J. C., and Best, N. (2006). A Bayesian toolkit for genetic association studies. *Genetic Epidemiology*, **30**(3), 231–247.

Mandelbrot, B. B. (1963). The variation of certain speculative prices. *Journal of Business*, **36**(4), 394–419.

Mascarenhas, B. and Aaker, D. A. (1989). Mobility barriers and strategic groups. *Strategic Management Journal*, **10**(5), 475–485.

Mata, J. and Freitas, E. (2012). Foreignness and exit over the life cycle of firms. *Journal of International Business Studies*, **43**(7), 615–630.

Matsumoto, M. and Nishimura, T. (1998). Mersenne twister: a 623-dimensionally equidistributed uniform pseudo-random number generator. *ACM Transactions on Modeling and Computer Simulation*, **8**(1), 3–30.

McCullagh, P. and Nelder, J. A. (1989). *Generalized Linear Models*. Boca Raton, FL: Chapman & Hall/CRC, 2nd edition.

McCullough, B. D. (1998). Assessing the reliability of statistical software: Part I. *American Statistician*, **52**(4), 358–366.

McCullough, B. D. (1999). Assessing the reliability of statistical software: Part II. *American Statistician*, **53**(2), 149–159.

McCutchen, W. W. Jr. (1993). Strategy changes as a response to alterations in tax policy. *Journal of Management*, **19**(3), 575–593.

McNamara, G., Aime, F., and Vaaler, P. M. (2005). Is performance driven by industry- or firm-specific factors? a response to Hawawini, Subramanian, and Verdin. *Strategic Management Journal*, **26**(11), 1075–1081.

Metropolis, N., Rosenbluth, A. W., Rosenbluth, M. N., Teller, A. H., and Teller, E. (1953). Equation of state calculations by fast computing machines. *Journal of Chemical Physics*, **21**(6), 1087–1092.

Metropolis, N. and Ulam, S. (1949). The Monte Carlo method. *Journal of the American Statistical Association*, **44**(247), 335–341.

Millar, R. B. (2009). Comparison of hierarchical Bayesian models for overdispersed count data using DIC and Bayes' factors. *Biometrics*, **65**(3), 962–969.

von Mises, R. (1928). *Probability, Statistics and Truth*. New York: Dover Publications, 2nd edition. 1981.

Molitor, N.-T., Best, N., Jackson, C., and Richardson, S. (2009). Using Bayesian graphical models to model biases in observational studies and to combine multiple sources of data: application to low birth weight and water disinfection by-products. *Journal of the Royal Statistical Society. Series A*, **172**(3), 615–637.

Murrell, P. (2011). *R Graphics*. Boca Raton, FL: Chapman and Hall/CRC Press, 2nd edition.

Nelder, J. A. and Wedderburn, R. W. M. (1972). Generalized linear models. *Journal of the Royal Statistical Society, Series A*, **135**(3), 370–384.

Norris, J. (1997). *Markov Chains*. Cambridge: Cambridge University Press.

Ntzoufras, I. (2002). Gibbs variable selection using BUGS. *Journal of Statistical Software*, **7**(7).

Ntzoufras, I. (2009). *Bayesian Modeling Using WinBUGS*. Hoboken, NJ: John Wiley & Sons, Inc.

O'Hara, R. and Sillanpää (2009). A review of Bayesian variable selection methods: what, how and which. *Bayesian Analysis*, **6**(1), 82–118.

Ott, R. L. (1993). *An Introduction to Statistical Methods and Data Analysis*. Belmont, CA: Duxbury, 4th edition.

Pham-Gia, T. and Turkkan, N. (1993). Bayesian analysis of the difference of two proportions. *Communications in Statistics–Theory and Methods*, **22**, 1755–1771.

Plummer, M., Best, N., Cowles, K., and Vines, K. (2006). CODA: convergence diagnosis and output analysis for MCMC. *R News*, **6**(1), 7–11, March.

Powers, D. A. and Xie, Y. (2000). *Statistical Methods for Categorical Data Analysis*. San Diego, CA: Academic Press.

Press, S. J. (2003). *Subjective and Objective Bayesian Statistics: Principles, Models, and Applications*. Hoboken, NJ: John Wiley & Sons, Inc.

Raftery, A. E. and Lewis, S. M. (1992). Comment: one long run with diagnostics: implementation strategies for Markov chain Monte Carlo. *Statistical Science*, **7**(4), 493–497.

Raudenbush, S. W. and Bryk, A. S. (2002). *Hierarchical Linear Models: Applications and Data Analysis Methods*. Thousand Oaks, CA: Sage, 2nd edition.

Ripley, B. D. (1987). *Stochastic Simulation*. New York: John Wiley & Sons, Inc.

Robert, C. and Casella, G. (2004). *Monte Carlo Statistical Methods*. New York: Springer-Verlag, 2nd edition.

Roberts, G. (1996). Markov chain concepts related to sampling applications. In *Markov Chain Monte Carlo in Practice*, W. R. Gilks, S. Richardson, and D. J. Spiegelhalter, eds., pp. 45–57. Boca Raton, FL: Chapman & Hall/CRC.

Roberts, G. O. and Sahu, S. K. (2001). Approximate predetermined convergence properties of the Gibbs sampler. *Journal of Computational and Graphical Statistics*, **10**(2), 216–229.

Robins, J. M., van der Vaart, A., and Ventura, V. (2000). Asymptotic distribution of p values in composite null values. *Journal of the American Statistical Associateion*, **95**(452), 1143–1156.

Rosenkopf, L. and Nerkar, A. (2001). Beyond local search: boundary-spanning, exploration, and impact in the optical disk industry. *Strategic Management Journal*, **22**(4), 287–306.

Rossi, P. E., McCulloch, R. E., and Allenby, G. M. (1996). The value of purchase history data in target marketing. *Marketing Science*, **15**(4), 321–340.

Rossi, P. E., Allenby, G. M., and McCulloch, R. E. (2005). *Bayesian Statistics and Marketing*. Hoboken, NJ: John Wiley & Sons, Inc.

Sanders, W. G. (2001). Behavioral responses of CEOs to stock ownership and stock option pay. *Academy of Management Journal*, **44**(3), 477–492.

Savage, L. (1954). *The Foundations of Statistics*. New York: Dover Publications, 2nd edition. 1972.

Schafer, J. L. (1997). *Analysis of Incomplete Multivariate Data*. Boca Raton, FL: Chapman & Hall/CRC.

Schulz, K. F. and Grimes, D. A. (2005). Multiplicity in randomised trials II: Subgroup and interim analyses. *Lancet*, **365**(9471), 1657–1661.

Schwarz, G. E. (1978). Estimating the dimension of a model. *Annals of Statistics*, **6**(2), 461–464.

Seaman, S. R., Bartlett, J. W., and White, I. R. (2012). Multiple imputation of missing covariates with non-linear effects and interactions: an evaluation of statistical methods. *BMC Medical Research Methodology*, **23**(46), http://www.biomedcentral.com/1471–2288/12/46, accessed Apr. 09, 2014.

Shamdasani, P. (2011). Smart money. South China Morning Post, p. 2, Jul. 4.

Shipper, F., Manz, C. C., Manz, K. P., and Harris, B. W. (2013). Collaboration that goes beyond co-op-eration: it's not just "if" but "how" sharing occurs that makes the difference. *Organizational Dynamics*, **42**(2), 100–109.

Short, J. C. Jr., Ketchen, D. J., Palmer, T. B., and Hult, G. T. M. (2007). Firm, strategic group, and industry influences on performance. *Strategic Management Journal*, **28**(2), 147–167.

Smith, A. F. M. and Roberts, G. O. (1993). Bayesian computation via the Gibbs sampler and related Markov chain Monte Carlo methods. *Journal of the Royal Statistical Society, Series B*, **55**(1), 3–23.

Spiegelhalter, D., Thomas, A., Best, N., and Gilks, W. (1996a). BUGS 0.5: Bayesian inference using Gibbs sampling manual (version ii). Cambridge: MRC Biostatistics Unit, Institute of Public Health, http://www.mrc-bsu.cam.ac.uk/bugs/documentation/contents.shtml, accessed Apr. 19, 2006.

Spiegelhalter, D., Thomas, A., Best, N., and Gilks, W. (1996b). BUGS 0.5: examples volume 1 (version i). Cambridge: MRC Biostatistics Unit, Institute of Public Health, http://www.mrc-bsu.cam.ac.uk/bugs/documentation/contents.shtml, accessed Apr. 19, 2006.

Spiegelhalter, D., Thomas, A., Best, N., and Gilks, W. (1996c). BUGS 0.5: examples volume 2 (version ii). Cambridge: MRC Biostatistics Unit, Institute of Public Health, http://www.mrc-bsu.cam.ac.uk/bugs/documentation/contents.shtml, accessed Apr. 19, 2006.

Spiegelhalter, D. J., Best, N. G., Carlin, B. P., and van der Linde, A. (2002). Bayesian measures of model complexity and fit. *Journal of the Royal Statistical Society, Series B (Statistical Methodology)*, **64**(4), 583–639.

Spiegelhalter, D. J., Freedman, L. S., and Parmar, M. K. B. (1994). Bayesian approaches to randomized trials. *Journal of the Royal Statistical Society, Series A (Statistics in Society)*, **157**(3), 357–416.

Spiegelhalter, D. J., Thomas, A., Best, N. G., and Lunn, D. (2003). WinBUGS user manual version 1.4. Available at http://www.mrc-bsu.cam.ac.uk/bugs/winbugs/manual14.pdf, accessed May 1, 2012.

Standard and Poor's (2013). Sovereigns rating list. www.standardandpoors.com/ratings/sovereigns/ratings-list/, accessed Nov. 19, 2013.

Stern, H. S. (2000). Comment. *Journal of the American Statistical Associateion*, **95**(452), 1157–1159.

Stone, M. and Springer, B. G. F. (1965). A paradox involving quasi prior distributions. *Biometrika*, **52**(3/4), 623–627.

Stuart, T. E. (2000). Interorganizational alliances and the performance of firms: a study of growth and innovation rates in a high-technology industry. *Strategic Management Journal*, **21**(8), 791–811.

Suzuki, Y. (1980). The strategy and structure of top 100 Japanese industrial enterprises 1950–1970. *Strategic Management Journal*, **1**(3), 265–291.

Tanner, M. A. and Wong, W. H. (1987). The calculation of posterior distributions by data augmentation. *Journal of the American Statistical Association*, **82**(398), 528–540.

Thisted, R. (1988). *Elements of Statistical Computing: Numerical Computing*. London: Chapman & Hall.

Tierney, L. (1994). Rejoinder: Markov chains for exploring posterior distributions. *Annals of Statistics*, **22**(4), 1758–1762.

Tierney, L. (1996). Introduction to general state-space Markov theory. In *Markov Chain Monte Carlo in Practice*, W. R. Gilks, S. Richardson, and D. J. Spiegelhalter, eds., pp. 59–74. Boca Raton, FL: Chapman & Hall/CRC.

Tukey, J. W. (1977). *Exploratory Data Analysis*. Reading, MA: Addison-Wesley.

Twitter, Inc. (2013, Oct. 3). Form S-1. EDGAR search, http://www.sec.gov/Archives/edgar/data/1418091/000119312513390321/d564001ds1.htm, accessed Oct. 14, 2013.

United Nations Conference on Trade and Development (2013). UNCTADstat. http://unctadstat.unctad.org/, accessed Sep. 2, 2013.

U.S. Bureau of the Census (1996). *Statistical Abstract of the United States: 1996*. Washington, DC: U.S. Bureau of the Census, 116th edition.

U.S. Food and Drug Administration (1999). Transscan T-Scan 2000–P970033, Apr. 16. http://www.accessdata.fda.gov/scripts/cdrh/cfdocs/cfTopic/pma/pma.cfm?num=p970033, accessed Jan. 17, 2012.

Venables, W. N. and Ripley, B. D. (2002). *Modern Applied Statistics with S*. New York: Springer, 4th edition.

Wade, J. B., Porac, J. F., and Pollock, T. G. (1997). Worth, words, and the justification of executive pay. *Journal of Organizational Behavior*, **18**, 641–664.

Whitfield, J. (2004). A hunt for the haphazard. Financial Times, p. 13, Jun. 11.

Wickham, H. (2009). *ggplot2: Elegant Graphics for Data Analysis*. Dordrecht: Springer.

Winglee, M., Kalton, G., Rust, K., and Kasprzyk, D. (2001). Handling item nonresponse in the U.S. component of the IEA reading literacy study. *Journal of Educational and Behavioral Statistics*, **26**(3), 343–359.

World Bank (2012). World development indicators. http://data.worldbank.org/data-catalog/world-development-indicators, accessed Aug. 23, 2012.

World Bank (2013). World development indicators. http://data.worldbank.org/indicator/NY.GDP.PCAP.CD, accessed Nov. 19, 2013.

Zellner, A. (1971). *An Introduction to Bayesian Inference in Econometrics*. New York: John Wiley & Sons, Inc.

Zellner, A. (1986). On assessing prior distributions and Bayesian regression analysis with g-prior distributions. In *Bayesian Inference and Decision Techniques: Essays in Honor of Bruno de Finetti*, P. Goel and A. Zellner, eds., pp. 233–243. New York: North Holland Publishing Co.

Zellner, A. (1995). Bayesian and non-Bayesian approaches to statistical inference and decision-making. *Journal of Computational and Applied Mathematics*, **64**(1-2), 3–10.

AUTHOR INDEX

Bayesian Methods for Management and Business: Pragmatic Solutions for Real Problems,
First Edition. Eugene D. Hahn.

SUBJECT INDEX

Bayesian Methods for Management and Business: Pragmatic Solutions for Real Problems,
First Edition. Eugene D. Hahn.

CPSIA information can be obtained
at www.ICGtesting.com
Printed in the USA
BVOW06*2156020118
504266BV00001B/3/P